VOLUME SEVEN HUNDRED AND FIFTEEN

METHODS IN
ENZYMOLOGY

Enzymes of Polyamine Metabolism

METHODS IN ENZYMOLOGY

Editors-in-Chief

KAREN N. ALLEN, Ph.D.

*Professor and Chair
Department of Chemistry
Professor of Material Science and Engineering
Boston University*

DAVID CHRISTIANSON

*Roy and Diana Vagelos Laboratories
Department of Chemistry
University of Pennsylvania
Philadelphia, PA*

Founding Editors

SIDNEY P. COLOWICK and NATHAN O. KAPLAN

VOLUME SEVEN HUNDRED AND FIFTEEN

METHODS IN ENZYMOLOGY

Enzymes of Polyamine Metabolism

Edited by

ROBERT A. CASERO, Jr.
Johns Hopkins University, School of Medicine, Baltimore

TRACY MURRAY STEWART
Johns Hopkins University, School of Medicine, Baltimore

ACADEMIC PRESS
An imprint of Elsevier

Academic Press is an imprint of Elsevier
50 Hampshire Street, 5th Floor, Cambridge, MA 02139, United States
125 London Wall, London, EC2Y 5AS, United Kingdom

First edition 2025

Copyright © 2025 Elsevier Inc. All rights are reserved, including those for text and data mining, AI training, and similar technologies.

For accessibility purposes, images in electronic versions of this book are accompanied by alt text descriptions provided by Elsevier. For more information, see https://www.elsevier.com/about/accessibility.

Publisher's note: Elsevier takes a neutral position with respect to territorial disputes or jurisdictional claims in its published content, including in maps and institutional affiliations.

No part of this publication may be reproduced or transmitted in any form or by any means, electronic or mechanical, including photocopying, recording, or any information storage and retrieval system, without permission in writing from the publisher. Details on how to seek permission, further information about the Publisher's permissions policies and our arrangements with organizations such as the Copyright Clearance Center and the Copyright Licensing Agency, can be found at our website: www.elsevier.com/permissions.

This book and the individual contributions contained in it are protected under copyright by the Publisher (other than as may be noted herein).

Notices
Knowledge and best practice in this field are constantly changing. As new research and experience broaden our understanding, changes in research methods, professional practices, or medical treatment may become necessary.

Practitioners and researchers must always rely on their own experience and knowledge in evaluating and using any information, methods, compounds, or experiments described herein. In using such information or methods they should be mindful of their own safety and the safety of others, including parties for whom they have a professional responsibility.

To the fullest extent of the law, neither the Publisher nor the authors, contributors, or editors, assume any liability for any injury and/or damage to persons or property as a matter of products liability, negligence or otherwise, or from any use or operation of any methods, products, instructions, or ideas contained in the material herein.

ISBN: 978-0-443-31784-2
ISSN: 0076-6879

For information on all Academic Press publications
visit our website at https://www.elsevier.com/books-and-journals

Publisher: Zoe Kruze
Editorial Project Manager: Saloni Vohra
Production Project Manager: James Selvam
Cover Designer: Bakyalakshmi S
Typeset by MPS Limited, India

Contents

Contributors	xxiii
Preface	xxix

1. Genetic analyses of Myc and hypusine circuits in tumorigenesis   1
Shima Nakanishi and John L. Cleveland

1. Introduction	2
2. Rationale	4
3. Analyses of roles of MYC and hypusinated eIF5A in tumor development	5
3.1 Genetic approaches to assess effects on tumor development	5
3.2 Isolation and analyses of primary, premalignant B cells	6
4. Analyses of tumor maintenance and progression	8
4.1 Tumorigenic potential – targeting Myc and the hypusine axis with small molecules	8
4.2 Tumorigenic potential – gene silencing studies	10
4.3 Assessment of tumor maintenance by subcutaneous tumor transplantation	12
5. Concluding remarks	15
Acknowledgements	16
References	16

2. Expression, purification, and crystallization of "humanized" *Danio rerio* histone deacetylase 10 "HDAC10", the eukaryotic polyamine deacetylase   19
Juana Goulart Stollmaier, Corey J. Herbst-Gervasoni, and David W. Christianson

1. Introduction	20
2. Construct design	23
3. Transformation, expression, and purification	23
3.1 Equipment	23
3.2 Materials	24
3.3 Transformation	25
3.4 Expression and purification	26
4. Activity assay using a colorimetric method	30
4.1 Equipment	30

4.2 Materials	31
4.3 Assay setup	31
4.4 Notes	32
5. Crystallization of HDAC10-inhibitor complexes	32
5.1 Equipment	34
5.2 Materials	34
5.3 Procedure	35
5.4 Notes	37
6. Summary	38
References	38

3. TR-FRET assay for profiling HDAC10 inhibitors and PROTACs   41
Kim Remans, Peter Sehr, Raphael R. Steimbach, Nikolas Gunkel, and Aubry K. Miller

1. Introduction	42
2. Synthesis of Tubastatin-AF647 FRET tracer molecule	43
2.1 Equipment and materials	43
2.2 Procedure	45
3. Baculovirus-mediated expression and purification of TwinStrep-GST-hHDAC10 in Sf21 cells	48
3.1 Equipment and materials	48
3.2 Procedure	51
4. Preparation of DTBTA-Eu^{3+}-labelled Strep-TactinXT®	59
4.1 Equipment and materials	59
4.2 Procedure	59
5. Performing and analyzing the HDAC10 TR-FRET ligand displacement assay	60
5.1 Equipment and materials	60
5.2 Procedure	61
References	62

4. Polyamine transport inhibitors: Methods and caveats associated with measuring polyamine uptake in mammalian cells   65
Alexandra Bunea and Otto Phanstiel IV

1. Introduction and rationale	66
1.1 Cell selection	68
1.2 Maximum tolerated concentration of the PTI	69
1.3 PTI solubility and the use of dimethylsulfoxide (DMSO)	69

- 2. Assay 1 and assay 2 — 70
  - 2.1 Assay 1 description and results — 70
  - 2.2 Assay 2 description and results — 72
  - 2.3 Equipment — 75
  - 2.4 Reagents — 75
  - 2.5 Procedure — 76
  - 2.6 Notes — 78
  - 2.7 Calculations — 79
- 3. Assay 3 — 79
  - 3.1 Description and results — 79
  - 3.2 Equipment — 81
  - 3.3 Reagents — 81
  - 3.4 Procedure — 82
  - 3.5 Notes — 85
  - 3.6 Calculations — 86
- 4. Assay 3 data analysis calculations — 88
- 5. Results and discussion — 88
- 6. Conclusions — 89
- Acknowledgments — 89
- References — 89

5. Evaluation of platinum drug toxicity resulting from polyamine catabolism — 93
Kamyar Zahedi, Sharon Barone, and Manoocher Soleimani

- 1. Introduction — 94
- 2. Rationale — 96
- 3. Cisplatin induced renal injury — 96
  - 3.1 Animals, reagents and equipment — 96
  - 3.2 Procedures — 97
  - 3.3 Notes — 98
- 4. Harvesting of biological samples — 99
  - 4.1 Reagents and equipment — 99
  - 4.2 Procedure — 100
- 5. Assessment of renal function — 100
  - 5.1 Reagents and equipment — 100
  - 5.2 Procedures — 100
- 6. RNA isolation, real-time quantitative RT-PCR (qRT-PCR) and northern blot analysis — 101
  - 6.1 Reagents and equipment — 101

		6.2	Procedures	102
		6.3	Notes	106
	7.	Preparation of kidney protein extracts and western blot analysis		106
		7.1	Reagents and equipment	106
		7.2	Procedures	107
		7.3	Notes	109
	8.	Histological staining of harvested kidneys for the evaluation of tubular injury and fibrosis		109
		8.1	Reagents and equipment	109
		8.2	Procedure	110
	9.	Immunofluoresence microscopic examination of kidney sections		111
		9.1	Reagents and equipment	111
		9.2	Procedure	111
		9.3	Notes	112
	10.	Measurement of tissue polyamine levels		112
		10.1	Reagents and equipment	112
		10.2	Procedure	112
		10.3	Notes	114
	11.	General equipment		114
	Acknowledgments			114
	References			114

6. Measurements of acrolein adducts resulting from polyamine catabolism   117

Keiko Kashiwagi and Kazuei Igarashi

1.	Introduction		118
2.	Acrolein production from spermine		119
	2.1	Measurement of acrolein	119
	2.2	Acrolein is produced from spermine by spermine oxidase	119
3.	Increase in polyamine oxidases and protein-conjugated acrolein in plasma of stroke patients		120
	3.1	Measurement of polyamine oxidases (AcPAO and SMOX) and protein-conjugated acrolein (PC-Acro) in plasma	120
	3.2	Increase in polyamine oxidases (AcPAO and SMOX) and protein-conjugated acrolein (PC-Acro) in plasma of stroke patients	121
4.	Plasma levels of PC-Acro, IL-6 and CRP as markers of silent brain infarction (SBI)		122
	4.1	Measurement of PC-Acro, IL-6 and CRP in plasma	122
	4.2	Imaging	122

4.3	Statistics	123
4.4	Relationship between biochemical markers (PC-Acro, IL-6 and CRP) and SBI, CA and WMH determined by imaging	123
5. Urinary amino acid-conjugated acrolein and taurine as new biomarkers for detection of dementia		124
5.1	Collection of urine and plasma samples	124
5.2	Measurement of urinary amino acid-conjugated acrolein and taurine contents, and plasma $A\beta_{40}$ and $A\beta_{42}$	125
5.3	Statistics	125
5.4	Correlation between MMSE and age, and the decrease in urinary AC-Acro and taurine	125
6. Inactivation of GAPDH by acrolein conjugation		127
6.1	Identification of glyceraldehyde-3-phosphate dehydrogenase (GAPDH) as an acrolein-conjugated protein	127
6.2	Determination of amino acid residues of GAPDH conjugated with acrolein	129
6.3	Translocation of acrolein-conjugated GAPDH to nuclei resulting in apoptosis	130
7. Inhibition of dendritic spine extension by acrolein conjugation with cytoskeleton proteins during brain infarction		130
7.1	Acrolein conjugation with cytoskeleton proteins	130
7.2	Induction of brain infarction in mice	131
7.3	Time-dependent decrease of dendritic spine extension in infarct brain	131
8. Summary and conclusions		132
Acknowledgments		132
References		132

7. Alterations in polyamine metabolism induced by the pathogen *Helicobacter pylori*: Implications for gastric inflammation and carcinogenesis   137

Alain P. Gobert, Caroline V. Hawkins, Kara M. McNamara, and Keith T. Wilson

1. Introduction		138
2. Culture of *H. pylori*		140
2.1	Equipment	140
2.2	Reagents	140
2.3	Procedure	141
2.4	Notes	142

3. Animal models of *H. pylori* infection	143
3.1 Equipment	143
3.2 Reagents	143
3.3 Procedures	143
3.4 Notes	144
4. Generation and infection of gastric organoids	145
4.1 Equipment	145
4.2 Reagents	145
4.3 Procedure	146
4.4 Notes	147
5. AcroleinRED assay with organoids	148
5.1 Equipment	148
5.2 Reagents	148
5.3 Procedure	148
5.4 Notes	149
6. Immunodetection of acrolein adducts	149
6.1 Equipment	149
6.2 Reagents	150
6.3 Procedure	150
Statements and Declarations	151
Competing interest	151
Grant Support	151
References	151

8. Yeast reconstituted translation assays for analysis of eIF5A function   155

Byung-Sik Shin and Thomas E. Dever

1. Introduction	156
2. Overview of in vitro translation assays	158
3. Preparation of initiation factors	160
4. Purification of eIF5A	160
5. Purification of elongation factors	162
5.1 Purification of eEF1A	162
5.2 Purification of eEF2 and eEF3	164
6. Purification of termination factors	165
6.1 Purification of eRF1	165
6.2 Purification of eRF3	166
7. Preparation of mRNA	166
7.1 Preparation of DNA template for T7 transcription	167
7.2 Preparation of mRNA by in vitro T7 transcription	168

8.	Preparation of tRNA^{Pro}	169
9.	Purification of tRNA synthetases and CCA-adding enzyme	170
	9.1 Purification of methionyl-tRNA synthetase (MetRS)	170
	9.2 Purification of Prolyl tRNA synthetase (PRS)	171
	9.3 Purification of Phenylalanyl and Lysyl tRNA synthetase	171
	9.4 Purification of CCA-adding enzyme	172
10.	Limited charging of tRNA$_i$ using ^{35}S methionine	173
11.	Aminoacylation of tRNA^{Phe}, tRNA^{Pro}, and tRNA^{Lys}	174
12.	Purification of 40S and 60S ribosomal subunits	175
13.	Peptide formation and release assays	176
	13.1 Preparation of initiation complexes	176
	13.2 Peptide formation assay	177
	13.3 Peptide release (termination) assay	179
	Acknowledgments	180
	References	180

9. Quantification of spermine oxidase (SMOX) activity in tissues by HPLC   183

Jackson R. Foley and Cassandra E. Holbert

1.	Introduction	184
2.	Rationale	187
3.	Tissue sample preparation	188
	3.1 Equipment	188
	3.2 Reagents	188
	3.3 Procedure	190
	3.4 Notes	190
4.	PAOX and SMOX reactions	190
	4.1 Equipment	190
	4.2 Reagents	190
	4.3 Procedure	191
	4.4 Notes	192
5.	HPLC analysis of polyamine oxidase activity	192
	5.1 Equipment	192
	5.2 Reagents	192
	5.3 Procedure	193
	5.4 Notes	196
	Acknowledgments	196
	References	196

10. Direct measurement of ATP13A2 polyamine-dependent ATPase activity following rapid purification of lysosomes   201
Christina Efthymiou, Sydney Drury, and Kenneth Lee

1. Introduction   202
2. LysoPure: rapid purification of lysosomes from mammalian cells   203
   - 2.1 Equipment   203
   - 2.2 Materials   203
   - 2.3 Procedure and expected results   204
   - 2.4 Considerations for the purification   205
3. Determination of lysosomal ATP13A2 spermine-dependent ATPase activity using a NADH-coupled ATPase assay   205
   - 3.1 Equipment   205
   - 3.2 Materials   205
   - 3.3 Procedure and expected results   206
   - 3.4 Considerations for the ATPase assay   208

Acknowledgments   208
References   208

11. Analysis of translational regulation using polysome profiling and puromycin incorporation   211
Quinton W. Wood and Teresa L. Mastracci

1. Introduction   212
2. Before you begin   215
3. Key resources table   216
4. Materials and equipment   217
5. Step-by-step method details   217
   - 5.1 Preparing the sucrose gradient   217
   - 5.2 Preparing cell and tissue samples (lysis)   218
   - 5.3 Ultracentrifugation of samples through sucrose gradient   219
   - 5.4 Collection of polysome profiles   219
6. Expected outcomes   221
7. Quantification and statistical analysis   222
8. Advantages   222
9. Limitations   222
10. Optimization and troubleshooting   223
    - 10.1 Improperly layered sucrose gradient   223
    - 10.2 Potential solution for an improperly layered sucrose gradient   223

		10.3 Insufficient cycloheximide treatment of cells and tissue	223
		10.4 Potential solution for an insufficient cycloheximide treatment of cells and tissue	223
	11.	Safety considerations and standards	224
	12.	Before you begin	224
	13.	Key resources table	224
	14.	Materials and equipment	225
	15.	Step-by-step method details	225
		15.1 Dissecting sample of interest	225
		15.2 Puromycin treatment	226
		15.3 Preparation of lysate for western	226
		15.4 Protein quantification and western blot	226
	16.	Expected outcomes	226
	17.	Quantification and statistical analysis	227
	18.	Advantages	227
	19.	Limitations	228
	20.	Optimization and troubleshooting	228
		20.1 Insufficient puromycin treatment of samples	228
		20.2 Potential solution to insufficient puromycin treatment of samples	228
	21.	Safety considerations and standards	228
	22.	Conclusion	228
	Acknowledgments		229
	Competing interests		229
	References		229

12. Targeting polyamine metabolism in an ex vivo prostatectomy model   231

Hayley C. Affronti, Aryn M. Rowsam, Spencer R. Rosario, and Dominic J. Smiraglia

1.	Introduction	232
2.	Materials and reagents	234
	2.1 Key materials	234
	2.2 Graphical overview	235
3.	Protocols	236
	3.1 Tissue explant culture	236
	3.2 Preparation for metabolic analysis	236
	3.3 Ex Vivo sample metabolomics analysis	237
	3.4 IHC analysis	238
References		238

13. Development and characterization of a *Drosophila* model of Snyder-Robinson syndrome
241

Xianzun Tao and R. Grace Zhai

1. Introduction	242
2. Rational	244
3. Generation of *dSms*-deficient flies	245
3.1 Equipment	245
3.2 Reagents	246
3.3 Fly lines	246
3.4 Procedure	246
3.5 Notes	246
4. *dSms* mRNA level measurement	248
4.1 Equipment	248
4.2 Reagents	248
4.3 Procedure	248
5. Polyamine levels measurement	248
5.1 Equipment	248
5.2 Reagents	248
5.3 Procedure	249
6. SMS re-expression	249
6.1 Fly lines	249
6.2 Procedure	249
6.3 Notes	249
7. Embryo viability measurement	250
7.1 Equipment	250
7.2 Procedure	250
8. Adult lifespan measurement	250
8.1 Equipment	250
8.2 Procedure	250
9. Climbing performance analysis	251
9.1 Equipment	251
9.2 Procedure	251
10. Eye morphology analysis	251
10.1 Equipment	252
10.2 Procedure	252
11. Eye activity analysis - electroretinography	253
11.1 Equipment	253
11.2 Procedure	253

12. Discussion	253
Acknowledgments	254
References	254

14. Monitoring ODC activity and polyamines in Bachmann-Bupp syndrome patient biological samples — 257
Chad R. Schultz, Elizabeth A. VanSickle, Caleb P. Bupp, and André S. Bachmann

1. Introduction	258
2. Rationale	259
3. Experimental section	260
3.1 RBC preparation	260
3.2 Protein quantification assay	260
3.3 ODC activity assay	262
3.4 Dansylating polyamines in the RBC preparations	263
3.5 Running samples on the HPLC	266
Acknowledgments	267
Competing interests	268
References	268

15. Gene replacement therapy to restore polyamine metabolism in a Snyder-Robinson syndrome mouse model — 271
Oluwaseun Akinyele, Krystal B. Tran, Marie A. Johnson, and Dwi U. Kemaladewi

1. Introduction	272
2. Materials	275
2.1 Adeno-associated viral vector (AAV)	275
2.2 Animals and genotyping	275
2.3 Temporal vein injection in neonatal mice	276
2.4 Tail vein injection in juvenile mice	276
2.5 Tissue isolation and biodistribution evaluation	276
2.6 Expression analysis	276
3. Methods	277
3.1 AAV design	277
3.2 Preparation of AAV aliquots	279
3.3 Mouse genotyping	281
3.4 Temporal vein injection in neonatal mice	283
3.5 Tail vein injection in juvenile mice	285

3.6 Tissue isolation and biodistribution evaluation	286
3.7 Analysis of SMS protein expression using western blot	287
4. Conclusion	289
Acknowledgments	289
References	290

16. Methods to study polyamine metabolism during osteogenesis 293
Amin Cressman and Fernando A. Fierro

1. Introduction	294
2. Materials and reagents	296
2.1 Isolation and expansion of MSCs	296
3. Osteogenic differentiation and quantification	296
4. Polyamine modulation	297
5. Methods	297
5.1 Isolation and expansion of MSCs	297
6. Characterization of MSCs	299
7. Osteogenic differentiation and quantification methods	300
8. Osteogenic differentiation	301
9. Quantification of mineralization	301
10. Quantification of alkaline phosphatase activity	302
11. Quantification of gene expression	302
12. Polyamine modulation to investigate role in osteogenesis	304
13. Supplementation with polyamines and pathway inhibitors	304
14. Reproducibility and statistical considerations	305
15. Conclusions and future directions	305
References	306

17. Collection, preparation, and biobanking of clinical specimens for analysis in polyaminopathies 309
Elizabeth A. VanSickle, Chad R. Schultz, André S. Bachmann, and Caleb P. Bupp

1. Introduction	310
2. Collection of biological samples	310
2.1 Identification of patients with polyaminopathies	310
2.2 Obtaining control samples	311
2.3 Informed consent	312
2.4 Remote sample collection, collection kits, and shipping	313
2.5 Collection timing, volume limitations, and other considerations	314
3. Preparation of biological samples	315

3.1	Blood	315
3.2	Urine	318
3.3	Skin	318
3.4	Tissue	319
4. Biobanking of biological samples		319
4.1	Storage considerations	319
4.2	Sample organization and tracking	320
Acknowledgments		320
Competing interests		320
References		321

18. An enhanced method for the determination of polyamine content in biological samples by dansyl chloride derivatization and HPLC — 323

Ashley Nwafor, Tracy Murray Stewart, and Robert A. Casero, Jr.

1. Introduction		324
2. Rationale		326
3. Polyamine standards preparation		327
3.1	Equipment	327
3.2	Reagents	327
3.3	Procedure	328
4. Cell, media, and urine sample lysate preparation		328
4.1	Equipment	328
4.2	Reagents	328
4.3	Procedure	328
4.4	Notes	328
5. Tissue sample lysate preparation		329
5.1	Equipment	329
5.2	Reagents	329
5.3	Procedure	329
5.4	Notes	330
6. Dansyl chloride labeling of polyamines for HPLC		330
6.1	Equipment	330
6.2	Reagents	330
6.3	Procedure	330
6.4	Notes	331
7. HPLC analysis of polyamine content		331
7.1	Equipment	331
7.2	Reagents	332

7.3 Procedure	332
7.4 Notes	333
8. Results	334
Acknowledgments	335
References	335

19. Cell-based determination of HDAC10-mediated polyamine deacetylase activity 337
Ishika Gupta, Ashley Nwafor, Robert A. Casero, Jr., and Tracy Murray Stewart

1. Introduction	338
2. Rationale	339
3. HDAC10-dependent growth rescue	340
3.1 Equipment and materials	340
3.2 Reagents	341
3.3 Procedure	341
3.4 Expected outcomes	346
3.5 Optimization and troubleshooting	346
4. Summary	347
References	348

20. *Drosophila melanogaster* imaginal disc assays to study the polyamine transport system 351
Shannon L. Nowotarski and Justin R. DiAngelo

1. Introduction	352
2. Rationale	353
3. Preparation of *Drosophila* imaginal discs	354
3.1 Equipment	354
3.2 Reagents	354
3.3 Procedure	355
3.4 Notes	356
4. Assessing disc development in the presence or absence of 20-hydroxyecdysone	357
4.1 Equipment	357
4.2 Reagents	357
4.3 Procedure	357
4.4 Notes	357
5. Imaginal disc assay to study the polyamine transport system	358
5.1 Equipment	358

5.2 Reagents	358
5.3 Procedure	358
5.4 Notes	359
6. Results	359
Acknowledgments	359
References	360

21. A fluorescence-based assay for measuring aminopropyltransferase activity — 363
Pallavi Singh, Jae-Yeon Choi, and Choukri Ben Mamoun

1. Introduction	364
2. Assay design	366
3. Things to keep ready beforehand	368
4. Key resources table	368
5. Preparation of reagents and standard curves for APT reactions	371
6. Setting up spermidine and spermine synthase reactions and detection using the DAB-APT assay	376
7. Validation of APT reactions using thin-layer chromatography (TLC) and mass spectrometry	380
8. Quantification and statistical analysis	382
9. Summary	384
References	386

22. Isolation of *Hafnia paralvei* FB315, a polyamine-high-producing bacterium, from aged cheese — 389
Yuta Ami, Emi Sugiwaka, and Shin Kurihara

1. Introduction	390
2. Isolation of bacteria from cheese	391
2.1 Equipment	391
2.2 Reagents	392
2.3 Procedure	392
3. Bacterial culture	393
3.1 Equipment	393
3.2 Reagents	393
3.3 Procedure	393
4. Purification of PuO derived from *Rhodococcus erythropolis* NCIMB 11540	393
4.1 Equipment	393
4.2 Reagents	393
4.3 Procedure	394

5. Simplified quantification of putrescine concentration in culture supernatants using the PuO-POD-4AA-TOPS method	395
5.1 Equipment	395
5.2 Reagents	395
5.3 Procedure	395
5.4 Note	397
6. Quantification of polyamine concentration by HPLC	397
6.1 Equipment	397
6.2 Reagents	397
6.3 Procedure	398
7. 16S rRNA gene analysis	398
7.1 Equipment	398
7.2 Reagents	398
7.3 Procedure	399
8. Cultivation in MRS medium supplemented with precursors of putrescine	399
8.1 Equipment	399
8.2 Reagents	399
8.3 Procedure	400
9. Results	400
9.1 Putrescine concentration in aged cheeses	400
9.2 Isolation of bacteria from aged cheeses and screening for polyamine-producing strain	400
9.3 Detailed analysis of the polyamine-producing candidate strain	401
9.4 Effects of putrescine precursors on putrescine production by *Hafnia paralvei* FB315	401
9.5 Growth and Putrescine Production of a High Putrescine-Producing Bacterium Isolated from Aged Cheese	401
10. Discussion	402
Appendix A. Supporting information	404
References	404

23. Methylated polyamines derivatives and antizyme-related effects   407

Maxim A. Khomutov, Arthur I. Salikhov, Olga A. Smirnova, Vladimir A. Mitkevich, and Alex R. Khomutov

1. Introduction	408
2. Rationale	408

2.1	MeSpds differentially modulate (mOAZ1)$_2$-polyamine complex formation	409
2.2	Electrophoresis of mOAZ1-polyamine complex	411
3.	Synthesis of research instruments (C-methylated spermidines)	411
3.1	Synthesis of N-(benzyloxycarbonyl)-aminoalcohols (2a-2c)	413
3.2	Synthesis of N-Cbz-protected C-methylated analogs of Spd (4a-4c)	414
3.3	Synthesis of C-methylated spermidine trihydrochloride (5-7)	416
4.	Detection of (mOAZ1)$_2$-polyamine complex formation using ITC	418
4.1	Equipment	418
4.2	Reagents	418
4.3	Procedure	418
4.4	Notes	419
5.	Electrophoresis of mOAZ1–polyamine complex	419
5.1	Equipment	419
5.2	Reagents	419
5.3	Procedure	419
6.	Conclusion	420
Acknowledgements		420
References		421

24. Novel LC-MS/MS assay to quantify D,L-alpha-difluoromethylornithine (DFMO) in mouse plasma — 423

Matthew A. Swanson, Carlye Szarowicz, Schuyler T. Pike, Chad R. Schultz, André S. Bachmann, and Thomas C. Dowling

1.	Introduction	424
2.	Rationale	425
3.	Materials and methods	425
3.1	Chemicals and reagents	425
3.2	Standard solutions preparation	426
3.3	Calibration curve, QCs and study sample preparation	426
3.4	Chromatographic and mass spectrometric conditions	426
3.5	Method validation	427
3.6	Application of the method to *in vivo* pharmacokinetic study	428
4.	Results	428
4.1	LC-MS/MS method development	428
4.2	Method validation	429
4.3	Assay stability	432
4.4	Application to *in vivo* pharmacokinetic study	433

5. Conclusion	433
Acknowledgments	433
Declaration of competing interest	433
CRediT authorship contribution statement	433
References	434

25. Polyamine quantitation by LC-MS using isobutyl chloroformate derivatives 437

Christine Isaguirre, Megan Gendjar, Kelsie M. Nauta, Nicholas O. Burton, and Ryan D. Sheldon

1. Introduction	438
2. Methods	439
2.1 Before you begin	439
2.2 Metabolite extraction	443
2.3 Polyamine derivatization	444
2.4 LC-MS analysis	446
2.5 Data analysis	449
3. Expected outcomes	450
3.1 Representative results	450
3.2 Quantitation	452
4. Advantages, limitations, and troubleshooting	452
4.1 Advantages	452
4.2 Limitations	455
4.3 Troubleshooting	456
Acknowledgments	456
References	456

Contributors

Hayley C. Affronti
Department of Biology, Vassar College, Poughkeepsie, NY, United States

Oluwaseun Akinyele
Division of Genetic and Genomic Medicine, Department of Pediatrics, University of Pittsburgh School of Medicine, UPMC Children's Hospital of Pittsburgh, Pittsburgh, PA, United States

Yuta Ami
Faculty of Biology-Oriented Science and Technology, Kindai University, Kinokawa, Wakayama, Japan

André S. Bachmann
International Center for Polyamine Disorders; International Center for Polyamine Disorders; Department of Pediatrics and Human Development, College of Human Medicine, Michigan State University, Grand Rapids, MI, United States

Sharon Barone
Division of Nephrology, Department of Internal Medicine, University of New Mexico Health Sciences Center Albuquerque; Research Services, New Mexico Veterans Health Care System, Albuquerque, NM, United States

Choukri Ben Mamoun
Department of Internal Medicine, Section of Infectious Diseases, Yale School of Medicine, New Haven, CT, United States

Alexandra Bunea
College of Medicine, University of Central Florida, Orlando, FL, United States

Caleb P. Bupp
Department of Pediatrics and Human Development, College of Human Medicine, Michigan State University; International Center for Polyamine Disorders; Division of Medical Genetics, Corewell Health/Helen DeVos Children's Hospital, Grand Rapids, MI, United States

Nicholas O. Burton
Department of Metabolism and Nutritional Programming, Van Andel Institute, Grand Rapids, MI, United States

Robert A. Casero, Jr.
Department of Oncology, Johns Hopkins University School of Medicine; Sidney Kimmel Comprehensive Cancer Center, Johns Hopkins School of Medicine, Baltimore, MD, United States

Jae-Yeon Choi
Department of Internal Medicine, Section of Infectious Diseases, Yale School of Medicine, New Haven, CT, United States

David W. Christianson
Roy and Diana Vagelos Laboratories, Department of Chemistry, University of Pennsylvania, Philadelphia, PA, United States

John L. Cleveland
Department of Tumor Microenvironment & Metastasis, Moffitt Cancer Center & Research Institute, Tampa, FL, United States

Amin Cressman
Stem Cell Program, Institute for Regenerative Cures, University of California Davis, Sacramento, CA, United States

Thomas E. Dever
Division of Molecular and Cellular Biology, Eunice Kennedy Shriver National Institute of Child Health and Human Development, National Institutes of Health, Bethesda, MD, United States

Justin R. DiAngelo
Division of Science, Penn State Berks, Reading, PA, United States

Thomas C. Dowling
Department of Pharmaceutical Sciences, College of Pharmacy, Ferris State University, Big Rapids, MI, United States

Sydney Drury
Department of Cellular and Molecular Physiology, The Pennsylvania State University, College of Medicine, PA, United States

Christina Efthymiou
Department of Cellular and Molecular Physiology, The Pennsylvania State University, College of Medicine, PA, United States

Fernando A. Fierro
Stem Cell Program, Institute for Regenerative Cures, University of California Davis, Sacramento; Department of Cell Biology and Human Anatomy, University of California Davis, Davis, CA, United States

Jackson R. Foley
Department of Oncology, Johns Hopkins University School of Medicine, Baltimore, MD, United States

Megan Gendjar
Mass Spectrometry Core, Van Andel Institute, Grand Rapids, MI, United States

Alain P. Gobert
Division of Gastroenterology, Hepatology, and Nutrition, Department of Medicine; Center for Mucosal Inflammation and Cancer; Program in Cancer Biology, Vanderbilt University Medical Center, Nashville, TN, United States

Juana Goulart Stollmaier
Roy and Diana Vagelos Laboratories, Department of Chemistry, University of Pennsylvania, Philadelphia, PA, United States

Nikolas Gunkel
German Cancer Research Center (DKFZ), Cancer Drug Development Group, Heidelberg, Germany

Ishika Gupta
Sidney Kimmel Comprehensive Cancer Center, Johns Hopkins School of Medicine, Baltimore, MD, United States

Caroline V. Hawkins
Division of Gastroenterology, Hepatology, and Nutrition, Department of Medicine, Vanderbilt University Medical Center, Nashville, TN, United States

Corey J. Herbst-Gervasoni
Roy and Diana Vagelos Laboratories, Department of Chemistry, University of Pennsylvania, Philadelphia, PA, United States

Cassandra E. Holbert
Department of Oncology, Johns Hopkins University School of Medicine; Department of Biology, Loyola University Maryland, Baltimore, MD, United States

Kazuei Igarashi
Amine Pharma Research Institute, Innovation Plaza at Chiba University, Chiba, Japan

Christine Isaguirre
Mass Spectrometry Core, Van Andel Institute, Grand Rapids, MI, United States

Marie A. Johnson
Division of Genetic and Genomic Medicine, Department of Pediatrics, University of Pittsburgh School of Medicine, UPMC Children's Hospital of Pittsburgh, Pittsburgh, PA, United States

Keiko Kashiwagi
Amine Pharma Research Institute, Innovation Plaza at Chiba University, Chiba, Japan

Dwi U. Kemaladewi
Division of Genetic and Genomic Medicine, Department of Pediatrics, University of Pittsburgh School of Medicine, UPMC Children's Hospital of Pittsburgh, Pittsburgh, PA, United States

Alex R. Khomutov
Engelhardt Institute of Molecular Biology, Russian Academy of Sciences, Moscow, Russia

Maxim A. Khomutov
Engelhardt Institute of Molecular Biology, Russian Academy of Sciences, Moscow, Russia

Shin Kurihara
Faculty of Biology-Oriented Science and Technology, Kindai University, Kinokawa, Wakayama, Japan

Kenneth Lee
Department of Cellular and Molecular Physiology, The Pennsylvania State University, College of Medicine, PA, United States

Teresa L. Mastracci
Department of Biology, Indiana University Indianapolis; Department of Biochemistry and Molecular Biology; Center for Diabetes and Metabolic Diseases, Indiana University School of Medicine, Indianapolis, IN, United States

Kara M. McNamara
Division of Gastroenterology, Hepatology, and Nutrition, Department of Medicine; Program in Cancer Biology, Vanderbilt University Medical Center, Nashville, TN, United States

Aubry K. Miller
German Cancer Research Center (DKFZ), Cancer Drug Development Group, Heidelberg, Germany

Vladimir A. Mitkevich
Engelhardt Institute of Molecular Biology, Russian Academy of Sciences, Moscow, Russia

Shima Nakanishi
Department of Tumor Microenvironment & Metastasis, Moffitt Cancer Center & Research Institute, Tampa, FL, United States

Kelsie M. Nauta
Department of Metabolism and Nutritional Programming, Van Andel Institute, Grand Rapids, MI, United States

Shannon L. Nowotarski
Division of Science, Penn State Berks, Reading, PA, United States

Ashley Nwafor
Department of Oncology, Johns Hopkins University School of Medicine; Sidney Kimmel Comprehensive Cancer Center, Johns Hopkins School of Medicine, Baltimore, MD, United States

Otto Phanstiel IV
College of Medicine, University of Central Florida, Orlando, FL, United States

Schuyler T. Pike
Department of Biological Sciences, College of Arts and Sciences, Ferris State University, Big Rapids, MI, United States

Kim Remans
European Molecular Biology Laboratory (EMBL), Protein Expression and Purification Core Facility, Heidelberg, Germany

Spencer R. Rosario
Department of Biostatistics and Bioinformatics, Roswell Park Comprehensive Cancer Center, Buffalo, NY, United States

Aryn M. Rowsam
Department of Biology, Hartwick College, Oneonta, NY, United States

Arthur I. Salikhov
Engelhardt Institute of Molecular Biology, Russian Academy of Sciences, Moscow, Russia

Chad R. Schultz
Department of Pediatrics and Human Development, College of Human Medicine, Michigan State University; International Center for Polyamine Disorders, Grand Rapids, MI, United States

Peter Sehr
European Molecular Biology Laboratory (EMBL), Chemical Biology Core Facility, Heidelberg, Germany

Ryan D. Sheldon
Mass Spectrometry Core, Van Andel Institute, Grand Rapids, MI, United States

Byung-Sik Shin
Division of Molecular and Cellular Biology, Eunice Kennedy Shriver National Institute of Child Health and Human Development, National Institutes of Health, Bethesda, MD, United States

Pallavi Singh
Department of Internal Medicine, Section of Infectious Diseases, Yale School of Medicine, New Haven, CT, United States

Dominic J. Smiraglia
Department of Cell Stress Biology, Roswell Park Comprehensive Cancer Center, Buffalo, NY, United States

Olga A. Smirnova
Engelhardt Institute of Molecular Biology, Russian Academy of Sciences, Moscow, Russia

Manoocher Soleimani
Division of Nephrology, Department of Internal Medicine, University of New Mexico Health Sciences Center Albuquerque; Research Services, New Mexico Veterans Health Care System, Albuquerque, NM, United States

Raphael R. Steimbach
German Cancer Research Center (DKFZ), Cancer Drug Development Group, Heidelberg, Germany

Tracy Murray Stewart
Department of Oncology, Johns Hopkins University School of Medicine; Sidney Kimmel Comprehensive Cancer Center, Johns Hopkins School of Medicine, Baltimore, MD, United States

Emi Sugiwaka
Faculty of Biology-Oriented Science and Technology, Kindai University, Kinokawa, Wakayama, Japan

Matthew A. Swanson
Department of Biological Sciences, College of Arts and Sciences, Ferris State University, Big Rapids, MI, United States

Carlye Szarowicz
Department of Biological Sciences, College of Arts and Sciences, Ferris State University, Big Rapids, MI, United States

Xianzun Tao
Department of Neurology, University of Chicago, Chicago, IL, United States

Krystal B. Tran
Division of Genetic and Genomic Medicine, Department of Pediatrics, University of Pittsburgh School of Medicine, UPMC Children's Hospital of Pittsburgh, Pittsburgh, PA, United States

Elizabeth A. VanSickle
Division of Medical Genetics, Corewell Health/Helen DeVos Children's Hospital; International Center for Polyamine Disorders, Grand Rapids, MI, United States

Keith T. Wilson
Division of Gastroenterology, Hepatology, and Nutrition, Department of Medicine; Center for Mucosal Inflammation and Cancer; Program in Cancer Biology; Department of Pathology, Microbiology, and Immunology, Vanderbilt University Medical Center; Veterans Affairs Tennessee Valley Healthcare System, Nashville, TN, United States

Quinton W. Wood
Department of Biology, Indiana University Indianapolis, Indianapolis, IN, United States

Kamyar Zahedi
Division of Nephrology, Department of Internal Medicine, University of New Mexico Health Sciences Center Albuquerque; Research Services, New Mexico Veterans Health Care System, Albuquerque, NM, United States

R. Grace Zhai
Department of Neurology, University of Chicago, Chicago, IL, United States

Preface

Although polyamines have be known to exist since Leeuwenhoek described spermine crystals using his newly improved microscope in 1678, the study of polyamines and the recognition of their importance to almost all critical biological processes has been hampered throughout the years by the lack of sufficient methodologies and instrumentation. We are now at a point in time where the availability of advanced technologies and protocols allow the necessary in-depth investigation into the functional, molecular roles that polyamines play in key processes in both normal and diseased cells and systems. In this volume, we have strived to include, in a single source, the most up-to-date, detailed protocols and methods that should allow those interested in studying polyamines to get started in their studies. Although many of the protocols included in this volume arose through the study of various diseases, including cancers and various polyaminopathies, they will also be invaluable in the study of normal processes that involve polyamines.

This volume comes at an opportune time when the recognition that polyamines are not only intimately involved in cell proliferation pathways but are also important in regulating the immune system in both normal and pathological conditions. This recognition has resulted in a resurgence of interest in polyamine biology across several disciplines.

Another recent finding that has changed the trajectory of the study of polyamines is the discovery that the regulation of polyamine metabolism is more complicated than was appreciated just a few short years ago. The discovery that histone deacetylase 10 (HDAC10) is the long sought after N^8-acetylspermidine deacetylase has opened completely new avenues of studies. HDAC10 activity has significant implications in normal physiology as well as in multiple pathologies that may serve as targets for therapy. Accordingly, this volume includes chapters dedicated to HDAC10 structure and enzymology, a chemical ligand-displacement assay for identification of HDAC10 inhibitors and PROTACS, as well as a cell-based method for detecting HDAC10-mediated polyamine deacetylase activity.

Although cancer-related studies have dominated the field of polyamine function and metabolism for the last 30 years, the continuing discovery of various pathologies and syndromes resulting from dysregulated polyamine metabolism has caused the field to become more circumspect in nature. Snyder-Robison syndrome, caused by reduced or lost spermine synthase activity, Bachmann/Bupp syndrome, resulting from excessive ornithine decarboxylase activity, and the various hypusine-related deficiencies all result in potentially debilitating sequalae in humans. Additionally, there are specific polyamine metabolism-related pathologies that result from various physiological stresses, including ischemia/reperfusion and response to specific drug treatments. Consequently, the protocols in this volume cover a broad spectrum of in vitro and in vivo studies, using both human-derived cells and other model organisms, including Drosophila, mice, and yeast, that will provide the research community the means to study many of these processes.

An emerging area of interest in the polyamine field is the interaction between certain pathogenic organisms and the host. These interactions can be the response to parasitic infections or even the carcinogenic transformation of effected tissues. These areas are also covered here.

Finally, as polyamine transport, function and metabolism continue to be rational therapeutic targets for an expanding array of pathologies, detailed synthetic methods have been included for some of the most recently reported polyamine-modulating compounds. The recent FDA approval of difluoromethylornithine (DFMO) as a maintenance therapy for high-risk neuroblastoma patients has strengthened its potential as a component of anticancer combination therapies as well as its repurposing for other polyamine-related conditions. Detailed methods for quantifying DFMO in plasma as well as improved techniques for measuring individual polyamines, their derivatives, and reaction products are presented that are applicable to a variety of sample types, including model systems and clinical samples.

In all, we hope this volume will be of assistance to those interested in pursuing studies in the field of polyamine metabolism and function, while also facilitating the adoption of updated methodologies by those already established in the field. We anticipate that by incorporating chapters on topics ranging from basic polyamine biochemistry to analysis of human patient samples, this compilation will aid in continuing the recent resurgent interest in this field.

Robert A. Casero, Jr., Ph.D.
Tracy Murray Stewart, Ph.D.

CHAPTER ONE

Genetic analyses of Myc and hypusine circuits in tumorigenesis

Shima Nakanishi and John L. Cleveland*

Department of Tumor Microenvironment & Metastasis, Moffitt Cancer Center & Research Institute, Tampa, FL, United States
*Corresponding author. e-mail address: John.Cleveland@moffitt.org

Contents

1. Introduction	2
2. Rationale	4
3. Analyses of roles of MYC and hypusinated eIF5A in tumor development	5
3.1 Genetic approaches to assess effects on tumor development	5
3.2 Isolation and analyses of primary, premalignant B cells	6
4. Analyses of tumor maintenance and progression	8
4.1 Tumorigenic potential – targeting Myc and the hypusine axis with small molecules	8
4.2 Tumorigenic potential – gene silencing studies	10
4.3 Assessment of tumor maintenance by subcutaneous tumor transplantation	12
5. Concluding remarks	15
Acknowledgements	16
References	16

Abstract

A prominent metabolic pathway induced by MYC family oncoproteins in cancer is the polyamine-hypusine circuit, which post-translationally modifies a specific lysine residue of eukaryotic translation initiation factor 5 A (eIF5A) with a unique amino acid coined hypusine [N$^\varepsilon$-(4-amino-2-hydroxybutyl)lysine]. This modification occurs in a two-step process, whereby the aminobutyl group of the polyamine spermidine is covalently linked to lysine-50 of eIF5A via deoxyhypusine synthase (DHPS) to form the intermediate deoxyhypusinated eIF5A, which is subsequently hydoxylated by deoxyhypusine hydroxylase (DOHH) to form the fully mature eIF5A^{Hyp}. As a result, eIF5A^{Hyp} is elevated in MYC-driven cancers.

Recently it has become evident that eIF5A^{Hyp} (i) plays key roles in the development, progression and maintenance of tumors; and (ii) eIF5A^{Hyp} functions are often tissue/cell context-specific. Thus, it is important to mechanistically assess how eIF5A^{Hyp} affects normal cells and tumorigenesis using suitable in vivo and ex vivo models. In this chapter, we describe the methods used in our laboratory to assess the effects of

MYC-polyamine-hypusine axis on the development and maintenance of MYC-driven B-cell lymphoma. The goals of this chapter are twofold. First, we discuss genetic and cell biological approaches that can be applied to assess roles of eIF5A^{Hyp} on lymphoma and normal B cell development. Second, we discuss methods that can be used to assess the roles of eIF5A^{Hyp} on the growth and maintenance of lymphoma. Collectively, these approaches provide a template that can be applied to evaluate roles of any putative regulator of the development and/or maintenance of lymphoma.

1. Introduction

MYC is a master regulatory basic helix-loop-helix leucine zipper (bHLH-Zip) transcription factor that directs cell growth (mass), cell division/proliferation and metabolism (Bello-Fernandez, Packham, & Cleveland, 1993; Nakanishi et al., 2023; Nilsson et al., 2005). Notably, via alterations in signaling circuits, amplification or chromosomal translocations, *MYC*, or its orthologs *MYCN* and *MYCL*, are overexpressed in more than 50 % of cancers, and this leads to dysregulation of MYC-downstream signaling pathways, and to the selection for malignant clones that evolve various means to bypass cell-cycle and apoptotic checkpoints during tumor development and/or progression. For example, malignant conversion in MYC-driven lymphoma is driven by events that include loss-of-function mutations in the ARF-p53 tumor suppressor pathway (Eischen, Weber, Roussel, Sherr, & Cleveland, 1999), and a myriad of mutations, copy number alterations (amplifications and deletions) and chromosome rearrangements (e.g., translocations, inversions) drive the expression of genes that cooperate with MYC in B cell transformation, such as the anti-apoptotic protein BCL2 and the transcription factor BCL6 that is required for the formation and the function of germinal centers (Ranuncolo et al., 2007).

A recent revelation is the role of MYC in post-transcriptional control of tumorigenesis. While this includes MYC-directed control of short-lived mRNAs via the induction of mRNA-destabilizing factors such as tristetraprolin (TTP) (Rounbehler et al., 2012), recent studies have shown that MYC also controls translation via its control of the polyamine-hypusine metabolic circuit, which is critical for the translational control of oncogenic circuits that are necessary for the development and maintenance of MYC-driven lymphoma (Nakanishi et al., 2023). First, MYC drives the biosynthesis of the polyamines putrescine, spermidine and spermine via transcriptional induction of all three biosynthetic enzymes, ornithine decarboxylase (ODC), spermidine synthase (SRM) and spermine synthase (SMS) (Pegg, 2016) (Fig. 1), leading to elevated levels of polyamines in cancer cells (Durie, Salmon, & Russell, 1977; Russell, 1971). Furthermore, MYC also induces the expression of two enzymes, deoxyhypusine synthase (DHPS) and deoxyhypusine hydroxylase (DOHH), that use the amino-butyl group of spermidine to post-translationally and covalently modify a single lysine residue (Lys-50 in humans) of eukaryotic translation initiation factor, eIF5A, with the amino acid hypusine, via a process coined hypusination (Park, Cooper, & Folk, 1981) (Fig. 1). Importantly, mechanistic studies have demonstrated that hypusinated eIF5A (eIF5A^{Hyp}) promotes translation elongation by alleviating stalled ribosomes at specific sequence motifs that present structural constraints (Gutierrez et al., 2013). Additionally, eIF5A^{Hyp} also facilitates translation start codon selection and termination (Manjunath et al., 2019; Schuller, Wu, Dever, Buskirk, & Green, 2017).

Recent studies investigating the roles of eIF5A^{Hyp} in various tissues, cell types, and physiological contexts have revealed surprising tissue and/or cell context specific roles of eIF5A^{Hyp} (Nakanishi & Cleveland, 2024). For example, eIF5A^{Hyp} is essential for the malignant conversion of MYC-overexpressing B cells and for the development and maintenance of MYC-driven lymphoma (Nakanishi et al., 2023). In contrast, eIF5A^{Hyp} activity is necessary to prevent chronic inflammation and

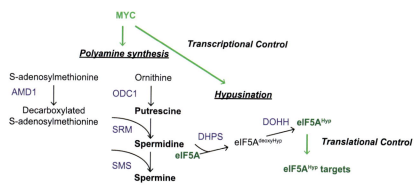

Fig. 1 Transcriptional regulation of the polyamine-hypusine circuit by MYC. Genes encoding the enzymes (blue font) involved in polyamines-hypusine circuit are frequently upregulated by MYC in various cancers. Moreover, levels of eIF5A is also elevated in cancers by MYC and/or other signaling pathways, resulting in increased levels of eIF5A^{Hyp}. eIF5A^{Hyp} acts as a downstream effector of MYC, modulating the translation of key regulators, including those controlling the cell cycle, thereby contributing to MYC-driven tumorigenesis.

carcinogenesis of intestinal epithelium cells, suggesting it plays a tumor suppressive role in this context (Gobert et al., 2023). Even more strikingly, eIF5A^{Hyp} has been reported to be necessary for the translation of TFEB, a master regulator of the autophagy pathway, in normal B cells (Zhang et al., 2019), yet MYC inhibits TFEB expression and functions in MYC-driven mouse and human B cell lymphoma where MYC augments expression of the entire polyamine-hypusine circuit.

2. Rationale

To define lymphoma and cell context specific-intrinsic roles of the polyamine-hypusine circuit, detailed genetic analyses using several in vivo and ex vivo models are necessary. For example, ex vivo depletion of ODC, the first enzyme of the polyamine biosynthesis, or knockdown or pharmacological inhibition of the end product of the circuit, eIF5A^{Hyp}, both significantly impair B-lymphoma cell growth, yet our analyses indicate that they likely do so through distinct mechanisms. Indeed, although genetic depletion studies indicate that ODC has intrinsic roles in contributing to malignant transformation, MYC-driven lymphoma can occur in the absence of ODC yet not in the absence of DHPS, which we have shown is essential for lymphomagenesis (Nakanishi et al., 2023).

As an in vivo tool to assess the roles of effector and modulators of MYC-driven lymphomagenesis we and others have used the Eμ-*Myc* transgenic mouse, which was generated as a model of human Burkitt lymphoma (BL) (Adams et al., 1985) where the *MYC* gene on chromosome 8 is translocated near the enhancers present in immunoglobulin heavy chain (*IgH*) on chromosome 14 (Adams et al., 1985) or near enhancers for the kappa (*IgK*) or lambda (*IgL*) light chain gene loci on chromosome 2 or 22, respectively. Hemizygous Eμ-*Myc* mice almost invariably develop lymphomas, with the majority of mice succumbing to the disease within the first 4–6 months of life. Indeed, the median survival of Eμ-*Myc* transgenic mice of 100–130 days is reproducible among different researchers, allowing a direct comparison of the results when assessing the effects of loss or gain of function of different targets on the rates of disease progression and metastasis. In this chapter, we describe the genetic and pharmacological approaches that can be applied to study roles of the polyamine-hypusine circuit in MYC driven lymphoma.

3. Analyses of roles of MYC and hypusinated eIF5A in tumor development

It is critical to choose the appropriate standard, conditional or knock-in mouse model, as well as the appropriate Cre recombinase-expressing transgenic or knock-in model to control tissue-specific knockout of *loxP*-flanked alleles, or to activate the expression of transgenes that are flanked by transcriptional stop cassettes in the desired tissue or cell types of interest. To understand tumor cell-intrinsic effects of the hypusine circuit on MYC-driven B cell lymphoma, we utilized the $Dhps^{fl/fl}$ conditional knockout mouse (Levasseur et al., 2019) crossed to Eμ-*Myc* transgenic mice that bear a knock-in of the *Cre recombinase* gene into the B cell-specific *Cd19* gene.

3.1 Genetic approaches to assess effects on tumor development

3.1.1 Equipment
1. CO_2 regulator system for CO_2 asphyxiation
2. Surgical knife
3. Forceps
4. Table scale

3.1.2 Reagents
1. Eμ-Myc transgenic mice (Adams et al., 1985)
2. $Dhps^{fl/fl}$ conditional knockout mouse (Levasseur et al., 2019)
3. CD19-Cre mice (Rickert, Roes, & Rajewsky, 1997) (The Jackson Laboratory, 006785)

3.1.3 Procedure
1. Backcross all mice to the C57BL/6J strain for at least 10 generations.
2. Cross floxed *Dhps* mice with hemizygous Eμ-Myc transgenic mice and knock-in CD19-Cre mice to generate Eμ-*Myc*;CD19-*Cre*;*Dhps*$^{fl/fl}$ mice, along with the littermates carrying only one floxed *Dhps* allele (Eμ-*Myc*;CD19-*Cre*;*Dhps*$^{fl/+}$) or wild-type *Dhps*.
3. Monitor mice of each cohort daily for signs of disease.
4. At an end point, document the age of the mice, the number and location of lymphomas per mouse and metastases of lymphoma to other organs, which for the Eμ-*Myc* transgenic model include the liver, lung and brain.
5. Harvest the lymphoma from moribund mice at an endpoint for expression and molecular analyses. Cut the lymphoma into small pieces and store them in an Eppendorf tube at −80 °C.

6. Analyze survival for each cohort using Kaplan-Meier survival curves and determine the median survival.
7. Determine statistical significance using the Mantel-Cox log-rank test.
8. In addition to the survival study, pool additional mice to follow several phenotypes during the course of disease. See 3.2 for the isolation of bone marrow (BM) cells and primary B-cells from each of the cohorts (aged 5-6 weeks) for the B-cell profiling analysis or analysis of gene and protein expression of interest (e.g. *Dhps*, *Myc*, *Cre*).
9. Measure the weight of the spleen to assess splenomegaly, which is caused by rampant bone marrow infiltration of lymphoma cells and extra-medullary hematopoiesis.

3.1.4 Notes
1. It is essential that all mice used are backcrossed onto the same genetic background as overall survival time (including median survival) can vary depending on the genetic background.
2. It is important to backcross mice under study to wild type mice (e.g., in our studies C57BL/6J) from a commercial source (such as Jackson Laboratories) to prevent genetic drift.

3.2 Isolation and analyses of primary, premalignant B cells
3.2.1 Equipment
1. 6-well plates (CytoOne, CC7682-7506)
2. 23-gauge needles (BD, PrecisionGlide-Needle)
3. 3-mL syringes (BD, 309557)
4. Cell strainer (Fisherbrand, 22-363-549)
5. Centrifuge (Eppendorf, 5810R and 5418)
6. QuadroMACS separator (Miltenyi Biotech, 130-091-051)
7. MACS Stand (Miltenyi Biotech, 130-042-303)
8. MACS Column (Miltenyi Biotech, 130-042-202)

3.2.2 Reagents
1. Mice generated in 3.1
2. Phosphate-buffered saline (PBS, Corning, 21-031-CV)
3. Anti-CD19 MicroBeads (Miltenyi Biotech, 130-121-301)

3.2.3 Procedure
1. Euthanize the mice using CO_2 asphyxiation, followed by cervical dislocation. Ensure that the euthanasia method complies with your institute's animal ethics regulations.

2. Place a 6-well plate on ice and add 2 mL of ice-cold PBS (or HBSS for primary cell culture) to a well.
3. Remove the femur and tibia from both legs, clean the connective tissue and muscle off, and place the bones in the well on ice.
4. Add 1 mL of ice-cold PBS (or HBSS for primary cell culture) to another well.
5. Flush the bone marrow out of the bones using 3-mL syringe with a 23-gauge needle and 1 mL of PBS, collecting the flush into the well.
6. Filter the bone marrow suspension through a 100 μm cell strainer into a 50-mL conical tube to remove debris.
7. (Optional) To improve cell yields, collect the vertebrae. Clean the spine and cut it into pieces, remove the spinal cord, and place vertebrae in ice-cold PBS in a mortar on ice. Crush the vertebrae using a pestle until the debris appears white, indicating that all bone marrow cells have been released (red). Combine this bone marrow suspension with the suspension from the femur and tibia.
8. Centrifuge at 300 × g for 5 min at 4 °C, and then discard the supernatant.
9. Treat the pellet with 5 mL 1X Red blood cell lysis solution (prepared from 10X stock: 8.02 g ammonium chloride, 0.84 g sodium bicarbonate, 0.37 g disodium EDTA in 100 mL water) at room temperature for no more than 5 min to lyse the red blood cells.
10. Add PBS to stop the lysis, centrifuge at 300 × g for 5 min at 4 °C, and discard supernatant.
11. Resuspend the pellet in 45 mL PBS, centrifuge again, and discard the supernatant (second wash).
12. Resuspend the pellet in 10 mL of PBS, count the cells and assess viability using the trypan blue exclusion method.
13. Centrifuge and discard the supernatant (third wash). Resuspend the pellets in PBS, and the cell concentration will be adjusted based on the previous count.
14. (Optional) Label the cells with specific sets of lymphocyte markers for profiling by flow cytometry (e.g., an assessment of the percentage of B cells in the BM sample).
15. For B-cell isolation using a MACS Separator and magnetic separation column, resuspend the pelleted cells with ice-cold MACS buffer (2 mM EDTA, 0.5 % of bovine serum albumin in PBS) after the second wash.
16. Filter the suspension through a 70 μm cell strainer into a 50 mL Falcon tube

17. Centrifuge, discard the supernatant, and resuspend the pellet in MACS buffer at 90 μL per 10 million cells.
18. Add MicroBeads conjugated with CD19 or B220 antibodies (Miltenyi Biotech, 10 μL per 10 million cells) and incubate for 15 min at 4 °C.
19. Place the column in the MACS separator, rinse with 3 mL MACS buffer.
20. Apply the suspension onto the column.
21. Wash the column with an additional 3 mL of MACS buffer three times.
22. Remove the column and elute the CD19$^+$ B cells with 5 mL of buffer.
23. Count the purified B-cells and aliquot them for downstream procedures, such as the isolation of genomic DNA, RNA and protein.

3.2.4 Notes
To assess the potential effects of the *Dhps* deficiency on the premalignant state, harvest bone marrow-derived B cells from the all cohorts of experimental mice following weaning, a time at which most Eμ-*Myc* mice lack signs of disease. However, note that some of Eμ-*Myc*;CD19-*Cre*;*Dhps*$^{fl/+}$ mice have early onset disease and develop the tumor within 6 weeks.

4. Analyses of tumor maintenance and progression

To study effects of targeting polyamine-hypusine circuit on lymphoma maintenance and progression *in* vivo, intravenous tumor transplantation (e.g., via tail vein injection) and subcutaneous tumor transplantation are two common methods used to introduce tumors into mice.

4.1 Tumorigenic potential – targeting Myc and the hypusine axis with small molecules

To assess the effects of pharmacologic inhibition of ODC or DHPS on lymphoma tumorigenic potential, lymphoma-bearing mice were treated with the ODC suicide inhibitor, DFMO (α-difluoromethylornithine) or with GC7 (N$_1$-guanyl-1,7-diaminoheptane), a spermidine analog and competitive inhibitor of DHPS (Jakus, Wolff, Park, & Folk, 1993).

4.1.1 Equipment
1. Microwave
2. Microwavable mug cup

3. Restraint device or 50 mL falcon tube with a hole fixed on the tube rack
4. Cotton swabs

4.1.2 Reagents
1. 6-7 week-old C57BL/6J mice (The Jackson Laboratory, 000664)
2. Fresh or frozen Eμ-*Myc* lymphoma, isolated from Eμ-*Myc* mice
3. PBS
4. α-difluoromethylornithine (DFMO, provided by Dr. Patrick Woster, University of South Carolina, MUSC)
5. N_1-guanyl-1,7-diaminoheptane (GC7, Biosearch Technologies, G1000-100)
6. Microwaved warm water
7. 27 mm gauge needles (BD, PrecisionGlide-Needle)
8. 1-mL syringes (BD)

4.1.3 Procedure
1. Thaw the frozen Eμ-*Myc* lymphoma cells and suspend the cells in PBS at 1×10^7 cells/mL.
2. Restrain the mouse securely, either using a restraint device or 50 mL falcon tube with a hole for the tail. Fix the falcon tube on the top of the tube rack using a rubber band so it will remain in place.
3. Using a cotton swab dipped into the microwaved water, warm the tail of the mouse. This helps dilate the tail veins and improve visibility of the tail vein.
4. Inject 1 million cells (100 μL of the lymphoma cell suspension) into the tail vein of each mouse (6-7 week-old C57BL/6J mice) using a 27 mm gauge needle.
5. Randomize the recipient mice into three cohorts: vehicle, DFMO, or GC7.
6. Administer DFMO in drinking water (1% wt/vol) starting one day following lymphoma transplant.
7. Three days after transplant, treat the cohorts with GC7 (4 or 20 mg/kg) or vehicle (PBS) via daily intraperitoneal injection until animals reach the endpoint.

4.1.4 Notes
In our past studies, treatment with GC7 as low as 4 mg/kg reduced tumorigenic potential and significantly extended survival of the recipient mice beyond 20 days (Nakanishi et al., 2023). Furthermore, treatment with 20 mg/kg GC7 did not cause any overt health issues.

4.2 Tumorigenic potential – gene silencing studies

4.2.1 Equipment
1. Equipment used for transplant experiments described in 4.1.1
2. Cell strainer (Fisherbrand, 22-363-549)
3. Centrifuge (Eppendorf, 5810R and 5418)
4. 24-well plate (Corning 3524)
5. 6-well plate (CytoOne, CC7682-7506)

4.2.2 Reagents
1. Eμ-*Myc* transgenic mice (Adams et al., 1985)
2. Rosa26-rt*TA*^{2} mice (Hochedlinger, Yamada, Beard, & Jaenisch, 2005)
3. NU/J, athymic nude mice (The Jackson Laboratory, 002019)
4. Eco Pack cells (Clontech 631507)
5. Doxycycline (Dox, ThermoFisher, NC0424034)
6. Doxycycline Chow Diet (Envigo, 200mg/kg)
7. ProFection Mammalian Transfection System (Promega, E1200)
8. Polybrene (Sigma Aldrich, H9268)
9. TET system approved FBS (Takara, 631106)
10. G418 (AdipoGen Life Sciences, AG-CN2-0030)
11. Anti-eIF5A (Abcam, ab32443)

4.2.3 Procedure
1. Cross Eμ-*Myc* transgenic mouse with the Rosa26-rt*TA*^{2} mouse that express a transgene encoding an optimized form of the reverse tetracycline-controlled transactivator (rtTA^2), which allows doxycycline (Dox)-inducible control of transgenes (Hochedlinger et al., 2005).
2. Harvest the lymphoma from the Eμ-*Myc*;Rosa26-rt*TA*^{2} mouse and establish the lymphoma cell lines.
3. To establish the rtTA-expressing lymphoma cell lines cut the tumors into small pieces.
4. Place these tumors in 100 μm cell strainer on the 50 mL falcon tube.
5. Add 1 mL of PBS and homogenize the tumors by gently pressing them using the back of 3 mL syringe.
6. Wash the tumor suspension using PBS, centrifuge at $300 \times g$ for 5 min, and discard the supernatant. Repeat the wash step three times.
7. Resuspend a single-cell suspension in 45% DMEM, 45% IMDM, supplemented with 10% FBS, 1% penicillin/streptomycin (100X), 4 mmol/L GlutaMax, 25 μmol/L β-mercaptoethanol, 1 mmol/L

sodium pyruvate, and 5 ng/mL mouse IL7 and place them in 24-well plate at different concentrations (0.5–10 million cells/mL).
8. Culture cells at 37 °C with 5% CO_2 in the tissue culture incubator. Depending on the cell expansion rate, gradually shift the plate size (12-, 6- well plates). This process may take up to ∼ 5–8 weeks.
9. To establish shRNA-expressing lymphoma cell lines, prepare the retrovirus or lentivirus via transfection of Eco Pack 293 or 293 T cells, respectively, with a retroviral or lentiviral vector that expresses the shRNA of interest (e.g., shRNA that targets mouse *Eif5a* or *Renilla* luciferase as a control (Nakanishi et al., 2023)) as well as the envelope (Env) and packaging plasmids. Note that Eco Pack cells express the MMLV Gag, Pol, and Env proteins. Various transfection methods (calcium phosphate-, lipofectamine-, or polyethylenimine-mediated transfection system) are widely used and are commercially available. Follow manufacturer's instructions.
10. For retroviral transduction, collect the virus 48–72 hr after transfection and filter them through 0.45 μm filter.
11. Place the 1.5 mL lymphoma cells at 0.5–1 million cells per mL and the 1.5 mL virus solution in the 6-well plate and add polybrene (8 μg/mL).
12. Perform a spinning infection by centrifugation of the plate at 2500 rpm for 80 min at room temperature.
13. Place infected cells in the tissue culture incubator.
14. Remove 1.5 mL media and add the fresh media next day as polybrene can be toxic to the cells.
15. Select the successfully infected cells either by cell sorting on a flow cytometer (BD, FACSArea) or by supplementation with antibiotics appropriate for the drug resistance gene (if any).
16. (Optional) Use the all-in-one inducible shRNA vector containing rtTA, instead of using the Rosa26-rtTA^2 mouse noted above.
17. Check the degree of knockdown of the target protein by immunoblots prior to performing subsequent ex vivo cell biological and in vivo lymphoma transplant studies.
18. (Pilot experiment) Determine the optimal amounts of the newly established shRNA-expressing lymphoma cells needed for the transplant experiments to induce a robust tumorigenic response. Inject 01, 0.25, 0.5, 0.75, 1 and 1.5 million cells into the tail veins of 6-7 week-old nude or C57BL/6J mice and monitor these mice daily for disease progression.

19. Purchase 5-week old nude mice ($n = 10$ per each cohort) and let them adjust to the environment of the mouse facility for at least a week. (Transplant experiment) Inject the lymphoma cells expressing Dox-inducible sh-*Eif5a* or sh-*Renilla* (control) construct via intravenous tail vein.
20. Three days after the transplant, give half of these mice a Dox chow diet daily until endpoint.
21. Harvest the lymphoma at the endpoint and store it in −80 °C.
22. Immunoblot the protein of interest to test if depletion of the target protein is sustained throughout the experiments.

4.2.4 Notes
1. Depending on the type of your study, immunodeficient mice may be necessary, and this is required when injecting human cancer cell lines or when transplanting highly immunogenic mouse tumor lines.
2. It is recommended to use an shRNA system in which GFP-shRNA polycistronic transcripts are expressed, which allows one to assess the retention of shRNA expression by monitoring GFP expression using flow cytometry.
3. Silencing the expression of shRNAs targeting oncogenes or essential genes is a common selection mechanism that allows tumors to override the effects of the shRNA. To predict this possibility, it is recommended to assess the expression GFP-shRNAs over time in the cell lines ex vivo prior to performing the in vivo experiments.

4.3 Assessment of tumor maintenance by subcutaneous tumor transplantation

To test the therapeutic potential of targeting eIF5A^{Hyp} in vivo, tumor maintenance and growth can be examined following subcutaneous tumor transplantation, which allows direct measurement of tumor growth.

4.3.1 Equipment
1. Aria flow cytometer (BD Biosciences)
2. IVIS spectrum 200 (PerkinElmer)
3. 24-well plate (Corning 3524)

4.3.2 Reagents
1. Eμ-*Myc* transgenic mouse (Adams et al., 1985)
2. Rosa26-LSL-*CreER*^{T2} mouse (Ventura et al., 2007) (The Jackson Laboratory, 008463)

3. $Dhps^{fl/fl}$ conditional knockout mouse (Levasseur et al., 2019)
4. Polybrene (Sigma Aldrich, H9268)
5. 4-hydroxytamoxifen (4-OHT, Sigma Aldrich, T5648)
6. Ethanol (Sigma Aldrich, 459844)
7. pRetroX-Tight-MCS-PGK-*OgNLuc* (Schaub et al., 2015)
8. Matrigel Matrix (Corning 354234)
9. Tamoxifen (Sigma Aldrich, T5648-1G)
10. Corn oil (Sigma Aldrich, C8267)
11. 2-furanylmethyl-deoxy-coelenterazine (furimazine) in the Nano-Glo Luciferase Assay System kit, (Promega, N1120)
12. Recombinant human Interleukin-7 (IL-7, Bio-techne, 407-ML)

4.3.3 Procedure

1. Cross the $Dhps^{fl/fl}$ conditional knockout mouse to Eμ-*Myc* and Rosa26-*CreER*^{T2} mice to generate the desired Eμ-*Myc*;Rosa26-*CreER*^{T2};$Dhps^{fl/fl}$ cohort along with control (Eμ-*Myc*;Rosa26-*CreER*^{T2};$Dhps^{+/+}$) littermate mice (Fig. 2).
2. Harvest the lymphoma from the Eμ-*Myc*;Rosa26-rt*TA*^{2}; $Dhps^{fl/fl}$ and the Eμ-*Myc*;Rosa26-*CreER*^{T2};$Dhps^{+/+}$ and establish lymphoma cell lines (See 4.2.3 for establishment of lymphoma cell lines).
3. To characterize the established inducible lymphoma cell lines, treat the cells ex vivo with increasing concentrations of 4-hydroxytamoxifen (4-OHT), which activates Cre-ER^{T2} recombinase activity or with ethanol (EtOH, the vehicle) to test and optimize the efficiency of *Dhps* deletion.
4. Confirm Cre-mediated deletion of *Dhps* at both the gene and protein levels as well as loss of hypusination of eIF5A by performing genomic DNA (gDNA)-PCR and immunoblotting.
5. Transduce the established cells with pRetroX-Tight-MCS-PGK-*OgNLuc* retrovirus, which expresses the LSSmOrange-NanoLuc reporter.
6. Sort the transduced LSSmOrange$^+$ cells using Aria flow cytometer.
7. (Transplant experiment) Suspend the cells in a 1:1 mixture of PBS and Matrigel (Corning).
8. Implant the cell suspension (1 million cells) subcutaneously into the right or left flanks of 6-week-old *nude* (or other immunocompromised) mice.
9. Monitor tumor growth of each mouse daily by measuring tumor volumes. Measure the greatest longitudinal diameter (L), the transverse diameter (W), and the height (H) using calipers, and calculate the volume using the formula, volume = $\pi/6$ (L × W × H) (Tomayko & Reynolds, 1989).

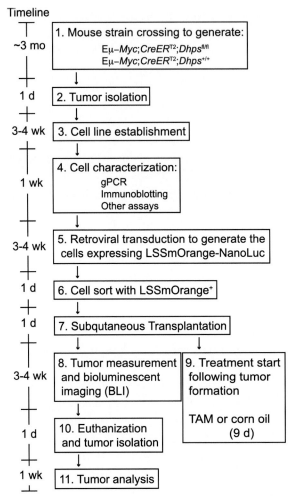

Fig. 2 Flow chart depicting the investigation of the effects of eIF5A^{Hyp} loss on the maintenance of lymphoma. The principal steps shown in the right panel and the corresponding timeline shown in the left panel that are necessary to achieve the goal are illustrated.

10. Monitor tumor growth twice weekly by bioluminescence imaging (BLI). Measure bioluminescence signals following intraperitoneal injection of furimazine using IVIS spectrum 200 and quantify them using Living Image software (PerkinElmer).
11. Once a palpable tumor is formed, randomize the mice by assigning each mouse alternatively to either tamoxifen or vehicle (corn oil) treatment cohort and administer the pre-determined dose daily by oral gavage (e.g., 100−145 mg/kg tamoxifen).

12. Treat Eμ-*Myc*;Rosa26-*CreER*^{T2};*Dhps*$^{fl/fl}$ lymphoma-bearing recipient mice with tamoxifen or corn oil for 9 consecutive days and terminate the treatment.
13. Treat the Control mice bearing Eμ-*Myc*;Rosa26-*CreER*^{T2};*Dhps*$^{+/+}$ lymphomas similarly and monitor them in parallel.
14. Harvest the lymphoma at the endpoint and store it in −80 °C.
15. Perform immunoblotting for the protein of interest (e.g., DHPS, eIF5A^{Hyp}) to test if depletion of the target protein is retained.

4.3.4 Notes

1. *Dhps* deletion should not be observed in the uninduced (EtOH-treated) Eμ-*Myc*;Rosa26-Cre-ER^{T2};*Dhps*$^{fl/fl}$ lymphoma cells, nor following 4-OHT or EtOH treatment of the control Eμ-*Myc*;Rosa26-*CreER*^{T2};*Dhps*$^{+/+}$ lymphomas.
2. It is important that oral gavage treatment should be stopped if the tumor lysis syndrome becomes severe.

5. Concluding remarks

The Eμ-*Myc* transgenic mouse is an invaluable tool to study MYC-driven lymphoma. To explore the polyamine-hypusine circuit as a potential therapeutic target for B-cell aggressive lymphoma with MYC involvement, this model, the conditional *Odc* or *Dhps* knockout mice, and pharmacological inhibitors were used to determine the contribution of polyamine biosynthesis versus the hypusine circuit on the development, progression and maintenance of lymphoma.

Ex vivo approaches of studies examining the effects of loss of function using various lymphoma cell lines are key tools to define the mechanism by which identified targets and effectors contribute to the development and maintenance of cancer. However, such ex vivo models have limitations in assessing long-term effects of targeting a suspected vulnerability or pathway. On the other hand, long term effects observed in genetically engineered mouse models can provide important platforms to affirm biological and clinical relevance. Therefore, both approaches are complementary, and both should be applied to provide a comprehensive mechanistic understanding of the therapeutic potential of high priority targets.

Acknowledgements

We thank members of the Cleveland laboratory, past and present, for their many contributions to all the MYC related studies. This work was supported by R01 grants CA100603 and CA241713 (J.L.C.), by NCI Comprehensive Cancer Center Grant P30 CA076292, by the Cortner-Couch Endowed Chair for Cancer Research from the University of South Florida School of Medicine (J.L.C.), and by monies from the State of Florida to the H. Lee Moffitt Cancer Center & Research Institute.

References

Adams, J. M., Harris, A. W., Pinkert, C. A., Corcoran, L. M., Alexander, W. S., Cory, S., ... Brinster, R. L. (1985). The c-myc oncogene driven by immunoglobulin enhancers induces lymphoid malignancy in transgenic mice. *Nature, 318*(6046), 533–538.

Bello-Fernandez, C., Packham, G., & Cleveland, J. L. (1993). The ornithine decarboxylase gene is a transcriptional target of c-Myc. *Proceedings of the National Academy of Sciences of the United States of America, 90*(16), 7804–7808.

Durie, B. G., Salmon, S. E., & Russell, D. H. (1977). Polyamines as markers of response and disease activity in cancer chemotherapy. *Cancer Research, 37*(1), 214–221.

Eischen, C. M., Weber, J. D., Roussel, M. F., Sherr, C. J., & Cleveland, J. L. (1999). Disruption of the ARF-Mdm2-p53 tumor suppressor pathway in Myc-induced lymphomagenesis. *Genes & Development, 13*(20), 2658–2669.

Gobert, A. P., Smith, T. M., Latour, Y. L., Asim, M., Barry, D. P., Allaman, M. M., ... Wilson, K. T. (2023). Hypusination maintains intestinal homeostasis and prevents colitis and carcinogenesis by enhancing aldehyde detoxification. *Gastroenterology, 165*(3), 656–669 e8.

Gutierrez, E., Shin, B. S., Woolstenhulme, C. J., Kim, J. R., Saini, P., Buskirk, A. R., & Dever, T. E. (2013). eIF5A promotes translation of polyproline motifs. *Molecular Cell, 51*(1), 35–45.

Hochedlinger, K., Yamada, Y., Beard, C., & Jaenisch, R. (2005). Ectopic expression of Oct-4 blocks progenitor-cell differentiation and causes dysplasia in epithelial tissues. *Cell, 121*(3), 465–477.

Jakus, J., Wolff, E. C., Park, M. H., & Folk, J. E. (1993). Features of the spermidine-binding site of deoxyhypusine synthase as derived from inhibition studies. Effective inhibition by bis- and mono-guanylated diamines and polyamines. *The Journal of Biological Chemistry, 268*(18), 13151–13159.

Levasseur, E. M., Yamada, K., Pineros, A. R., Wu, W., Syed, F., Orr, K. S., ... Mirmira, R. G. (2019). Hypusine biosynthesis in beta cells links polyamine metabolism to facultative cellular proliferation to maintain glucose homeostasis. *Science Signaling, 12*(610).

Manjunath, H., Zhang, H., Rehfeld, F., Han, J., Chang, T. C., & Mendell, J. T. (2019). Suppression of ribosomal pausing by eIF5A Is necessary to maintain the fidelity of start codon selection. *Cell reports, 29*(10), 3134–3146.e6.

Nakanishi, S., & Cleveland, J. L. (2024). The many faces of hypusinated eIF5A: Cell context-specific effects of the hypusine circuit and implications for human health. *International Journal of Molecular Sciences, 25*(15).

Nakanishi, S., Li, J., Berglund, A. E., Kim, Y., Zhang, Y., Zhang, L., ... Cleveland, J. L. (2023). The polyamine-hypusine circuit controls an oncogenic translational program essential for malignant conversion in MYC-driven lymphoma. *Blood Cancer Discov, 4*(4), 294–317.

Nilsson, J. A., Keller, U. B., Baudino, T. A., Yang, C., Norton, S., Old, J. A., ... Cleveland, J. L. (2005). Targeting ornithine decarboxylase in Myc-induced lymphomagenesis prevents tumor formation. *Cancer Cell, 7*(5), 433–444.

Park, M. H., Cooper, H. L., & Folk, J. E. (1981). Identification of hypusine, an unusual amino acid, in a protein from human lymphocytes and of spermidine as its biosynthetic precursor. *Proceedings of the National Academy of Sciences of the United States of America, 78*(5), 2869–2873.

Pegg, A. E. (2016). Functions of polyamines in mammals. *The Journal of Biological Chemistry, 291*(29), 14904–14912.

Ranuncolo, S. M., Polo, J. M., Dierov, J., Singer, M., Kuo, T., Greally, J., ... Melnick, A. (2007). Bcl-6 mediates the germinal center B cell phenotype and lymphomagenesis through transcriptional repression of the DNA-damage sensor ATR. *Nature Immunology, 8*(7), 705–714.

Rickert, R. C., Roes, J., & Rajewsky, K. (1997). B lymphocyte-specific, Cre-mediated mutagenesis in mice. *Nucleic Acids Research, 25*(6), 1317–1318.

Rounbehler, R. J., Fallahi, M., Yang, C., Steeves, M. A., Li, W., Doherty, J. R., ... Cleveland, J. L. (2012). Tristetraprolin impairs myc-induced lymphoma and abolishes the malignant state. *Cell, 150*(3), 563–574.

Russell, D. H. (1971). Increased polyamine concentrations in the urine of human cancer patients. *Nature: New Biology, 233*(39), 144–145.

Schaub, F. X., Reza, M. S., Flaveny, C. A., Li, W., Musicant, A. M., Hoxha, S., ... Amelio, A. L. (2015). Fluorophore-nanoluc BRET reporters enable sensitive in vivo optical imaging and flow cytometry for monitoring tumorigenesis. *Cancer Research, 75*(23), 5023–5033.

Schuller, A. P., Wu, C. C., Dever, T. E., Buskirk, A. R., & Green, R. (2017). eIF5A functions globally in translation elongation and termination. *Molecular Cell, 66*(2), 194–205 e5.

Tomayko, M. M., & Reynolds, C. P. (1989). Determination of subcutaneous tumor size in athymic (nude) mice. *Cancer Chemotherapy and Pharmacology, 24*(3), 148–154.

Ventura, A., Kirsch, D. G., McLaughlin, M. E., Tuveson, D. A., Grimm, J., Lintault, L., ... Jacks, T. (2007). Restoration of p53 function leads to tumour regression in vivo. *Nature, 445*(7128), 661–665.

Zhang, H., Alsaleh, G., Feltham, J., Sun, Y., Napolitano, G., Riffelmacher, T., ... Simon, A. K. (2019). Polyamines control eIF5A hypusination, TFEB translation, and autophagy to reverse B cell senescence. *Molecular Cell, 76*(1), 110–125.e9.

CHAPTER TWO

Expression, purification, and crystallization of "humanized" *Danio rerio* histone deacetylase 10 "HDAC10", the eukaryotic polyamine deacetylase

Juana Goulart Stollmaier, Corey J. Herbst-Gervasoni, and David W. Christianson[*]

Roy and Diana Vagelos Laboratories, Department of Chemistry, University of Pennsylvania, Philadelphia, PA, United States
*Corresponding author. e-mail address: chris@sas.upenn.edu

Contents

1. Introduction	20
2. Construct design	23
3. Transformation, expression, and purification	23
3.1 Equipment	23
3.2 Materials	24
3.3 Transformation	25
3.4 Expression and purification	26
4. Activity assay using a colorimetric method	30
4.1 Equipment	30
4.2 Materials	31
4.3 Assay setup	31
4.4 Notes	32
5. Crystallization of HDAC10-inhibitor complexes	32
5.1 Equipment	34
5.2 Materials	34
5.3 Procedure	35
5.4 Notes	37
6. Summary	38
References	38

Abstract

The class IIb histone deacetylase HDAC10 is responsible for the deacetylation of intracellular polyamines, in particular N^8-acetylspermidine. HDAC10 is emerging as an attractive target for drug design owing to its role as an inducer of autophagy, and high-resolution crystal structures enable structure-based drug design efforts. The only crystal structure available to date is that of HDAC10 from *Danio rerio* (zebrafish), but a construct containing the A24E and D94A substitutions yields an active site contour that more closely resembles that of human HDAC10. The use of this "humanized" construct has advanced our understanding of HDAC10-inhibitor structure-activity relationships. Here, we outline the preparation, purification, assay, and crystallization of humanized zebrafish HDAC10-inhibitor complexes. The plasmid containing the humanized zebrafish HDAC10 construct for heterologous expression in *Escherichia coli* is available through Addgene (#225542).

1. Introduction

Polyamines such as putrescine, spermidine, and spermine are small polycations essential for cell growth and proliferation (Tabor & Tabor, 1984). Since aberrant polyamine metabolism is often observed in cancer cells, enzymes of polyamine metabolism can serve as attractive targets for cancer chemotherapy (Casero, Murray Stewart, & Pegg, 2018). Polyamine acetylation plays a prominent role in eukaryotic polyamine metabolism, and the generation and degradation of N^1-acetylated forms of spermine and spermidine have been studied extensively. An isomer of N^1-acetylspermidine, N^8-acetylspermidine, also sustains polyamine metabolism in the cell cytosol (Desiderio, 1992; Libby, 1980).

The hydrolysis of N^8-acetylspermidine to yield spermidine and acetate in cytosolic extracts from rat liver was reported nearly 50 years ago (Blankenship, 1978; Libby, 1978). This activity was specific for the hydrolysis of N^8-acetylspermidine but not N^1-acetylspermidine (Marchant, Abu Manneh, & Blankenship, 1986), and selective inhibition of N^8-acetylspermidine deacetylase activity in mice and HeLa cells was reported a few years later (Marchant et al., 1989); importantly, this activity was distinct from histone deacetylase activity. It was not until 2017 that the zinc metalloenzyme histone deacetylase 10 (HDAC10) was identified as the eukaryotic N^8-acetylspermidine deacetylase (Fig. 1) (Hai, Shinsky, Porter, & Christianson, 2017); this activity was subsequently confirmed in human neuroblastoma cell line BE(2)-C (Steimbach et al., 2022).

Since the biological substrate of HDAC10 is a small molecule and not a protein, it is clear that the "histone deacetylase" nomenclature is inadequate to describe the greater family of HDAC enzymes. While first discovered to play a role in reversible histone acetylation-deacetylation cycles (Allfrey, Faulkner, & Mirsky, 1964; Gallwitz, 1968; Inoue & Fujimoto, 1969), these

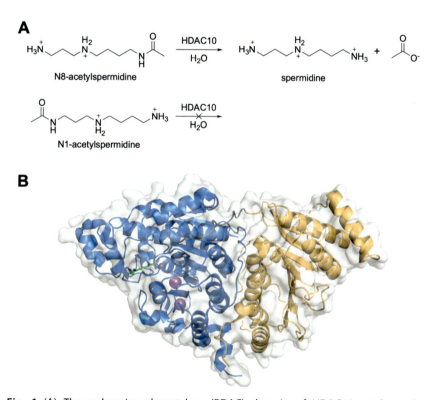

Fig. 1 (A) The polyamine deacetylase (PDAC) domain of HDAC10 catalyzes the hydrolysis of N^8-acetylspermidine, but not N^1-acetylspermidine, to form spermidine and acetate. (B) Crystal structure of H137A HDAC10 complexed with the intact substrate, N^8-acetylspermidine (PDB 7KUS). The substrate bound to the active site in the PDAC domain (blue) is represented by a green stick-figure; the catalytic Zn^{2+} ion is a small gray sphere and structural K^+ ions are purple spheres. The pseudo-deacetylase (ΨDAC) domain (dark yellow) has no known catalytic function.

enzymes were later shown to deacetylate myriad protein and small-molecule substrates in various cellular compartments (Hai et al., 2017; Narita, Weinert, & Choudhary, 2019; Patel, Pathak, & Mujtaba, 2011; Toro & Watt, 2020). The first zinc-dependent human histone deacetylase (HDAC1) was identified in Schreiber's laboratory (Taunton, Hassig, & Schreiber, 1996), and ten additional isozymes were subsequently discovered and classified by phylogenetic analysis (Gregoretti, Lee, & Goodson, 2004): class I (HDAC1, HDAC2, HDAC3, and HDAC8), class IIa (HDAC4, HDAC5, HDAC7, and HDAC9), class IIb (HDAC6 and HDAC10), and class IV (HDAC11). These enzymes exhibit substantial diversity in terms of subcellular localization, substrate specificity, and catalytic activity. For example, HDAC6 serves as a tubulin deacetylase in the cytosol and a lamin decrotonylase in the nucleus (Hubbert et al., 2002; Zhang et al., 2022). In another example, HDAC11 is not a lysine deacetylase, but instead is a lysine-fatty acid deacylase (Cao et al., 2019; Kutil et al., 2018; Moreno-Yruela, Galleano, Madsen, & Olsen, 2018). Lacking a better name for this enzyme family that encompasses such broad diversity of function, we cautiously retain use of the historical term "histone deacetylase" despite it being a misnomer.

Owing to their role in transcriptional regulation and epigenetics, HDACs are validated targets for cancer chemotherapy (Barneda-Zahonero & Parra, 2012; Cress & Seto, 2000; Glozak & Seto, 2007; Lane & Chabner, 2009), and more recently Duchenne muscular dystrophy (Mercuri et al., 2024; Mullard, 2024). At present, five pan-HDAC inhibitors (i.e., inhibitors that do not selectively inhibit one isozyme over another) have been approved by the U.S. Food and Drug Administration for therapeutic use: Givinostat, Vorinostat, Panobinostat, Belinostat, and Romidepsin (Bondarev et al., 2021; Rim, Karas, Barada, Dean, & Levitsky, 2024). Similar structural features shared by HDAC active sites challenge the design of selective inhibitors targeting one isozyme or another. However, the detailed structural characterization of HDAC-inhibitor complexes illuminates active site features that can be effectively exploited to this end.

As the cytosolic polyamine deacetylase, HDAC10 is unique among HDACs in that its native substrate is a small molecule and not a protein (Hai et al., 2017). HDAC10 is also unique because it contains a catalytically active polyamine deacetylase (PDAC) domain and a non-functional, vestigial pseudo-deacetylase (ΨDAC) domain (Fig. 1) (Hai et al., 2017). Two structural features in the PDAC active site confer narrow substrate specificity: (1) the $P^{23}(E,A)CE^{26}$ ("PEACE") motif helix, which constricts the active site so as to favor the binding of a slender substrate, and (2) the carboxylate group of E274, which serves as a gatekeeper by forming water-mediated hydrogen bonds with the secondary ammonium group of N^8-acetylspermidine but not N^1-acetylspermidine (Hai et al., 2017; Herbst-Gervasoni & Christianson, 2021; Shinsky & Christianson, 2018).

HDAC10 has been identified in multiple organisms, including model organisms such as zebrafish (*Danio rerio*) (The UniProt Consortium, 2023). Although zebrafish HDAC10 shares only 46% sequence identity with human HDAC10, these orthologues exhibit identical catalytic activity and polyamine substrate specificity (Hai et al., 2017). Zebrafish HDAC10 is readily crystallized and is thus established as a reference point for understanding structure-function relationships (Hai et al., 2017). Additionally, a "humanized" version of zebrafish

HDAC10 containing the A24E and D94A substitutions yields an active site contour that better mimics that of human HDAC10, and this variant has advanced X-ray crystallographic studies of HDAC10-inhibitor complexes (Christianson, 2024; Herbst-Gervasoni, Steimbach, Morgen, Miller, & Christianson, 2020; Steimbach et al., 2022; Zeyen et al., 2022).

In this chapter, we outline the preparation of "humanized" zebrafish HDAC10 (hereafter simply "HAC10"), the measurement of catalytic activity, and the general approach to cocrystallization of HDAC10-inhibitor complexes. We have been unable to prepare the PDAC domain as a standalone domain, likely due to the stabilization afforded by its interaction with the ΨDAC domain. However, the interdomain linker can be nicked with trypsin without disrupting PDAC-ΨDAC association, and this proteolysis step induces crystallization.

2. Construct design

The plasmid encoding zebrafish HDAC10 (Uniprot Q803K0, residues 2–675) was initially synthesized and codon-optimized by Genscript and subcloned into a modified pET28a(+) vector (a gift from Dr Scott Gradia, University of California, Berkeley; Addgene plasmid 29656) in-frame with a TEV-cleavable (TEV, tobacco etch virus) N-terminal 6xHis-MBP-tag (MBP, maltose binding protein) (Fig. 2) (Hai et al., 2017). The "humanized" construct was generated by introducing the A24E and D94A substitutions into this plasmid (Herbst-Gervasoni et al., 2020). The plasmid encoding humanized *D. rerio* HDAC10 can be obtained from Addgene (plasmid 225542) and transformed into *E. coli* BL21(DE3) for protein expression.

3. Transformation, expression, and purification
3.1 Equipment
1. Eppendorf™ Research™ Plus Pipettors
2. Milli-Q® Integral Water Purification System for Ultrapure Water (EMD Millipore; #ZRXQ003WW)
3. Isotemp™ incubator (Fisher Scientific)
4. Digital dry heat bath (USA Scientific #2510-1102)
5. 12 × 2-L and 2 × 250-mL baffled flasks
6. AMSCO® 250LS Small Steam Sterilizer (STERIS Healthcare)

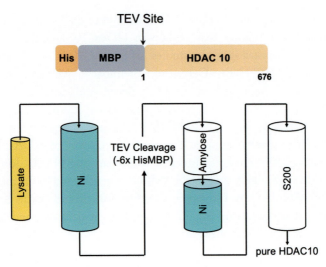

Fig. 2 (Top) HDAC10 expression construct: His, 6xHis purification tag; MBP, maltose binding protein; TEV, tobacco etch virus protease. The "humanized" construct contains S24E and D94A mutations. (Bottom) General purification scheme.

7. New Brunswick™ Innova® 40 R benchtop incubator shaker (Eppendorf #M1299-0094)
8. CO8000 Cell Density Meter (Biochrom #80-3000-45)
9. Revco® Ultima PLUS −80°C freezer (or comparable; Thermo Scientific)
10. Sorvall™ LYNX 6000 Superspeed Centrifuge (Thermo Fisher #75006590)
11. Transilluminator FBTIV-88 (Fisher Scientific)
12. XCell SureLock Mini-Cell (EI0001)
13. Polar Ware 250B Stainless Steel Griffin Style Beaker 250 mL Capacity (Stoelting #1526Q30EA)
14. Magnetic Stirrer RT Basic (Thermo Fisher Scientific)
15. Q700 Sonicator (Qsonica #Q700-110)
16. ÄKTAprime Plus FPLC (GE Healthcare #11001313)
17. Cytiva HisTrap™ HP Prepacked Column (5 mL) (GE Healthcare)
18. Cytiva MBPTrap™ HP Prepacked Column (5 mL) (GE Healthcare)
19. HiLoad™ Superdex™ 26/600 200 pg column (GE Healthcare)
20. PowerPac™ basic gel machine (Bio-Rad)
21. 50-mL and 150-mL Superloop (GE Healthcare)

3.2 Materials

1. Thermo Scientific™ S1 Pipet Fillers
2. MiniSpin® plus (Eppendorf)

3. BL21 (DE3) competent *E. coli* (New England BioLabs #C2527H)
4. Fisher BioReagents™ 2xYT (Fisher Scientific, #BP97435)
5. Kanamycin monosulfate, USP grade (GoldBio #K-120-5)
6. Isopropyl β-D-1-thiogalactopyranoside (IPTG) (GoldBio #I2481C)
7. Sodium chloride (Fisher Chemical #S271-500)
8. TCEP HCl (GoldBio #TCEP1)
9. Imidazole (Fisher Chemical #03196-500)
10. HEPES, sodium salt (GoldBio #H-401)
11. D-(+)-Maltose monohydrate (Fisher BioReagents, #BP684)
12. Aluminum foil
13. Fisher BioReagents™ Glycerol (Fisher Scientific #BP229-4)
14. Fisherbrand™ Disposable PES Bottle Top Filters (Fisher Scientific #SCGPT05RE)
15. Lysozyme, from egg white (GoldBio #L-040-1)
16. Benzonase® Nuclease (MilliporeSigma #E1014)
17. cOmplete™, Mini, EDTA-free Protease Inhibitor Cocktail Tablets (MilliporeSigma #04693159001)
18. NuPAGE™ 4-12% Bis-Tris Gel (Invitrogen #NP0323)
19. Spectra/Por® 1 Dialysis Membrane 6-8 kD (Repligen #132665)
20. Millex™ Sterile 0.22 μM PVDF Syringe Filter (MilliporeSigma #SLGVM33RS)
21. 1x MES-SDS Running Buffer, provided as a 20X stock (Invitrogen #NP0002)
22. Amicon® Ultra Centrifugal Filters – 30 kD molecular weight cut-off (EMD Millipore #UFC901024)
23. MilliporeSigma™ Ultrafree™ -MC Centrifugal Filters (0.22 μM)

3.3 Transformation

1. Retrieve an aliquot containing 50 μL of BL21 (DE3) Competent *E. coli* (NEB C2527H) cells from storage at −80 °C and thaw on ice for 10 min.
2. Add 1 μL of the HDAC10 gene inserted into pET28a(+) plasmid DNA to the thawed *E. coli* cells. Gently flick the tube several times and place back on ice. Incubate for 30 min.
3. Using the digital dry bath, heat shock the cells by placing the tube at 42 °C for 10 s and immediately place the cells back on ice for 5 min.
4. Add 950 μL of SOC medium (included with the NEB BL21 (DE3) *E. coli* cells).
5. Incubate the cells at 37 °C with shaking at 250 rpm for 1 h.

6. Ensure that the cells are well suspended and then dilute 1:100 in SOC medium and apply 50 μL of the diluted cells onto a pre-warmed LB-agar plate (supplemented with 50 μg/mL kanamycin); using aseptic technique, spread the cells evenly onto the plate and then place into a 37 °C incubator upside down.
7. Incubate the plated cells at 37 °C for 16–18 h and then store by sealing the plate with Parafilm® and placing at 4 °C for storage until needed for inoculation of overnight cultures.

3.4 Expression and purification

1. Remove the plates from the incubator and place them in the cold room with parafilm.
2. Separate glassware and media for autoclave. Prepare two 500-mL Fernbach flasks with 250 mL of 2xYT medium for the starter culture, and six 2-L Fernbach flasks for 1-L cultures.
3. Autoclave the glassware with media. Let it cool down before adding the antibiotic stock and the inoculation (50 mg/L kanamycin).
4. Pick one BL21 DE3 colony from the previously transformed agar plate and inoculate each Fernbach flask to generate starter cultures. Place flasks in the shaker overnight at 37 °C, 250 rpm.
5. Add 5 mL of the starter culture to each 1-L of 2xYT.
6. Grow at 37 °C at 250 rpm until OD_{600} reaches 0.6–0.8 and then lower the shaker temperature to 16 °C; stop shaking while samples cool to 16 °C.
7. Once the temperature reaches 16 °C, add 250 mM $ZnSO_4$ and 200 mM IPTG, in this order, to 1 L of culture. Grow at 16 °C at 250 rpm for 18 h.
8. Centrifuge growth medium and cells at $5422\,g$ at 4 °C for 15 min.
9. Collect cell pellets, determine mass, and store at −80 °C until purification. Approximately 30 g of cell pellet is expected for a typical 6-L growth.
10. The general purification procedure is summarized in Fig. 2. Prepare buffer solutions (Table 1: Loading Buffer A, Elution Buffer B, Elution Buffer C, and Size Exclusion (SE) Buffer) using ultrapure water. Make 5 L of A, 0.5 L of B, 0.5 L of C and 2 L of SE. Make only 80 % of the volume and do not add glycerol yet. Let the solutions sit at 4 °C overnight. The pH and volume will be adjusted on the following day when the buffers are at 4 °C.

Table 1 Buffer compositions for HDAC10 purification.

Loading Buffer A	Elution Buffer B	Elution Buffer C	Size Exclusion Buffer
50 mM HEPES	50 mM HEPES	50 mM HEPES	50 mM HEPES
300 mM KCl	300 mM KCl	300 mM KCl	300 mM KCl
2 mM TCEP•HCl	2 mM TCEP•HCl	2 mM TCEP•HCl	1 mM TCEP•HCl
10 µM ZnSO$_4$	10 µM ZnSO$_4$	10 µM ZnSO$_4$	5 % glycerol
30 mM imidazole	500 mM imidazole	500 mM imidazole	
10 % glycerol	10 % glycerol	10 % glycerol 10 mM maltose	

11. Adjust the pH of the buffer solutions to 7.5 while cold and add the remaining glycerol and water to correct the final volume. Vacuum filter each buffer through a separate bottle-top filter, except for the 4 L of buffer A. Store the buffer bottles at 4 °C.
12. Retrieve the frozen cell pellet and allow it to thaw in a beaker with water. Add 3 mL/g of loading buffer A to the cell pellet and pour the suspension into a metal beaker.
13. Add 2 protease inhibitor tablets, 0.4 mg/mL lysozyme, and 2.8 units/mL benzonase nuclease and stir for 45 min at 4 °C, 260 rpm.
14. Remove magnetic stir bar and put metal beaker containing cell pellet suspension on ice.
15. Sonicate for 8 min (amplitude = 30, 1 s on/2 s off cycle).
16. Centrifuge lysate at 18,000 rpm for 1 h at 4 °C. In the meantime, perform a system wash of the FPLC and equilibrate a 5-mL HisTrap column with loading buffer (buffer A).
17. Load lysate supernatant into the 150-mL superloop and inject lysate supernatant onto the HisTrap column (1 mL/min). Collect 12-mL fractions.
18. When the entire volume of lysate is injected onto the column, switch the FPLC to load and allow the A280 plateau to drop to approximately 0 (Fig. 3; this will take approximately 40 min).
19. Begin a 120-mL gradient with a maximum concentration of elution buffer (buffer B) to reach 100 % (remain at 1 mL•min-1). Collect 5-mL fractions.

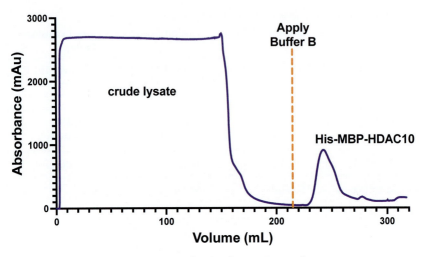

Fig. 3 Representative chromatogram for the first HisTrap column.

20. Collect and pool appropriate fractions under the A280 peak (Fig. 3) and confirm the presence of MBP-HDAC10 with SDS-PAGE.
21. Add 6 mg of TEV protease to the pooled fractions.
22. Dialyze in 6–8 kDa dialysis tubing in 2 L of loading buffer A while stirring in a 2-L graduated cylinder overnight.
23. Dispose of dialysis buffer and replace with 2 L of the remaining loading buffer and allow to dialyze for at least another 5 h. In the meantime, perform a FPLC system wash and equilibrate the size exclusion column with SE buffer.
24. After dialysis, centrifuge the protein solution at 18,000 rpm for 20 min then load it into the 50-mL superloop.
25. Equilibrate a tandem MBPTrap-HisTrap column (both 5 mL, Cytiva) with loading buffer A. Be sure that the MPBTrap column is above the HisTrap column.
26. Inject the TEV-digested protein solution onto tandem MBPTrap-HisTrap column and collect the flowthrough (1 mL/min).
27. Once the entire volume of the protein solution is injected onto the column, switch the FPLC to load. Allow for the A280 to reach approximately 0.
28. Set elution buffer C to 100 % and elute off proteins from tandem column (Fig. 4). This can be done at 3 mL/min to save time.
29. The fractions are stored at 4 °C overnight to be pooled on the following day.

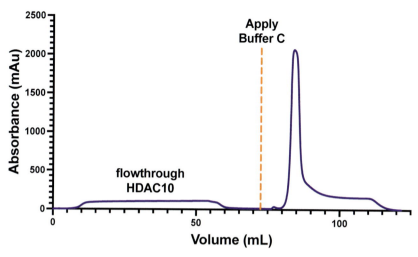

Fig. 4 Representative chromatogram for the MBPTrap-HisTrap tandem setup.

30. Pool the flowthrough and concentrate to at least 13 mL. Use Amicon Ultra centrifugal filter 30 kDa, centrifuge it at 4000 rpm during 15–20 min cycles.
31. Filter protein solution into a 50-mL superloop pre-rinsed with water and then SE buffer. Note: at this point, the protein will easily crash out of solution if there is a dramatic change in salt concentration. Not all protein will crash out, but there will be a noted "string"-like substance that is likely protein if the superloop is only rinsed with water and not SE buffer.
32. Inject concentrated flowthrough onto a HighLoad Superdex 200 column (1 mL/min). Switch to load after and set flow to 2.3 mL/min.
33. Collect and pool appropriate fractions under A280 (Fig. 5) and confirm HDAC10 purity by SDS-PAGE.
34. Concentrate protein to around 15 mg/mL. Don't filter the protein solution.
35. Flash-cool protein samples in liquid N2 (in individual PCR tubes) in volumes that will result in a 10 mg/mL HDAC10 solution when diluted to 30 μL. For example, if concentrated to 12.5 mg/mL, add 24 mL of the HDAC10 solution to the PCR tube so that 6 μL of SE buffer can be added to achieve 10 mg/mL HDAC10 concentration.
36. Store at −80 °C for further use.

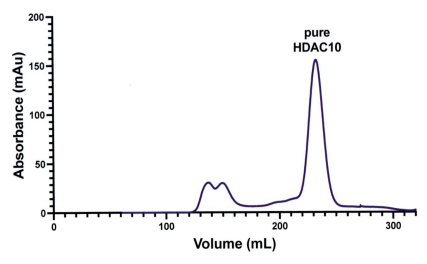

Fig. 5 Representative chromatogram for the size exclusion column.

4. Activity assay using a colorimetric method

The HDAC10 sample prepared as described in the previous section is suitable for use in a variety of biochemical and biophysical assays, e.g., isothermal titration calorimetry measurements of inhibitor binding (Herbst-Gervasoni et al., 2020), thermal shift assay (Goulart Stollmaier, Watson, & Christianson, 2024; Herbst-Gervasoni et al., 2020), and measurements of catalytic activity (Hai et al., 2017). An acetylpolyamine hydrolysis assay using liquid chromatography-mass spectrometry has been used to measure HDAC10 activity with non-fluorogenic substrates (Herbst-Gervasoni & Christianson, 2019). Below, we outline an alternative colorimetric method for measuring HDAC10 activity using a peroxidase enzymatic cascade (Fig. 6). This method was adapted from a published protocol (Holt & Palcic, 2006). We previously utilized this assay to study a bacterial polyamine deacetylase (Lombardi et al., 2011).

4.1 Equipment
1. Infinite® M1000Pro plate reader (Tecan)
2. Eppendorf™ Research™ Plus Pipettors
3. Isotemp™ incubator (Fisher Scientific)
4. Sorvall ST 8 Small Benchtop Centrifuge (ThermoFisher #75007203)

Expression, purification, and crystallization of HDAC10

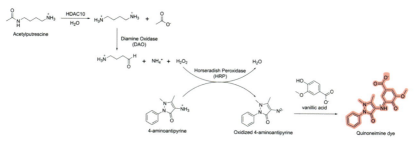

Fig. 6 Reaction scheme of the peroxidase enzymatic cascade and formation of the quinoneimine dye indicator to quantify polyamine deacetylase activity.

4.2 Materials

1. Thermo Scientific™ S1 Pipet Fillers
2. HEPES, sodium salt (GoldBio #H-401)
3. Sodium chloride (Fisher Chemical #S271-500)
4. Potassium chloride (Thermo Scientific Chemicals #AA1159530)
5. Calcium chloride dihydrate (Fisher BioReagents #BP510-500)
6. Magnesium chloride hexahydrate (Fisher BioReagents #BP214-500)
7. N-Acetylputrescine hydrochloride (Sigma-Aldrich A8784-25MG)
8. 4-Aminoantipyrene (Supelco #06800-25G)
9. Vanillic acid (Sigma-Aldrich #94770-10G)
10. Peroxidase from horseradish (Sigma-Aldrich #P8250-5KU)
11. Diamine oxidase from porcine kidney (Sigma-Aldrich #D7876-250MG)
12. 96-well Microplate (Greiner Bio-One #655101)
13. HDAC10 protein stored in SEC buffer
14. Tubastatin A (MedChemExpress #HY-13271A)

4.3 Assay setup

1. Prepare the assay buffer by making a solution with final concentration of 100 mM HEPES, 10 mM KCl, 4.0 mM CaCl$_2$ and 2.8 mM MgCl$_2$. Adjust the buffer pH to 7.4 with NaOH and calculate the amount of NaOH used (usually 40 mM) before adding NaCl so the final sodium concentration in solution is 280 mM.
2. Prepare a stock solution of 150 µM N-acetylputrescine hydrochloride in water.
3. For inhibition control: Solubilize Tubastatin A in DMSO to a final concentration of 40 mM. Use this DMSO stock for further dilutions in water. Make a dilute stock solution of 100 µM Tubastatin A in water.
4. Prepare the chromogenic solution by dissolving 4-aminoantipyrine (500 µM) and vanillic acid (1 mM) in assay buffer. Add horseradish

peroxidase (4 U/mL) and diamine oxidase (1 U/mL). Mix the solution carefully; diamine oxidase tends to foam when mixed vigorously.
5. Dilute HDAC10 to 0.5 μM in assay buffer.
6. On the 96-well microplate, add in the following order:
   i. 50 μL of 0.5 μM HDAC10 in assay buffer,
   ii. 50 μL of 100 μM Tubastatin (for inhibitor control; if not using inhibitor, add only 50 μL water),
   iii. 50 μL chromogenic solution in assay buffer, and
   iv. 50 μL of 150 μM N-acetylputrescine hydrochloride in water (substrate).
7. Quickly spin the microplate in the benchtop centrifuge to remove bubbles (30 s at 300 rpm).
8. Cover the microplate and place it in the incubator at 37 °C for 1 h.
9. Run absorbance readings at 498 nm on the plate reader. For wells exhibiting HDAC10 activity, a faint pink/red color should be evident. Colorless wells indicate no HDAC10 activity.

4.4 Notes

Always test the chromogenic solution before setting up the microplate by taking a small aliquot and mixing it with 30% hydrogen peroxide. If the solution quickly turns to a deep red color, then the chromogenic solution is active and can be used for the assay. If air bubbles are present in the wells, the absorbance readings will be incorrect and inconsistent.

Assays should be run in triplicate. It is best to run each trial with an inhibitor control, protein control, and buffer control.

5. Crystallization of HDAC10-inhibitor complexes

To date, 37 crystal structures of HDAC10 have been deposited and released in the Protein Data Bank (PDB). The resolutions of these structures range from 2.0 Å (PDB accession code 6WBQ) to 2.90 Å (PDB accession code 6UIJ). Of note, residues in the ΨDAC domain typically exhibit higher thermal B factors compared to residues in the PDAC domain. Increased thermal motion manifests as less well-defined electron density for the ΨDAC domain as compared to the PDAC domain.

A 48-residue linker connects the PDAC and ΨDAC domains of HDAC10 (Hai et al., 2017). It is postulated that this linker exhibits a high degree of flexibility which inhibits crystallization. Trypsin is added in a

1:1000 (trypsin:HDAC10) molar ratio to HDAC10 just prior to crystallization in order to cleave the linker. Varying lengths of linker remnants are observed in crystal structures; however, the entire linker generally cannot be modeled in electron density maps, suggesting that some degree of linker cleavage is necessary for crystallization.

The HDAC10 protein crystallizes exclusively in trigonal space group $P3_121$. The propensity of HDAC10 to crystallize only in this space group hints at specific interactions that must occur between HDAC10 monomers in the unit cell as well as in adjacent unit cells to initiate nucleation of a crystal. This construct of HDAC10 crystallizes exclusively in phosphate buffers supplemented with PEG 3350 (Fig. 7), and phosphate anions are observed to bridge HDAC10 monomers in the crystal. These phosphate-mediated interactions seem to be pH-dependent, as phosphate buffer solutions will yield crystals at some but not all pH values. A range of pH values is sampled by altering the ratio of monobasic to dibasic phosphate as well as changing the counterion in the formulation of precipitant buffer solutions.

Despite having developed a robust understanding of the conditions required for HDAC10 crystallization, crystals will not always form, even with a condition known to yield crystals. Many times, the same condition must be repeated multiple times to yield HDAC10 crystals. Due to this uncertainty, the HDAC10 Redundancy Screen has been developed (discussed below). The HDAC10 Redundancy Screen samples phosphate buffers at different pH values with different cations and repeated conditions, thereby increasing the overall probability of crystal formation.

If the HDAC10 Redundancy Screen is unsuccessful yielding crystals, seeding drops with HDAC10 crystal seeds can be effective. Generally, HDAC10-inhibitor/substrate structures that yield the highest resolution

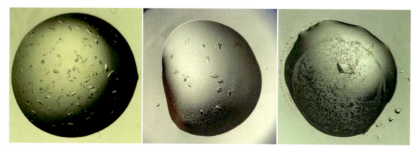

Fig. 7 Examples of diamond-shaped crystals of HDAC10-inhibitor complexes.

structures provide the best seed stock. The general practice for collecting HDAC10 crystals seed stock is to cocrystallize HDAC10 with Tubastatin A (Herbst-Gervasoni, et al., 2020) in the HDAC10 Redundancy Screen and utilize a Crystal Crusher (Hampton Research) to create crystal fragments. The crystal fragments are collected with precipitant buffer from the well and can be stored at −80 °C (see Note 1). Direct addition of this seed stock to each drop yields superior results compared to streak seeding.

5.1 Equipment
1. Eppendorf™ Research™ Plus Pipettors
2. MiniSpin® plus centrifuge (Eppendorf #022620207)
3. Mosquito® crystallization robot (TTP Labtech)
4. 4 °C cold room or EchoTherm™ Benchtop Incubator (Torrey Pines Scientific #IN55)
5. S8APO KL300 LED Microscope (Leica Microsystems)

5.2 Materials
1. HDAC10 in size exclusion buffer
2. Size exclusion buffer (50 mM HEPES (pH 7.5), 300 mM KCl, 1 mM TCEP, 5% glycerol)
3. Ultrafree®-MC centrifugal filter units (EMD Millipore UFC30GV00)
4. UV-transmissible polymer MRC 2-drop crystallization plates (Molecular Dimensions #MD1100U0100)
5. 1 mg/mL Trypsin (1 mM HCl, 2 mM $CaCl_2$ dihydrate) (Hampton Research #HR2-429-02)
6. HDAC10 Redundancy Screen comprised of:
    a. 0.2 M sodium phosphate monobasic monohydrate, 20% w/v polyethylene glycol 3350 (Hampton Research #HR2-922-39)
    b. 0.2 M sodium phosphate dibasic dihydrate, 20% w/v polyethylene glycol 3350 (Hampton Research #HR2-922-40)
    c. 0.2 M potassium phosphate monobasic, 20% w/v polyethylene glycol 3350 (Hampton Research #HR2-922-41)
    d. 0.2 M potassium phosphate dibasic, 20% w/v polyethylene glycol 3350 (Hampton Research #HR2-922-42)
    e. 0.2 M ammonium phosphate monobasic, 20% w/v polyethylene glycol 3350 (Hampton Research #HR2-922-43)
    f. 0.2 M ammonium phosphate dibasic, 20% w/v polyethylene glycol 3350 (Hampton Research #HR2-922-44)
7. Ethylene glycol (Hampton Research #HR2-621)

8. 5-μL and 2-μL micro-reservoir strips (sptlabtech #4150-03100 and #4150-03110)
9. Heavy duty packing clear tape

5.3 Procedure

5.3.1 Crystallization tray
1. Remove frozen aliquot of purified HDAC10 from −80 °C freezer and quickly thaw by holding tube in gloved hand just before all ice melts.
2. Put aliquot in ice bucket.
3. Increase volume to 30 μL by adding SE buffer (yields 10 mg/mL HDAC10 solution).
4. Add inhibitor or substrate to desired final concentration (generally, a 40–50 mM stock solution in DMSO or water is utilized) and let incubate for 1 h on ice.
5. Add 0.518 μL of 2 mg/mL trypsin in 1 mM HCl and 2 mM $CaCl_2$ to the 30 μL aliquot ensuring sufficient mixing by aspirating HDAC10 solution into pipette tip used to dispense trypsin solution multiple times while stirring with the pipette tip (see Section 5.4, Note 2 below).
6. Allow the HDAC10/trypsin solution to sit at room temperature for 1 h.
7. Filter protein solution utilizing 0.22 μm spin filters centrifuged at 10,000 rpm at 4 °C for 3 min.
8. Keep protein solution on ice until ready.
9. Pipette 80 μL of HDAC10 Redundancy Screen (described below) into appropriate wells of sitting drop crystallization plates.
10. Load crystal tray into appropriate position on Mosquito crystallization robot (or similar liquid handling instrument) and pipette protein into micro-reservoir strips (as well as seed stock if applicable to program).
11. Initiate crystallization tray program.
12. Seal crystal tray with clear tape.
13. Store crystal tray at 4 °C.
14. If successful, crystals should form in approximately one day.

5.3.2 HDAC10 redundancy screen
1. Obtain 24×50 mL disposable conical vials.
2. Number vials sequentially 1–24.
3. To each vial add reagents as outlined in Table 2.
4. Mix each vial well to ensure homogeneity (see Note 4).

Table 2 HDAC10 redundancy screen for crystallization trials.

Precipitant Solution	0.2 M Sodium Phosphate Monobasic Monohydrate, 20 % w/v Polyethylene Glycol 3350 (mL)	0.2 M Sodium Phosphate Dibasic Dihydrate, 20 % w/v Polyethylene Glycol 3350 (mL)
1	40.0	0.0
2	35.0	5.0
3	30.0	10.0
4	25.0	15.0
5	20.0	20.0
6	15.0	25.0
7	10.0	30.0
8	5.0	35.0
	0.2 M Potassium Phosphate Monobasic, 20 % w/v Polyethylene Glycol 3350 (mL)	0.2 M Potassium Phosphate Dibasic, 20 % w/v Polyethylene Glycol 3350 (mL)
9	40.0	0.0
10	35.0	5.0
11	30.0	10.0
12	25.0	15.0
13	20.0	20.0
14	15.0	25.0
15	10.0	30.0
16	5.0	35.0
	0.2 M Ammonium Phosphate Monobasic, 20 % w/v Polyethylene Glycol 3350 (mL)	0.2 M Ammonium Phosphate Dibasic, 20 % w/v Polyethylene Glycol 3350 (mL)
17	40.0	0.0
18	35.0	5.0
19	30.0	10.0

20	25.0	15.0
21	20.0	20.0
22	15.0	25.0
23	10.0	30.0
24	5.0	35.0

Table 3 Example of a 96-well crystallization tray using conditions outlined in Table 2.

	1	2	3	4	5	6	7	8	9	10	11	12
A	1	1	1	1	9	9	9	9	17	17	17	17
B	2	2	2	2	10	10	10	10	18	18	18	18
C	3	3	3	3	11	11	11	11	19	19	19	19
D	4	4	4	4	12	12	12	12	20	20	20	20
E	5	5	5	5	13	13	13	13	21	21	21	21
F	6	6	6	6	14	14	14	14	22	22	22	22
G	7	7	7	7	15	15	15	15	23	23	23	23
H	8	8	8	8	16	16	16	16	24	24	24	24

5. Pipette 80 μL of each condition into a 96-well crystallization tray in quadruplicate (Table 3).
6. Store remaining HDAC10 Redundancy Screen stock at 4 °C (see Note 5).

5.4 Notes

1. Generally, seed stocks are generated by adding 10 μL of precipitant buffer to the drop, crushing the crystals, then collecting the crushed crystals by aspiration. An additional 10 μL of precipitant buffer is added to the drop again to ensure complete transfer of seed crystals. Precipitant buffer is added at a 10 μL volume at least two more times. It is not advised to mix seed crystals obtained from conditions of different monobasic phosphate:dibasic phosphate ratios. Seed stock concentration is to be optimized each time new seeds are obtained.

2. It is important to add 0.518 μL of trypsin solution in the buffer described. Adding more may lower the pH of the HDAC10 solution, possibly resulting in an altered set of crystallization conditions.
3. By utilizing varying ratios of monobasic and dibasic phosphate salts, the HDAC10 Redundancy Screen allows precipitant buffers to sample the range of pH values within the buffering range of monobasic/dibasic phosphate salts (approximately pH 5.7–8.0).
4. Each condition of the HDAC10 Redundancy Screen must be thoroughly mixed to ensure crystal formation. Failure to do so will reduce crystal growth or even inhibit it entirely.
5. The HDAC10 Redundancy Screen expiry has not been determined, however, use within one year of formulation is generally advised.

6. Summary

HDAC10 is established as the eukaryotic polyamine deacetylase responsible for the hydrolysis of N^8-acetylspermidine in the cell cytosol, and the crystal structure of zebrafish HDAC10 now enables the structure-based design of isozyme-selective inhibitors (Hai et al., 2017; Steimbach et al., 2022). The detailed protocols outlined in this chapter can be used to generate a humanized variant of zebrafish HDAC10 containing the A24E and D94A substitutions so as to more closely mimic the active site of human HDAC10. This variant is suitable for cocrystallization trials and assays to assess inhibitory potency and selectivity. We expect that the methodologies documented herein will find broad use and accelerate the discovery of novel inhibitors.

References

Allfrey, V. G., Faulkner, R., & Mirsky, A. E. (1964). Acetylation and methylation of histones and their possible role in the regulation of RNA synthesis. *Proceedings of the National Academy of Sciences of the United States of America, 51*, 786–794.

Barneda-Zahonero, B., & Parra, M. (2012). Histone deacetylases and cancer. *Molecular Oncology, 6*, 579–589.

Blankenship, J. (1978). Deacetylation of N^8-acetylspermidine by subcellular fractions of rat tissue. *Archives of Biochemistry and Biophysics, 189*, 20–27.

Bondarev, A. D., Attwood, M. M., Jonsson, J., Chubarev, V. N., Tarasov, V. V., & Schiöth, H. B. (2021). Recent developments of HDAC inhibitors: Emerging indications and novel molecules. *British Journal of Clinical Pharmacology, 87*, 4577–4597.

Cao, J., Sun, L., Aramsangthenchai, P., Spiegelman, N. A., Zhang, X., Huang, W., ... Lin, H. (2019). HDAC11 regulates type I interferon signaling through defatty-acylation of SHMT2. *Proceedings of the National Academy of Sciences of the United States of America, 116*, 5487–5492.

Casero, R. A., Murray Stewart, T., & Pegg, A. E. (2018). Polyamine metabolism and cancer: Treatments, challenges and opportunities. *Nature Reviews. Cancer, 18*, 681–695.

Christianson, D. W. (2024). Chemical versatility in catalysis and inhibition of the class IIb histone deacetylases. *Accounts of Chemical Research, 57*, 1135–1148.

Cress, W. D., & Seto, E. (2000). Histone deacetylases, transcriptional control, and cancer. *Journal of Cellular Physiology, 184*, 1–16.

Desiderio, M. A. (1992). Opposite responses of nuclear spermidine N^8-acetyltransferase and histone acetyltransferase activities to regenerative stimuli in rat liver. *Hepatology (Baltimore, Md.), 15*, 928–933.

Gallwitz, D. (1968). Acetylation of histones by a kinase from rat liver nuclei. *Biochemical and Biophysical Research Communications, 32*, 117–121.

Glozak, M. A., & Seto, E. (2007). Histone deacetylases and cancer. *Oncogene, 26*, 5420–5432.

Goulart Stollmaier, J., Watson, P. R., & Christianson, D. W. (2024). Design, synthesis, and structural evaluation of acetylated phenylthioketone inhibitors of HDAC10. *ACS Medicinal Chemistry Letters* (in press).

Gregoretti, I., Lee, Y.-M., & Goodson, H. V. (2004). Molecular evolution of the histone deacetylase family: Functional implications of phylogenetic analysis. *Journal of Molecular Biology, 338*, 17–31.

Hai, Y., Shinsky, S. A., Porter, N. J., & Christianson, D. W. (2017). Histone deacetylase 10 structure and molecular function as a polyamine deacetylase. *Nature Communications, 8*, 15368.

Herbst-Gervasoni, C. J., & Christianson, D. W. (2019). Binding of N^8-acetylspermidine analogues to histone deacetylase 10 reveals molecular strategies for blocking polyamine deacetylation. *Biochemistry, 58*, 4957–4969.

Herbst-Gervasoni, C. J., & Christianson, D. W. (2021). X-ray crystallographic snapshots of substrate binding in the active site of histone deacetylase 10. *Biochemistry, 60*, 303–313.

Herbst-Gervasoni, C. J., Steimbach, R. R., Morgen, M., Miller, A. K., & Christianson, D. W. (2020). Structural basis for the selective inhibition of HDAC10, the cytosolic polyamine deacetylase. *ACS Chemical Biology, 15*, 2154–2163.

Holt, A., & Palcic, M. (2006). A peroxidase-coupled continuous absorbance plate-reader assay for flavin monoamine oxidases, copper-containing amine oxidases and related enzymes. *Nature Protocols, 1*, 2498–2505.

Hubbert, C., Guardiola, A., Shao, R., Kawaguchi, Y., Ito, A., Nixon, A., ... Yao, T. P. (2002). HDAC6 is a microtubule-associated deacetylase. *Nature, 417*, 455–458.

Inoue, A., & Fujimoto, D. (1969). Enzymatic deacetylation of histone. *Biochemical and Biophysical Research Communications, 36*, 146–150.

Kutil, Z., Novakova, Z., Meleshin, M., Mikesova, J., Schutkowski, M., & Barinka, C. (2018). Histone deacetylase 11 is a fatty-acid deacylase. *ACS Chemical Biology, 13*, 685–693.

Lane, A. A., & Chabner, B. A. (2009). Histone deacetylase inhibitors in cancer therapy. *Journal of Clinical Oncology: Official Journal of the American Society of Clinical Oncology, 27*, 5459–5468.

Libby, P. R. (1978). Properties of an acetylspermidine deacetylase from rat liver. *Archives of Biochemistry and Biophysics, 188*, 360–363.

Libby, P. R. (1980). Rat liver nuclear N-acetyltransferases: Separation of two enzymes with both histone and spermidine acetyltransferase activity. *Archives of Biochemistry and Biophysics, 203*, 384–389.

Lombardi, P. M., Angell, H. A., Whittington, D. A., Flynn, E. F., Rajashankar, K. R., & Christianson, D. W. (2011). Structure of prokaryotic polyamine deacetylase reveals evolutionary functional relationships with eukaryotic histone deacetylases. *Biochemistry, 50*, 1808–1817.

Marchant, P., Abu Manneh, V., & Blankenship, J. (1986). N^1-Acetylspermidine is not a substrate for N-acetylspermidine deacetylase. *Biochimica et Biophysica Acta (BBA) – General Subjects, 881,* 297–299.

Marchant, P., Dredar, S., Manneh, V., Alshabanah, O., Matthews, H., Fries, D., & Blankenship, J. (1989). A selective inhibitor of N^8-acetylspermidine deacetylation in mice and HeLa cells without effects on histone deacetylation. *Archives of Biochemistry and Biophysics, 273,* 128–136.

Mercuri, E., et al. (2024). Safety and efficacy of Givinostat in boys with Duchenne muscular dystrophy (EPIDYS): A multicentre, randomised, double-blind, placebo-controlled, phase 3 trial. *Lancet, 23,* 393–403.

Moreno-Yruela, C., Galleano, I., Madsen, A. S., & Olsen, C. A. (2018). Histone deacetylase 11 is an ε-N-myristoyllysine hydrolase. *Cell Chemical Biology, 25,* 849–856.

Mullard, A. (2024). FDA approves an HDAC inhibitor for Duchenne muscular dystrophy. *Nature Reviews Drug Discovery, 23,* 329.

Narita, T., Weinert, B. T., & Choudhary, C. (2019). Functions and mechanisms of non-histone protein acetylation. *Nature Reviews. Molecular Cell Biology, 20,* 156–174.

Patel, J., Pathak, R. R., & Mujtaba, S. (2011). The biology of lysine acetylation integrates transcriptional programming and metabolism. *Nutrition and Metabolism, 8,* 12.

Rim, M. H., Karas, B. L., Barada, F., Dean, C., & Levitsky, A. M. (2024). Recent and anticipated novel drug approvals (Q2 2024 through Q1 2025). *American Journal of Health-System Pharmacy: AJHP: Official Journal of the American Society of Health-System Pharmacists, 81,* 733–738.

Shinsky, S. A., & Christianson, D. W. (2018). Polyamine deacetylase structure and catalysis: Prokaryotic acetylpolyamine amidohydrolase and eukaryotic HDAC10. *Biochemistry, 57,* 3105–3114.

Steimbach, R. R., Herbst-Gervasoni, C. J., Lechner, S., Murray Stewart, T., Klinke, G., Ridinger, J., ... Miller, A. K. (2022). Aza-SAHA derivatives are selective histone deacetylase 10 chemical probes that inhibit polyamine deacetylation and phenocopy HDAC10 knockout. *Journal of the American Chemical Society, 144,* 18861–18875.

Tabor, C. W., & Tabor, H. (1984). Polyamines. *Annual Review of Biochemistry, 53,* 749–790.

Taunton, J., Hassig, C. A., & Schreiber, S. L. (1996). A mammalian histone deacetylase related to the yeast transcriptional regulator Rpd3p. *Science (New York, N. Y.), 272,* 408–411.

The UniProt Consortium. (2023). UniProt: The universal protein knowledgebase in 2023. *Nucleic Acids Research, 51*(D1), D523–D531.

Toro, T. B., & Watt, T. J. (2020). Critical review of non-histone human substrates of metal-dependent lysine deacetylases. *The FASEB Journal, 34,* 13140–13155.

Zeyen, P., Zeyn, Y., Herp, D., Mahmoudi, F., Yesiloglu, T. Z., Erdmann, F., ... Sippl, W. (2022). Identification of histone deacetylase 10 (HDAC10) inhibitors that modulate autophagy in transformed cells. *European Journal of Medicinal Chemistry, 234,* 114272.

Zhang, D., Tang, J., Xu, Y., Huang, X., Wang, Y., Jin, X., ... Liu, P. (2022). Global crotonylome reveals hypoxia-mediated lamin A crotonylation regulated by HDAC6 in liver cancer. *Cell Death & Disease, 13,* 717.

CHAPTER THREE

TR-FRET assay for profiling HDAC10 inhibitors and PROTACs

Kim Remans[a], Peter Sehr[b], Raphael R. Steimbach[c], Nikolas Gunkel[c], and Aubry K. Miller[c,*]

[a]European Molecular Biology Laboratory (EMBL), Protein Expression and Purification Core Facility, Heidelberg, Germany
[b]European Molecular Biology Laboratory (EMBL), Chemical Biology Core Facility, Heidelberg, Germany
[c]German Cancer Research Center (DKFZ), Cancer Drug Development Group, Heidelberg, Germany
*Corresponding author. e-mail address: aubry.miller@dkfz.de

Contents

1. Introduction — 42
2. Synthesis of Tubastatin-AF647 FRET tracer molecule — 43
   2.1 Equipment and materials — 43
   2.2 Procedure — 45
3. Baculovirus-mediated expression and purification of TwinStrep-GST-hHDAC10 in Sf21 cells — 48
   3.1 Equipment and materials — 48
   3.2 Procedure — 51
4. Preparation of DTBTA-Eu^{3+}-labelled Strep-TactinXT® — 59
   4.1 Equipment and materials — 59
   4.2 Procedure — 59
5. Performing and analyzing the HDAC10 TR-FRET ligand displacement assay — 60
   5.1 Equipment and materials — 60
   5.2 Procedure — 61
References — 62

Abstract

Quantitative biochemical characterization of the binding/inhibitory properties of investigative substances against their protein targets and anti-targets is a necessary step in modern drug discovery campaigns. The histone deacetylase family of proteins comprises eleven Zn^{2+} dependent enzymes, members of which are regularly investigated as therapeutic drug targets. The binding of histone deacetylases by small molecule inhibitors or PROTACs is typically measured in enzymatic assays that use acylated lysine-containing peptides as substrates. Histone deacetylase 10, however, is unique within the family in that it recognizes acetylated small molecule polyamines, as opposed to peptides, as substrates. We have therefore adapted a TR-FRET ligand displacement assay for histone deacetylase 10, which does not rely on enzymatic turnover of a substrate. In this chapter, we describe the preparation of the three different assay components: a small molecule dye conjugate "tracer", a TwinStrep-GST-HDAC10 fusion protein, and Eu^{3+}-labelled Strep-TactinXT®. Lastly, we describe how to combine these reagents and perform dose-response measurements of investigational HDAC10-binding molecules to produce IC_{50} values.

1. Introduction

Histone deacetylases (HDACs) are a family of eleven Zn^{2+} dependent hydrolases that catalyze the hydrolysis of N-acyl post-translational modifications on lysine side-chains of proteins as well as on small molecule primary metabolites (Yang & Seto, 2008). While it is currently understood that HDACs mediate myriad physiological processes on a wide variety of substrates, their name derives from their original discovery as acting on histones with effects on DNA packaging and gene regulation (Taunton, Hassig, & Schreiber, 1996; Yoshida, Kijima, Akita, & Beppu, 1990). Over the past three decades, histone deacetylase (HDAC) inhibitors have been shown to modulate numerous pathologies and have been widely developed in both academic and industrial settings, culminating in six different approved drugs, primarily for oncology indications (Ho, Chan, & Ganesan, 2020).

The HDAC family of enzymes is divided into four classes based on sequence homology: Class I (HDACs 1, 2, 3, and 8), Class IIa (HDACs 4, 5, 7, and 9), Class IIb (HDACs 6 and 10) and Class IV (HDAC11). It is considered best practice, when developing HDAC inhibitors, to measure the inhibitory capacity of each substance against a subset, or preferably all, of the 11 HDACs. This is typically performed by treating a lysine-containing peptide that has been appropriately acylated, with recombinant HDAC enzyme and inhibitor (Moreno-Yruela & Olsen, 2021). The extent to which the inhibitor prevents hydrolysis of the acyl group can then be quantified as an IC_{50} or K_D value.

We initiated a project to develop selective HDAC10 inhibitors and degraders (i.e. PROTACs) on the basis of reports of a critical role for HDAC10 in chemotherapeutic resistance in neuroblastoma (Oehme et al., 2013). HDAC10 is unique within the HDAC family in that it does not appear to accept acylated lysines as substrates; instead, it has been described as a polyamine deacetylase, acting on small molecule polyamines like N^8-acetylspermidine, acetylputrescine, and acetylcadaverine. As we were unable to use peptidic substrates in an enzymatic assay, we turned our attention to a published time-resolved fluorescence enery transfer (TR-FRET) ligand displacement assay for

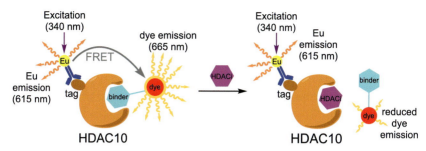

Fig. 1 Schematic depicting the concept of the HDAC10 TR-FRET assay.

HDACs (Marks, Fakhoury, Frazee, Eliason, & Riddle, 2011). In this assay, an HDAC10 fusion protein is combined with an a Eu^{3+}-labelled antibody and a "tracer" molecule, which is a dye-conjugated HDAC binder (Fig. 1). When all three components are present, excitation of Europium at 340 nm produces fluorescent emission at 615 nm. Productive FRET results in excitation of the dye molecule, which fluoresces at 665 nm. Disruption of FRET by the addition of a non-labelled competitive HDAC binder is measured as a change in the 615/665 emissions ratio. While this assay was commercially available as a kit, it was prohibitively expensive and the structure of the HDAC binder-dye conjugate was not disclosed. We, therefore, set out to develop a similar assay using "home-made" versions of all three components. This has enabled us to profile many different compound classes against HDAC10 (Chen et al., 2020; Chen et al., 2023; Géraldy et al., 2019; Herp et al., 2022; Lechner et al., 2022; Lechner et al., 2023; Morgen et al., 2020; Steimbach et al., 2022), and the following protocol provides the details for it to be reliably repeated in other laboratories.

2. Synthesis of Tubastatin-AF647 FRET tracer molecule

The originally published TR-FRET assay did not reveal the chemical structures of the tracer molecules. We therefore set out to synthesize one that would be compatible with Class IIb isozymes, using the highly potent Tubastatin A scaffold (Fig. 2).

2.1 Equipment and materials
2.1.1 Permanent equipment
- Temperature controlled stir plate (MR Hei-Standard, Heidolph, Germany)
- Rotary evaporator (Rotavapor R-200, Büchi, Switzerland)

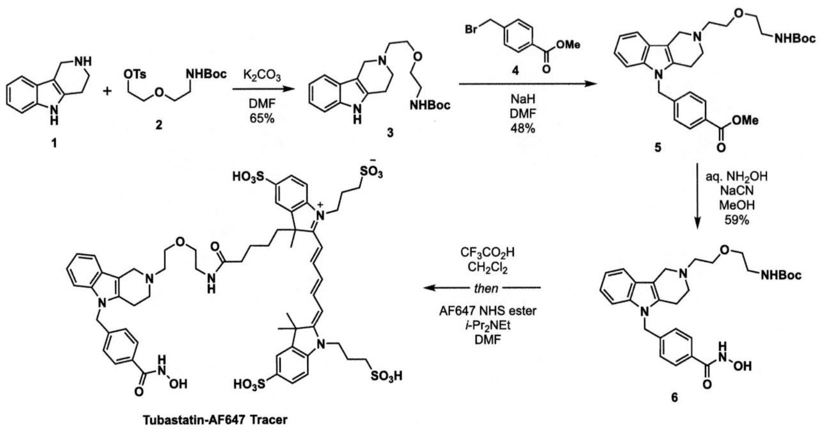

Fig. 2 Chemical synthesis of the Tubastatin-AF647 FRET tracer molecule from commercially available precursors.

- NMR spectrometer (Avance III, Bruker, Germany)
- Analytical LC/MS (6120 Series Single Quadrupole Electrospray, Agilent, USA) and preparative HPLC (1260 Infinity, Agilent, USA) systems
- Precision analytical balance (ML204, Mettler Toledo, USA)
- Lyophilizer (Alpha 2-4 LD Plus, Christ, Germany)
- UV-Vis spectrophotometer (Cary 60, Agilent, USA)

2.1.2 Disposable equipment

- Glassware (round bottom flasks, separatory funnels, funnels, chromatography columns)
- Syringes and needles
- Filter paper
- TLC plates (Silica gel 60 F_{254}, Merck, Germany) and chambers
- Disposable glass pipettes
- Graduated cylinders
- Preparative HPLC Column (Gemini® 5 μm C18 110 Å AXIA column 250 × 21.2 mm, Phenomenex, USA)
- Quartz cuvette (for UV-Vis measurements)

2.1.3 Materials

- 2,3,4,5-Tetrahydro-1H-pyrido[4,3-*b*]indole (**1**) (CAS: 6208-60-2; Sigma)
- 2-(2-((tert-Butoxycarbonyl)amino)ethoxy)ethyl 4-methylbenzenesulfonate (**2**) (CAS: 192132-77-7; BLD Pharm)
- Methyl 4-(bromomethyl)benzoate (**4**) (CAS: 2417-72-3; Sigma)

- AF647 NHS ester triethylammonium salt (CAS: 407627-61-6; MedChemExpress)
- K_2CO_3
- Dimethylformamide (anhydrous)
- EtOAc
- $MgSO_4$
- MeOH
- Concentrated aqueous NH_4OH
- CH_2Cl_2
- $CDCl_3$
- NaH (60 % in oil)
- Argon
- Saturated aqueous NaCl solution (brine)
- 50 % aqueous NH_2OH
- NaCN
- MeOH
- Trifluoroacetic acid
- i-Pr_2NEt
- MeCN (HPLC grade)
- H_2O (HPLC grade)

2.2 Procedure

2.2.1 Synthesis of tert-butyl (2-(2-(1,3,4,5-tetrahydro-2H-pyrido[4,3-b]indol-2-yl)ethoxy)ethyl)carbamate (3)

- Add K_2CO_3 (1.47 g, 10.64 mmol) to a solution of indole **1** (914 mg, 5.31 mmol) and tosylate **2** (1.91 g, 5.31 mmol) in dimethylformamide (15 mL) and heat to 70 °C for 16 h.
- Cool to room temperature, and dilute the reaction mixture with H_2O (400 mL) and EtOAc (100 mL).
- Separate the two layers using a separatory funnel and further extract the aqueous layer with EtOAc (2 × 100 mL).
- Combine the organic layers and dry them using $MgSO_4$.
- Remove the $MgSO_4$ by filtration, wash the filter cake with EtOAc, and concentrate the filtrate using a rotary evaporator.
- Purify the product via column chromatography using 5 % MeOH, 0.25 % NH_4OH in CH_2Cl_2 to obtain indole **3** (1.25 g, 65 % yield) as a beige solid, after concentration of the product fractions using a rotary evaporator.

2.2.2 Product characterization
Verify the identity and purity of the product by TLC, LC-MS, and NMR analysis.
TLC R_f 0.35 (10 % MeOH, 0.5 % NH$_4$OH in CH$_2$Cl$_2$).
LC/MS-ESI (*m*/*z*): [M+H$^+$] 360.2.
^{1}H NMR (400 MHz, CDCl$_3$) δ 7.82 (br s, 1 H), 7.40 (d, *J* = 7.5 Hz, 1 H), 7.29 (d, *J* = 7.5 Hz, 1 H), 7.13 (dt, *J* = 7.5, 1.3 Hz, 1 H), 7.07 (dt, *J* = 7.5, 1.3 Hz, 1 H), 5.16 (br s, 1 H), 3.84 (s, 2 H), 3.74 (t, *J* = 5.7 Hz, 2 H), 3.56 (t, *J* = 5.7 Hz, 2 H), 3.33 (br q, *J* = 5.1 Hz, 2 H), 3.01 (m, 2 H), 2.94–2.87 (m, *J* = 5.8 Hz, 4 H), 1.43 (s, 9 H) ppm.

2.2.3 Synthesis of methyl 4-((2-(2-(2-((tert-butoxycarbonyl)amino)ethoxy)ethyl)-1,2,3,4-tetrahydro-5H-pyrido[4,3-b]indol-5-yl)methyl)benzoate (5)

- Add NaH (167 mg, 4.18 mmol, 60 % suspension in oil) to a solution of indole **3** (1.247 g, 3.469 mmol) in anhydrous dimethylformamide (12.0 mL) at 0 °C under argon.
- Allow the reaction mixture to warm to room temperature.
- After 1 h (or after gas evolution ceases), cool the mixture to 0 °C and add bromide **4** (875 mg, 3.82 mmol).
- Allow the mixture to warm to room temperature and monitor reaction progress by thin later chromatography (5 % MeOH, 0.25 % NH$_4$OH in CH$_2$Cl$_2$).
- After 16 h, dilute the reaction mixture with H2O (500 mL) and EtOAc (200 mL).
- Separate the two layers using a separatory funnel and further extract the aqueous layer with EtOAc (3 × 100 mL).
- Combine the organic layers and wash them in a separatory funnel with H$_2$O (2 × 30 mL) and brine (30 mL).
- Dry the organic layer using MgSO$_4$.
- Remove the MgSO$_4$ by filtration, wash the filter cake with EtOAc, and concentrate the filtrate using a rotary evaporator.
- Purify the product via column chromatography using 5 % MeOH, 0.25 % NH$_4$OH in CH$_2$Cl$_2$ to obtain ester **5** (847 mg, 48 % yield) as a dark yellow oil, after concentration of the product fractions using a rotary evaporator.

2.2.4 Product characterization
Verify the identity and purity of the product by TLC, LC-MS, and NMR analysis.
TLC R_f 0.49 (10 % MeOH, 0.5 % NH$_4$OH in CH$_2$Cl$_2$).
LC/MS-ESI (*m*/*z*): [M+H$^+$] 508.3.

¹H NMR (400 MHz, CDCl₃) δ 7.93 (d, *J* = 8.6 Hz, 2 H), 7.47–7.44 (m, 1 H), 7.18–7.05 (m, 5 H), 5.29 (s, 2 H), 5.09 (br s, 1 H), 3.88 (s, 3 H), 3.86 (br s, 2 H), 3.72 (t, *J* = 5.6 Hz, 2 H), 3.56 (t, *J* = 5.2 Hz, 2 H), 3.33 (br q, *J* = 5.0 Hz, 2 H), 2.98 (m, 2 H), 2.88 (m, 2 H), 2.76 (m, 2 H), 1.42 (s, 9 H) ppm.

2.2.5 Synthesis of tert-butyl (2-(2-(5-(4-(hydroxycarbamoyl)benzyl)-1,3,4,5-tetrahydro-2H-pyrido[4,3-b]indol-2-yl)ethoxy)ethyl)carbamate (6)

- Add 50 % aqueous NH₂OH (4.0 mL, avoid using metal needles), followed by NaCN (4.6 mg, 0.094 mmol) to a solution of ester **5** (193 mg, 0.38 mmol) in MeOH (3.6 mL).
- After 24 h, dilute the reaction mixture with H₂O (50 mL) and CH₂Cl₂ (25 mL).
- Separate the layers using a separatory funnel and further extract the aqueous layer with CH₂Cl₂ (2 × 25 mL).
- Combine the organic layers and dry them using MgSO₄.
- Remove the MgSO₄ by filtration, wash the filter cake with CH₂Cl₂, and concentrate the filtrate using a rotary evaporator.
- Purify the product via column chromatography using 10 % MeOH, 0.5 % NH₄OH in CH₂Cl₂ to obtain hydroxamic acid **6** (114 mg, 59 %) as a light orange foam, after concentration of the product fractions using a rotary evaporator.

2.2.6 Product characterization

Verify the identity and purity of the product by LC-MS and NMR analysis.
LC/MS-ESI (*m/z*): [M+H]⁺ 509.3.
¹H NMR (400 MHz, CDCl₃) δ 7.43 (br d, *J* = 6.4 Hz, 1 H), 7.28 (m, 2 H), 7.13–7.05 (m, 3 H), 6.76 (m, 2 H), 5.31 (br s, 1 H), 5.10 (br s, 2 H), 3.97 (br s, 2 H), 3.76 (br s, 2 H), 3.52 (m, 2 H), 3.27 (br s, 2 H), 3.07 (br s, 2 H), 3.00 (br s, 2 H), 2.75 (br s, 2 H), 1.39 (s, 9 H) ppm.

2.2.7 Synthesis of "Tubastatin-AF647 tracer"

- Add trifluoroacetic acid (0.4 mL) to a solution of hydroxamic acid **6** (108.4 mg, 0.213 mmol) in CH₂Cl₂ (4.0 mL) at room temperature.
- After 2 h, concentrate the reaction mixture thoroughly on a rotary evaporator to give a sticky, brown oil.
- Prepare a solution of 16 mg of this oil in DMF (1.0 mL).
- Add 0.1 mL of this solution to a solution of AF647 NHS ester triethylammonium salt (1.0 mg) in DMF (0.1 mL).
- Add *i*-Pr₂NEt (10 μL).

- After 30 min, purify the product by reverse phase HPLC (Temperature: ambient; Solvent A = H$_2$O, 0.1 % NH$_3$; Solvent B = MeCN; Flow Rate = 15.0 mL/min; Gradient: 5 % B to 30 % B [over 4 min], then 30 % B to 60 % B [over 8 min], then 0 % B to 95 % B [over 4 min]) to provide Tubastatin-AF647 Tracer (2.0 mg) as a blue solid, after concentration of the product fractions using a rotary evaporator and then lyophilizer.

2.2.8 Product characterization
Verify the identity and purity of the product by LC-MS analysis.
LC/MS-ESI (m/z): [M–2H$^+$]$^{2-}$ 623.4; [M–3H$^+$]$^{3-}$ 415.3.

2.2.9 Preparation of quantified stock solution
- Dissolve the purified Tubastatin-AF647 Tracer in a small amount (<1 mL) of ultrapure water to prepare a master stock. Store between −80 °C and −20 °C.
- From this master stock take a small aliquot and prepare a serial dilution in ultrapure water.
- Use UV–VIS spectrophotometry to measure the absorbance of the diluted tracer samples and calculate the concentration of the master stock using the Lambert–Beer law and the extinction coefficient of the tracer dye (ε650 nm = 255,000 cm^{-1}M^{-1}).
- Prepare 20 μM work-dilutions of the tracer from the master stock to use in the assay.

3. Baculovirus-mediated expression and purification of TwinStrep-GST-hHDAC10 in Sf21 cells

The original publication (Marks et al., 2011) used a GST-HDAC10 fusion protein for the TR-FRET assay. While both homemade N-terminal and C-terminal GST-HDAC10 fusion protein bound by a Eu-anti-GST antibody performed quite poorly in the TR-FRET assay in our hands, we found that a TwinStrep-GST-HDAC10 fusion protein bound by a DTBTA-Eu-labelled Strept-Actin XT donor performed satisfactorily. A schematic overview of the entire human HDAC10 (hHDAC10) expression and purification workflow can be found in Fig. 3.

3.1 Equipment and materials
3.1.1 Permanent equipment
- Biological safety cabinet (Thermo MSCAdvantage 1.2 class II biological safety cabinet)

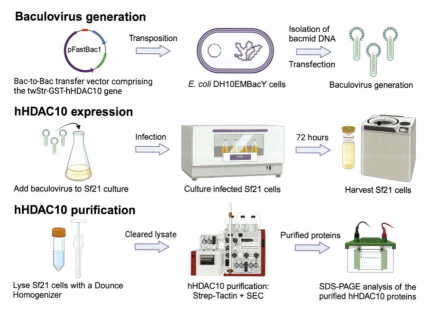

Fig. 3 Overview of the hHDAC10 expression and purification workflow. Baculovirus generation: the pFastBac1_TwStr-GST-hHDAC10 is used to transform *E. coli* DH10EMBacY cells. The bacmid DNA is isolated and transfected into Sf9 insect cells for generation of the recombinant baculoviruses. hHDAC10 expression: Sf21 insect cell cultures are infected with the recombinant baculoviruses carrying the TwStr-GST-hHDAC10 gene. After 72 h of culturing, the infected Sf21 cells are harvested by centrifugation. hHDAC10 purification: the Sf21 cells are lysed with a Dounce Homogenizer. The cleared lysate is loaded onto a Strep-Tactin affinity chromatography column. The elution fractions containing the TwStr-GST-hHDAC10 protein are further purified by size exclusion chromatography. The purified TwStr-GST-hHDAC10 protein is analyzed by SDS-PAGE. *Figure created with BioRender.com; https://BioRender.com/h86m437.*

- Shaking incubator(s) that can maintain 27 °C and 37 °C (Infors Multitron)
- Static incubator that can maintain 27 °C
- Electroporation device (Bio-Rad MicroPulser Electroporator)
- Cell counter (Bio-Rad TC20 automated cell counter)
- Benchtop centrifuges for Eppendorf tubes and falcons (Eppendorf 5430 R and Eppendorf 5804 R centrifuges, respectively)
- UV–VIS spectrophotometer for the determination of plasmid DNA and protein concentration (NanoDrop One Spectrophotometer)
- Cell culture microscope with Diascopic and Epi-fluorescence illumination options to assess the insect cells and check the YFP fluorescence (Nikon Eclipse Ts2-FL)

- Centrifuge, centrifuge rotor and centrifuge buckets suitable for harvesting larger scale expression cultures (1–6 L scale) (Beckman Coulter Avanti JXN-26 with JLA-8.100 rotor and J-LITE 1000 mL Polypropylene bottles)
- Dounce homogenizer (a microfluidizer device can also be used)
- Ultracentrifuge (Beckman Coulter Optima L-100 XP ultracentrifuge with 45 TI rotor)
- Automated chromatography system (Åkta Pure (Cytiva) or NGC Quest (Bio-Rad))
- Protein gel electrophoresis equipment (XCell Mini-Cell Electrophoresis System or Bio-Rad Mini-PROTEAN Electrophoresis Cells)

3.1.2 Disposable equipment
- Counting slides
- Sterile glass or plastic cell culture flasks
- Electroporation cuvettes (e.g., Bio-Rad Gene Pulser/MicroPulser Electroporation 0.2 cm gap Cuvettes)
- Sterile bacterial culturing tubes
- Bacmid selection plates: sterile TYE-agar plates (1 % Tryptone, 0.5 % Yeast extract, 0.8 % NaCl, 1.5 % agar) supplemented with 30 μg/mL kanamycin, 10 μg/mL tetracycline, 7 μg/mL gentamycin, 0.5 mM IPTG and 400 μg/mL X-Gal
- Sterile 6-well cell culture plates (e.g., Nunclon™ Delta Surface 6-well plates)
- Parafilm
- Aluminium foil
- 0.22 μm PVDF membrane filter units (Merck Millipore)
- 0.45 μm nitrocellulose membrane filters (Cytiva)
- Pre-packed 5 mL Strep-TactinXT® Superflow® high capacity column (IBA)
- HiLoad 16/600 Superdex 200 pg size exclusion chromatography column (Cytiva)
- 5 mL sample loop
- Fraction collector tubes and/or 96-well deepwell plate

3.1.3 Materials
- Insect cell lines: Sf9 and Sf21
- Insect cell culture medium: Sf-900™ III SFM (Thermo Fischer Scientific)
- *E. coli* DH10EMBacY cells (Geneva Biotech) [Note: as an alternative to electrocompetent *E. coli* DH10EMBacY cells, one can also prepare and use chemically competent *E. coli* DH10EMBacY cells and perform a heat shock (45 s at 42 °C) for the transformation process]

- pFastBac1_TwStr-GST-hHDAC10 plasmid DNA (see Fig. 4 and Vector Map file)
- SOC medium: 2% Tryptone, 0.5% Yeast extract, 10 mM NaCl, 2.5 mM KCl, 10 mM MgCl$_2$, 10 mM MgSO$_4$ and 20 mM glucose (autoclaved and stored at 4 °C)
- LB medium: 10 g Bactotryptone, 5 g Yeast extract and 5 g NaCl in 1 l H$_2$O (autoclaved and stored at 4 °C)
- Kanamycin: sterile 30 mg/mL stock in H$_2$O (stored at −20 °C)
- Tetracycline: sterile 10 mg/mL stock in H$_2$O (stored at −20 °C)
- Gentamycin: sterile 7 mg/mL stock in H$_2$O (stored at −20 °C)
- Isopropyl ß-D-1-thiogalactopyranoside (IPTG): sterile 1 mM stock in H$_2$O (stored at −20 °C)
- X-Gal: 20 mg/mL stock in DMSO
- Qiagen buffers P1, P2 and N3. These buffers can be purchased individually or as part of the Qiagen plasmid DNA preparation kits
- 100% i-PrOH (stored at room temperature)
- 70% EtOH (stored at room temperature)
- Transfection reagent: XtremeGENE HP (Sigma) (other suitable transfection reagents (e.g., FuGENE® (Promega)) can also be used)
- Liquid N$_2$
- Lysis/wash buffer: 100 mM Tris/HCl pH 8.0, 150 mM NaCl, 1 mM EDTA, 1 mM DTT (filtered)
- Elution buffer: 100 mM Tris/HCl pH 8.0, 150 mM NaCl, 1 mM EDTA, 1 mM DTT, 50 mM biotin (filtered)
- SEC buffer: 50 mM HEPES/NaOH pH 7.5, 300 mM NaCl, 1 mM EDTA, 2 mM DTT (filtered)
- cOmplete EDTA-free protease inhibitors (Roche)
- Benzonase® nuclease
- 1 M MgCl$_2$ stock (filtered and stored at room temperature)
- 100% glycerol
- Protein gels and SDS running buffer
- Protein ladder
- SDS sample buffer
- Coomassie staining solution

3.2 Procedure

3.2.1 Routine cell culture maintenance

We culture both our Sf9 and Sf21 insect cell lines in Sf-900™ III SFM. We generally maintain 20–25 mL cell culture stocks in 125 mL unbaffled glass or plastic culture flasks. We routinely split our Sf9 and Sf21 stocks to 0.5×10^6 cells/mL on Mondays and Wednesdays and to 0.3×10^6 cells/mL on Fridays. For large scale expression experiments, we expand our stocks as necessary. We always culture our insect cells stocks at 27 °C and with a shaking speed of 100–120 rpm (25 mm throw).

3.2.2 Transposition and bacmid preparation

For transposition and bacmid preparation, we use protocols based on the methods developed by the group of Imre Berger, which are all described in detail in the MultiBac manual (Berger et al., 2013; Gorda, Toelzer, Aulicino, & Berger, 2021).

- Thaw the electrocompetent *E. coli* DH10EMBacY cells on ice.
- Add ~50-100 ng of the pFastBac1_TwStr-GST-hHDAC10 plasmid (Fig. 4 and Vector Map File) to 50 µL of electrocompetent *E. coli* DH10EMBacY cells and transfer the mixtures to a pre-cooled electroporation cuvette.
- Incubate the samples on ice for 15 min

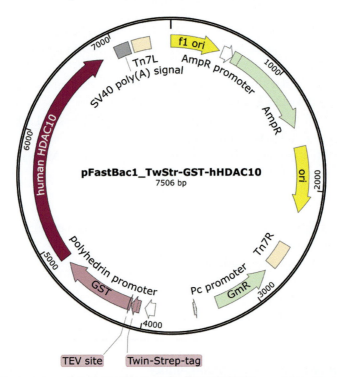

Fig. 4 Vector map of the pFastBac1_TwStr-GST-hHDAC10 construct. *Figure created with SnapGene.*

- Perform the electroporation process according to the manufacturer's recommendations.
- After the electroporation pulse, immediately add 1 mL of sterile SOC medium to the cuvette.
- Transfer the mixture to a sterile bacterial culturing tube and shake the culture overnight at 37 °C.
- The next day, prepare serial 10-fold dilutions (three are usually sufficient) of the transformed *E. coli* DH10EMBacY cells in sterile SOC medium.
- Plate 135 µL of all dilutions on the bacmid selection plates and incubate the plates overnight at 37 °C.
- The next day, check the bacmid selection plates for the presence of blue and white colonies. In the white colonies, the transposition should have been successful (meaning insertion of the TwStr-GST-hHDAC10 gene into the EMBacY baculoviral DNA). Re-streak 7 white colonies and 1 blue colony on fresh bacmid selection plates to confirm the absence of mixed clones. Incubate the bacmid selection plates overnight at 37 °C.
- The next day, pick 2 fully white clones and inoculate 2 mL LB cultures supplemented with 30 µg/mL kanamycin, 10 µg/mL tetracycline and 7 µg/mL gentamycin. Shake these liquid cultures overnight at 37 °C.
- The next day, centrifuge the overnight cultures (10 min, 3500 x g) [Note: after centrifugation, you can also freeze the bacterial cell pellets and store these at −20 °C if you want to continue with the bacmid preparation at a later point in time.] Discard the supernatant and resuspend the cell pellets in 250-300 µL buffer P1. Transfer the resuspended cell pellets to clean Eppendorf tubes.
- Add 250-300 µL of buffer P2. If LyseBlue has been added to buffer P1, your sample will turn blue. Gently invert the Eppendorf tubes until the samples are homogeneously blue.
- Add 300 µL of buffer N3. A white precipitate will appear. Gently invert the Eppendorf tubes until the blue colour has disappeared fully in case LyseBlue has been added to buffer P1.
- Centrifuge the samples for 10 min at ∼20,000 x g.
- Transfer the supernatants to clean Eppendorf tubes in one go (be careful not to disturb the pellet).
- Add 700 µL of 100 % isopropanol (final concentration ∼40 % isopropanol) and mix by inverting the Eppendorf tubes.
- Centrifuge the samples for 10 min at ∼20,000 x g at 4 °C.
- Carefully remove the supernatants.
- Gently add 500 µL of 70 % ethanol to the pellets (drop-by-drop).

- Centrifuge the samples for 5 min at ~20,000 × g at 4 °C.
- Carefully remove the supernatants.
- Gently add 50 µL of 70 % ethanol to the pellets (drop-by-drop).
- Centrifuge the samples for 5 min at ~20,000 × g at 4 °C.
- Move the samples to a biological safety cabinet and carefully remove most of the 70 % ethanol.
- Dry the pellets containing the recombinant bacmid DNA inside the biological safety cabinet.
- Once the bacmid DNA pellets are dry, add 40 µL of sterile water and resuspend the pellets by tapping the Eppendorf tubes approximately 10 times on the bench [Note: we strongly prefer the use of freshly isolated bacmid DNA for the transfections in Sf9 cells rather than bacmid DNA that has been frozen and stored at −20 °C.]
- Measure the concentration of each bacmid prep. We will use ~10 µg of bacmid DNA per transfection.
- If you wish to verify the insertion of the TwStr-GST-hHDAC10 gene into the baculoviral DNA via PCR, set aside 1-2 µL of the bacmid DNA for a bacmid PCR [Note: you can verify insertion of the TwStr-GST-hHDAC (2829 bps) gene into the EMBacY bacmid DNA by performing a bacmid PCR. To do so, we follow the bacmid PCR protocol described in the Thermo Fischer Scientific Bac-to-BacTM user guide, which makes use of pUC/M13-fw and pUC/M13-rev primers that hybridize to sites flanking the mini-attTn7 site in the bacmid DNA].

3.2.3 Transfection of the bacmid DNA into Sf9 cells and baculovirus amplification

For the transfection of the bacmid DNA into Sf9 cells and the subsequent baculovirus amplification, we use protocols based on the methods developed by the group of Imre Berger, which are all described in detail in the MultiBac manual (Berger et al., 2013; Gorda et al., 2021).

- Split your Sf9 insect cell culture to a density between 0.5-1 × 10^6 cells/mL in a biological safety cabinet. Prepare 5-10 mL of this new stock. After splitting, count the cells again to make sure that they are at the desired cell density.
- Label the lid of the 6-well plate that you will use for the transfections. 2 wells will serve as controls: 1 well will be used for non-transfected Sf9 cells and 1 well will be filled with medium only (to check for contamination issues). 2 other wells will be used for the generation of the recombinant baculovirus carrying the TwStr-GST-hHDAC10 gene. We will work with 2 clones (i.e., 2 wells) per construct.

- In the 6-well plate, add 2 mL of Sf-900™ III SFM medium per well for clones 1 and 2 and for the non-infected control cells. In the fourth well, add 3 mL of Sf-900™ III SFM medium.
- Gently add 1 mL of cells with a density of 0.5–1.0×10^6 cells/mL to all wells except for the "medium only" negative control well. Do this in a drop-by-drop manner with the pipet boy on a low setting.
- Incubate the 6-well plate for 15 min inside the biological safety cabinet to allow the Sf9 cells to settle. Afterwards, check whether the cells have settled properly using a standard cell culture microscope.
- During the 15 min incubation period, prepare the transfection cocktail:
  a. Add 100 µL of Sf-900™ III SFM medium to each Eppendorf tube containing ~10 µg of bacmid DNA from Section 3.2.2.
  b. Mix 200 µL of Sf-900™ III SFM medium with 20 µL of XtremeGENE HP transfection reagent.
  c. Add 100 µL of the XtremeGENE mixture to each Eppendorf tube with the bacmid DNA.
  d. Incubate this transfection cocktail for 10 min inside the biological safety cabinet.
- After both incubation steps have been completed, add the transfection cocktails to the individual wells of Sf9 cells in the 6-well plate. Do this in a drop-by-drop manner using a spiral movement.
- Carefully put parafilm around the sides of the 6-well plates to reduce evaporation.
- Take the plates out of the biological safety cabinet and wrap them in aluminium foil to shield them from light.
- Incubate the plates without shaking for 72 h at 27 °C.
- After 72 h, check the Sf9 insect cells in your 6-well plate under the microscope. The infected Sf9 cells should appear to be larger, more irregular in their shape and less confluent compared to the non-infected control cells. You should also be able to observe the YFP fluorescence, as the EMBacY baculoviral DNA contains a YFP gene under control of the polH promoter, allowing expression of YFP in the insect cells.
- Carefully remove the 3 mL cell culture supernatant from each well. This is your V_0 baculovirus stock, which will be further amplified to generate the V_1 baculovirus.
- Split the Sf9 insect cell culture to a density of 1.0×10^6 cells/mL. Prepare 25 mL of Sf9 insect cells at a density of 1.0×10^6 cells/mL per baculovirus [Note: start expanding the Sf9 insect cell culture stock in a timely

manner, so that you'll be able to prepare at least 50-100 mL of Sf9 cells at a density of 1.0×10^6 cells/mL on the day that you will start the V1 baculovirus amplification.]
- Add 3 mL of the V_0 baculovirus to 25 mL of Sf9 cells at a density of 1.0×10^6 cells/mL in a 125 mL shaking flask.
- Shake the Sf9 insect cell cultures for 72 h at 27 °C.
- After 72 h, transfer the infected Sf9 insect cell cultures to sterile 50 mL falcons and centrifuge the samples for 10 min at $500 \times g$ at room temperature [Note: your infected Sf9 insect cell cultures should have a slightly yellow shade due to the presence of the YFP gene on the EMBacY bacmid DNA. Normally, your infected Sf9 cells will have a larger diameter than non-infected Sf9 cells.]
- Take the supernatant, filter it using a 0.22 μm sterile filter and transfer it to a clean sterile 50 mL falcon. This is your V_1 baculovirus stock [Note: optionally, you can add 1–5 % FBS to increase virus stability for longer-term storage at 4 °C.]

3.2.4 Expression of TwinStrep-GST-hHDAC10 in Sf21 cells
- Split your Sf21 insect cell culture to a density of 1×10^6 cells/mL in a biological safety cabinet. For large scale expression of TwStr-GST-hHDAC10, we generally use 1-2 L of Sf21 cells [Note: start expanding your Sf21 insect cell culture stock in a timely manner, so that you'll be able to prepare 1-2 L of Sf21 cells at a density of 1.0×10^6 cells/mL on the day that you will start the large-scale expression.] After splitting, count the cells again to make sure that they are at the desired cell density.
- Add 10 mL of the recombinant V_1 baculovirus to 1 L of Sf21 cells at a density of 1×10^6 cells/mL in a 5 L cell culture flask inside a biological safety cabinet.
- Shake the infected Sf21 insect cell cultures for 72 h at 27 °C.
- After 72 h, take a 1 mL sample of the infected Sf21 cultures, count the cells to assess your final cell density and viability and check the YFP fluorescence.
- Harvest the infected Sf21 insect cell cultures by centrifugation (15 min, $600 \times g$) [Note: if you don't plan to continue with the protein purification immediately, the Sf21 cell pellets can be flash-frozen in liquid N_2 and stored at -20 °C until the start of the purification.]

3.2.5 Purification of TwinStrep-GST-hHDAC10
All purification steps are performed at 4 °C and the samples are kept on ice at all times throughout the purification process.

- For a 2 L Sf21 expression culture, resuspend the cell pellet in 50 mL of ice cold lysis/wash buffer and add benzonase® nuclease, 2 mM MgCl$_2$ and 1x cOmplete EDTA-free protease inhibitors.
- Incubate the sample for 10–15 min on ice.
- Lyse the cells using a Dounce Homogenizer (~20 strokes).
- Dilute the lysed sample to a total volume of 100 mL with ice cold lysis/wash buffer and add benzonase® nuclease again.
- Incubate the lysate for 10–15 min on ice.
- Take a sample of the total lysate for SDS-PAGE analysis (30 μL sample + 10 μL 4x SDS sample buffer).
- Centrifuge the lysate for 30 min at 4 °C at 125,000 x g in an ultracentrifuge.
- After centrifugation, carefully remove the supernatant and filter it through a 0.45 μm nitrocellulose membrane filter into a clean flask. Take a sample of the cleared supernatant for SDS-PAGE analysis (30 μL sample + 10 μL 4x SDS sample buffer).
- Attach a 5 mL Strep-TactinXT® Superflow® high capacity column to your chromatography system and equilibrate the column first with 5 column volumes of filtered ultrapure water and then with 5 column volumes of lysis/wash buffer at a speed of 1 mL/min.
- Load the cleared lysate onto the 5 mL Strep-TactinXT® Superflow® high capacity column with a speed of 1 mL/min.
- Collect the flow through fraction in a clean flask and take a sample for SDS-PAGE analysis (30 μL sample + 10 μL 4x SDS sample buffer).
- After loading, wash the 5 mL Strep-TactinXT® Superflow® high capacity column with lysis/wash buffer until the A280 signal returns to baseline.
- After washing, elute the 5 mL Strep-TactinXT® Superflow® high capacity column with at least 5 column volumes of elution buffer. Collect individual elution fractions and take samples for SDS-PAGE analysis (30 μL sample + 10 μL 4x SDS sample buffer).
- Analyse samples of the total lysate (before ultracentrifugation), the cleared lysate (supernatant after ultracentrifugation), the flow through fraction and the individual elution fractions via SDS-PAGE (Fig. 5).
- After SDS-PAGE analysis, pool the elution fractions containing the TwStr-GST-hHDAC10 (102.8 kDa) protein.
- Concentrate the pooled elution fractions to a volume of maximum 5 mL.
- Centrifuge the concentrated sample for 10 min at ~20,000 x g at 4 °C and transfer the supernatant to a clean tube.

- Attach a HiLoad 16/600 Superdex 200 pg size exclusion chromatography column to your chromatography system and equilibrate the column first with 2 column volumes of filtered ultrapure water and then with 2 column volumes of SEC buffer at a speed of 1 mL/min.
- Inject the centrifuged TwStr-GST-hHDAC10 sample using a 5 mL sample loop.
- Elute the HiLoad 16/600 Superdex 200 pg size exclusion chromatography column with at least 1 column volume of SEC buffer at a speed of 1 mL/min. Collect 2 mL elution fractions in individual tubes or in a 96-well deepwell plate.
- Analyze the elution fractions of the size exclusion chromatography via SDS-PAGE (Fig. 5).
- After SDS-PAGE analysis, pool the elution fractions containing the cleanest TwStr-GST-hHDAC10 (102.8 kDa) protein.
- Measure the concentration of the pooled size exclusion chromatography fractions based on the absorbance at 280 nm ($\varepsilon_{\text{TwStr-GST-hHDAC10}}$ = 122,730 M^{-1} cm^{-1}).
- Concentrate the pooled fractions from the size exclusion chromatography run to the desired concentration, centrifuge (10 min, ~20,000 x g, 4 °C) and aliquot.
- Flash-freeze the aliquots in liquid N_2 and store at −80 °C.

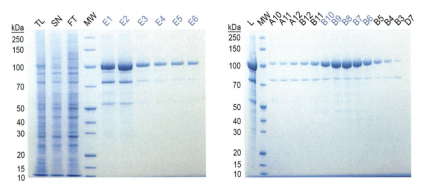

Fig. 5 SDS-PAGE analysis of the affinity and size exclusion chromatography steps of the purification of the recombinant TwStr-GST-hHDAC10 protein. Strep-Tactin (left) and size exclusion chromatography purification (right) of TwStr-GST-hHDAC10 (102.8 kDa). TL: total lysate; SN: cleared supernatant; FT: flow through; MW: molecular weight marker; E1-E6: Strep-Tactin elution fractions; L: load SEC; A10-A12, B12-B3, D7: SEC elution fractions. Fractions indicated in blue were pooled.

4. Preparation of DTBTA-Eu^{3+}-labelled Strep-TactinXT®
4.1 Equipment and materials

4.1.1 Permanent equipment
- Precision balance (TE214S, Sartorius, Germany)
- Benchtop vortexer (Vortex Genie 2, Scientific Industries, USA)
- Electronic tube roller (CAT RM5, M. Zipperer GmbH, Germany)
- Centrifuge (Centrifuge 5415 R, Eppendorf, Germany)
- Rotary mixer (Test tube rotator 34528, Snijders Scientific, Holland)
- NanoDrop spectrophotometer (NANODROP 8000, Thermo Scientific, USA)
- Pipette (PIPETMAN, Gilson, USA)

4.1.2 Disposable equipment
- Eppendorf tubes
- Pipette tips
- PD10 desalting column (17-0851-01; GE Healthcare)

4.1.3 Materials
- ATBTA-Eu^{3+} (CAS: 601494-52-4; TCI)
- Aqueous NaOAc solution (100 mM, pH 4.9)
- Cyanuric chloride (CAS: 108-77-0; Sigma)
- Acetone
- Unconjugated Strep-TactinXT® (purchased upon request from IBA)
- Na$_2$CO$_3$
- NaHCO$_3$
- 10 % BSA solution (1 g BSA in 10 mL H$_2$O)
- TBS buffer (50 mM Tris/HCl pH 7.5, 150 mM NaCl, 0.01 % NaN$_3$)

4.2 Procedure
- Prepare 100 mM Na$_2$CO$_3$ solution
- Prepare 100 mM NaHCO$_3$ solution
- Mix 0.9x NaHCO$_3$ solution and 0.1x Na$_2$CO$_3$ solution to make a 100 mM Carbonate buffer with ~pH 9.3.
- Dissolve cyanuric chloride (12.0 mg, 0.0651 mmol) in acetone (697 µL) to prepare a 93 mM solution.
- Dissolve ATBTA-Eu^{3+} (2.0 mg, 2.3 µmol) in aqueous NaOAc (60 µL, 100 mM, pH 4.9) to prepare a 39 mM solution.
- Add a portion of the cyanuric chloride solution (25 µL, 93 mM in acetone, 2.3 µmol) to the solution of ATBTA-Eu^{3+} (60 µL, 39 mM in aqueous NaOAc, 2.3 µmol).

- Briefly vortex the mixture and place on a tube roller at room temperature for 30 min
- Add the mixture dropwise to acetone (1 mL) and centrifuge (10,000 rpm, 3 min) in a 1.5 mL Eppendorf tube.
- Carefully remove the supernatant.
- Resuspend in acetone (1 mL), briefly vortex, centrifuge (10,000 rpm, 3 min), and remove the supernatant (perform this step a total of three times).
- Air dry the pellet at 37 °C with the Eppendorf lid open for 1 h.
- Dissolve the pellet in carbonate buffer (400 μL, 100 mM, pH 9.3).
- Add 100 μL of this solution to a solution of Strep-TactinXT® (5.0 mg, 94 nmol, IBA Life Sciences) in Carbonate buffer (400 μL, 100 mM, pH 9.3). This is roughly 4 equivalents of Eu^{3+} complex per Strep-TactinXT® tetramer.
- Briefly vortex the mixture and place on a rotary mixer at 4 °C overnight.
- The next day equilibrate a PD10-column with 25 mL TBS buffer.
- Add 10 % BSA solution (0.5 mL).
- Add all 0.5 mL of the DTBTA-Eu^{3+}-labelled Strep-TactinXT® reaction mixture, followed by TBS buffer (2.0 mL). Discard the flow through.
- Add TBS buffer (3.3 mL) and collect the flow through, which can be split into smaller aliquots if desired.
- Measure absorbance at 335 nm (Abs_{335}) and 280 nm (Abs_{280}) with a NanoDrop spectrophotometer (1 mm path length).
- Calculate the concentration of DTBTA-Eu^{3+} using the following equation: $C = \frac{Abs_{335}}{3100}$. We measured $Abs_{335} = 0.258$ and a concentration of 83.2 μM.
- Calculate the concentration of Strep-TactinXT® tetramer using the following equation: $C = \frac{Abs_{280} - 0.8 \cdot (Abs_{335})}{14240}$. We measured $Abs_{280} = 0.556$ and a concentration of 24.6 μM, giving a labelling ratio of 3.4 DTBTA-Eu^{3+} chelates per Strep-TactinXT® tetramer.
- Make a 100 nM stock solution in TBS buffer based on the concentration of the Strep-TactinXT® tetramer and store at −80 °C.

5. Performing and analyzing the HDAC10 TR-FRET ligand displacement assay

5.1 Equipment and materials

5.1.1 Permanent equipment
- Plate reader (CLARIOstar, BMG Labtech, Germany) equipped with TR-FRET filters

- Digital dispenser (D300e, Tecan, Switzerland). Note: this is optional and manual serial dilutions can also be performed
- Orbital shaker (Orbit P2, Labnet, USA)
- Centrifuge equipped for well plates (Thermo Scientific, Megafuge 16 R, USA)
- Calibrated multichannel pipette for low µL volumes (VIAFLO, Integra, Switzerland)
- Software for performing non-linear regression analysis (Prism, GraphPad, USA)

5.1.2 Disposable equipment
- White low-volume 384-well plates (Corning 4512)
- Transparent 384-well plate sealing film (Starlab E2796-0100)
- Pipette tips
- Low residual volume reagent reservoir for multichannel pipette

5.1.3 Materials
- Base buffer (50 mM HEPES pH 8.0, 150 mM NaCl, 10 mM $MgCl_2$, 1 mM EGTA, and 0.01 % Brij-35)
- Tubastatin-AF647 FRET tracer solution (See Section 2)
- TwinStrep-GST-HDAC10 protein (See Section 3)
- DTBTA-Eu^{3+}-labelled Strep-TactinXT® (See Section 4)
- SAHA (Vorinostat)
- Experimental HDAC inhibitors or PROTACs
- DMSO

5.2 Procedure
- To prepare 6260 µL of TR-FRET Assay Solution for one 384-well plate, combine Base Buffer (6206 µL), Tubastatin-AF647 FRET tracer (7.97 µL of a 19.6 µM stock solution in H_2O; see Section 2), TwinStrep-GST-HDAC10 (40.22 µL of a 778.21 nM stock solution; see Section 3), and DTBTA-Eu^{3+}-labelled Strep-TactinXT® (6.26 µL of a 100 nM stock solution; see Section 4) to prepare 6260 µL of TR-FRET Assay Buffer. Final concentrations in assay wells: Tubastatin-AF647 FRET tracer: 25 nM, TwinStrep-GST-HDAC10: 5.00 nM, DTBTA-Eu^{3+}-labelled Streptactin: 0.10 nM.
- Using a multichannel pipette, transfer 15 µL of TR-FRET Assay Buffer to each well of a 384-well assay plate.
- Transfer experimental substances (up to 14 per plate) from DMSO stock solutions of 10 mM and 0.1 mM in triplicate using a digital dispenser (e.g. D300e from Tecan) according to the plate map in Fig. 6. Tested concentrations range from 50 µM to 87 pM.

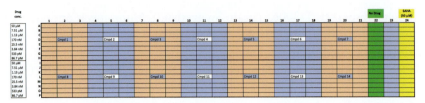

Fig. 6 Plate map to be used for determining IC$_{50}$ values of experimental HDAC10 inhibitors or degraders.

- Transfer SAHA (positive control) from a 10 mM DMSO stock solution using a digital dispenser (e.g. D300e from Tecan) according to the plate map in Fig. 6.
- Cover the plate with sealing film.
- Shake the plate at 800 rpm on an orbital shaker for 30 s
- Centrifuge the plate at 300 g for 1 min
- Incubate at room temperature in the dark for 90 min
- Carefully remove the sealing film and insert the plate in a plate reader. Excite wells with 100 flashes and detect fluorescence emission at 665 nm and 620 nm with appropriately adjusted gain and focal height.
- Export data to a data processing software (e.g. GraphPad Prism) and calculate the ratio of 665 nm and 620 nm fluorescence data for each well on the plate.
- Calculate the mean of all ratios for the 16 SAHA positive controls to define the background.
- Subtract this background from all calculated ratios on the plate, including each individual SAHA-containing well.
- Fit the experimental data to a 4-parameter non-linear dose-response analysis, where 0 % is defined as 0 % and 100 % is defined as the mean of the 16 background-substracted "no-drug" treatments.

References

Berger, I., Garzoni, F., Chaillet, M., Haffke, M., Gupta, K., & Aubert, A. (2013). The MultiBac protein complex production platform at the EMBL. *Journal of Visualized Experiments: JoVE*(77), e50159.

Chen, X., Chen, X., Steimbach, R. R., Wu, T., Li, H., Dan, W., & Zhu, Y. (2020). Novel 2, 5-diketopiperazine derivatives as potent selective histone deacetylase 6 inhibitors: Rational design, synthesis and antiproliferative activity. *European Journal of Medicinal Chemistry, 187*, 111950.

Chen, X., Wang, J., Zhao, P., Dang, B., Liang, T., Steimbach, R. R., & Gao, J. (2023). Tetrahydro-β-carboline derivatives as potent histone deacetylase 6 inhibitors with broad-spectrum antiproliferative activity. *European Journal of Medicinal Chemistry, 260*, 115776.

Géraldy, M., Morgen, M., Sehr, P., Steimbach, R. R., Moi, D., Ridinger, J., & Miller, A. K. (2019). Selective inhibition of histone deacetylase 10: Hydrogen bonding to the gatekeeper residue is implicated. *Journal of Medicinal Chemistry, 62*(9), 4426–4443.

Gorda, B., Toelzer, C., Aulicino, F., & Berger, I. (2021). Chapter six—The MultiBac BEVS: Basics, applications, performance and recent developments. In In. W. B. O'Dell, & Z. Kelman (Vol. Eds.), *Methods in Enzymology: 660*, (pp. 129–154). Academic Press.

Herp, D., Ridinger, J., Robaa, D., Shinsky, S. A., Schmidtkunz, K., Yesiloglu, T. Z., & Jung, M. (2022). First fluorescent acetylspermidine deacetylation assay for HDAC10 identifies selective inhibitors with cellular target engagement. *Chembiochem: A European Journal of Chemical Biology, 23*(14), e202200180.

Ho, T. C. S., Chan, A. H. Y., & Ganesan, A. (2020). Thirty years of HDAC inhibitors: 2020 Insight and hindsight. *Journal of Medicinal Chemistry, 63*(21), 12460–12484.

Lechner, S., Malgapo, M. I. P., Grätz, C., Steimbach, R. R., Baron, A., Rüther, P., & Médard, G. (2022). Target deconvolution of HDAC pharmacopoeia reveals MBLAC2 as common off-target. *Nature Chemical Biology, 18*(8), 812–820.

Lechner, S., Steimbach, R. R., Wang, L., Deline, M. L., Chang, Y.-C., Fromme, T., & Kuster, B. (2023). Chemoproteomic target deconvolution reveals histone deacetylases as targets of (R)-lipoic acid. *Nature Comm, 14*(1), 3548.

Marks, B. D., Fakhoury, S. A., Frazee, W. J., Eliason, H. C., & Riddle, S. M. (2011). A substrate-independent TR-FRET histone deacetylase inhibitor assay. *SLAS Discovery, 16*(10), 1247–1253.

Moreno-Yruela, C., & Olsen, C. A. (2021). High-throughput screening of histone deacetylases and determination of kinetic parameters using fluorogenic assays. *STAR Protocols, 2*(1), 100313.

Morgen, M., Steimbach, R. R., Géraldy, M., Hellweg, L., Sehr, P., Ridinger, J., & Miller, A. K. (2020). Design and synthesis of dihydroxamic acids as HDAC6/8/10 inhibitors. *ChemMedChem, 15*(13), 1163–1174.

Oehme, I., Linke, J.-P., Böck, B. C., Milde, T., Lodrini, M., Hartenstein, B., & Witt, O. (2013). Histone deacetylase 10 promotes autophagy-mediated cell survival. *Proceedings of the National Academy of Sciences of the United States of America, 110*(28), E2592–E2601.

Steimbach, R. R., Herbst-Gervasoni, C. J., Lechner, S., Stewart, T. M., Klinke, G., Ridinger, J., & Miller, A. K. (2022). Aza-SAHA derivatives are selective histone deacetylase 10 chemical probes that inhibit polyamine deacetylation and phenocopy HDAC10 knockout. *Journal of the American Chemical Society, 144*(41), 18861–18875.

Taunton, J., Hassig, C. A., & Schreiber, S. L. (1996). A mammalian histone deacetylase related to the yeast transcriptional regulator Rpd3p. *Science, 272*(5260), 408–411.

Yang, X.-J., & Seto, E. (2008). The Rpd3/Hda1 family of lysine deacetylases: From bacteria and yeast to mice and men. *Nature Reviews. Molecular Cell Biology, 9*(3), 206–218.

Yoshida, M., Kijima, M., Akita, M., & Beppu, T. (1990). Potent and specific inhibition of mammalian histone deacetylase both in vivo and in vitro by trichostatin A. *The Journal of Biological Chemistry, 265*(28), 17174–17179.

CHAPTER FOUR

Polyamine transport inhibitors: Methods and caveats associated with measuring polyamine uptake in mammalian cells

Alexandra Bunea and Otto Phanstiel IV*

College of Medicine, University of Central Florida, Orlando, FL, United States
*Corresponding author. e-mail address: otto.phanstiel@ucf.edu

Contents

1. Introduction and rationale	66
1.1 Cell selection	68
1.2 Maximum tolerated concentration of the PTI	69
1.3 PTI solubility and the use of dimethylsulfoxide (DMSO)	69
2. Assay 1 and assay 2	70
2.1 Assay 1 description and results	70
2.2 Assay 2 description and results	72
2.3 Equipment	75
2.4 Reagents	75
2.5 Procedure	76
2.6 Notes	78
2.7 Calculations	79
3. Assay 3	79
3.1 Description and results	79
3.2 Equipment	81
3.3 Reagents	81
3.4 Procedure	82
3.5 Notes	85
3.6 Calculations	86
4. Assay 3 data analysis calculations	88
5. Results and discussion	88
6. Conclusions	89
Acknowledgments	89
References	89

Abstract

Combination therapies which target both polyamine biosynthesis and polyamine transport have shown promise as anti-cancer strategies and as potentiators of the immune response. While polyamine biosynthesis inhibitors like difluoromethylornithine (DFMO) exist, cancers often escape via upregulated polyamine import. As a result, polyamine transport inhibitors (PTIs) are needed to inhibit polyamine uptake and create a 'full-court press' on polyamine metabolism. As new PTIs are developed, they need to be ranked for their ability to inhibit polyamine uptake. This paper describes three polyamine transport assays to evaluate polyamine transport inhibition. The first tests the ability of the PTI to inhibit the uptake of an anthracene-containing polyamine poison (Ant44). The second assay evaluates the ability of the PTI to inhibit the uptake of a rescuing dose of spermidine into DFMO-treated cells. The final assay is the gold standard for the field and involves determining the concentration of PTI needed to inhibit 50 % of the uptake of each of the radiolabeled native polyamines: ^{3}H-putrescine, ^{3}H-spermidine or ^{14}C-spermine. These assays provide EC_{50} and IC_{50} values which allow a formal ranking of transport inhibition potency to aid in PTI selection.

1. Introduction and rationale

The native polyamines (putrescine, spermidine and spermine) play critical roles in translation, transcription, chromatin remodeling and immune privilege (Fig. 1). Certain tumor types are addicted to polyamines and contain higher polyamine levels than normal tissues (Phanstiel, 2018). Polyamine blocking therapy (PBT) is an anti-cancer strategy which inhibits polyamine metabolism by inhibiting both polyamine biosynthesis and uptake (Alexander, Mariner, Donnelly, Phanstiel IV, & Gilmour, 2020; Alexander, Minton, Peters, Phanstiel, & Gilmour, 2017; Hayes, Burns, & Gilmour, 2014; Hayes, Shicora, et al., 2014). Inhibition of polyamines has a profound effect on tumor growth and enables an enhanced immune response in several tumor types (Alexander, Mariner, Donnelly, Phanstiel, & Gilmour, 2020; Gitto et al., 2018; Nakkina, Gitto, Beardsley, et al., 2021; Nakkina, Gitto, Pandey, et al., 2021). For example, difluoromethylornithine (DFMO) inhibits ornithine decarboxylase (ODC), which is the rate determining enzyme in polyamine biosynthesis. DFMO is FDA approved for the treatment of high-risk neuroblastoma (Yang, 2024), but typically results in decreased intracellular levels of putrescine and spermidine and a cytostatic effect. Many cancer cell types escape DFMO-induced polyamine depletion by upregulating polyamine transport to replenish their intracellular polyamine pools (Gitto et al., 2018; Madan et al., 2016). Therefore, polyamine transport inhibitors (PTIs) are needed to block this tumor escape pathway (Muth et al., 2014; Phanstiel & Archer, 2016).

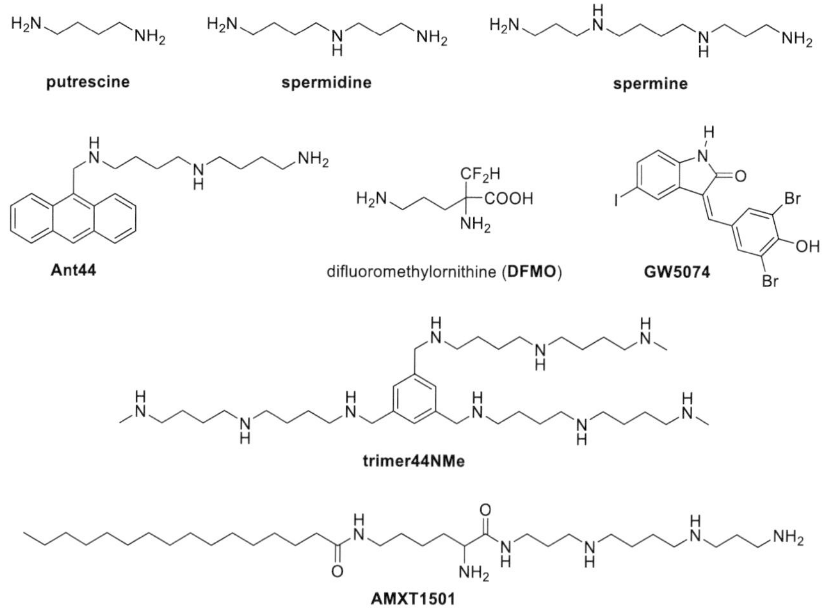

Fig. 1 Structures of the native polyamines, Ant44, DFMO, GW5074, trimer44NMe, and AMXT1501.

The design of PTIs has been challenging without knowledge of the precise molecular target. Several candidate proteins have been suggested as polyamine transporters and these have recently been reviewed (Poulin, Casero, & Soulet, 2012).

Our group and others have demonstrated that the P5B-type ATPase, ATP13A3, is the primary polyamine transport protein in mammalian cells. For example, we showed that L3.6pl human pancreatic cancer cells (Bruns, Harbison, Kuniyasu, Eue, & Fidler, 1999) have high full-length ATP13A3 expression and high polyamine import activity (Madan et al., 2016). Using CRISPR methods, we generated a functionally-dead, in-frame deletion mutant of ATP13A3 in this cell line (Sekhar, Andl, & Phanstiel IV, 2022). We then demonstrated that ATP13A3 is associated with the import of spermidine and spermine, but not putrescine in L3.6pl cells (Sekhar, Andl, & Phanstiel, 2022). This feature is especially helpful as PTI compounds which target ATP13A3 should inhibit spermidine and spermine, but not putrescine uptake in L3.6pl cells (Sekhar, Andl, & Phanstiel, 2022). This unique phenotype opens the door for screening of future ATP13A3 inhibitors. We note that polyamine transport inhibition could occur either through direct or indirect mechanisms involving ATP13A3 because the precise signaling and ATP13A3 regulatory pathway(s) have yet to be defined (Fig. 2).

This Methods paper details three assays to evaluate PTI candidates.

1.1 Cell selection

For PTI screening purposes, the selected cell line matters. Obviously, one needs to have a cell line with measurable polyamine transport activity to screen for small molecules which inhibit this activity. Until recently, this property was difficult to assess a priori. In our studies of pancreatic cancer cell lines we found that the selected cell line should have high full-length ATP13A3 and low Cav-1 protein expression (Madan et al., 2016). For example, we found that L3.6pl pancreatic cancer cells had the highest expression of full length ATP13A3 protein and low Cav1 expression and high polyamine import activity. In contrast, AsPC1 cells had only trace expression of full-length ATP13A3 and had low polyamine uptake activity (Madan et al., 2016). The basal polyamine uptake activity also reflected the cell line's obligate transport activity as measured by their ability to be rescued from DFMO treatment with exogenous spermidine. DFMO-treated L3.6pl cells were readily rescued back to 100% relative growth with exogenous spermidine, however, AsPC1 cells could not be rescued by exogenous spermidine from DFMO treatment (Madan et al., 2016).

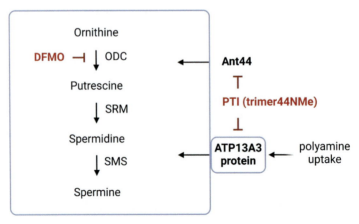

Fig. 2 Polyamine metabolism, DFMO, Ant44, and the polyamine transport protein ATP13A3.

Alternatively, Chinese hamster ovary (CHO-K1) cells are also highly responsive and were used in a high throughput screening effort to identify new PTI structures (Dobrovolskaite, Madan, Pandey, Altomare, & Phanstiel IV, 2021). Consistent with our findings in human cells, recent studies by the Vangheluwe group demonstrated that ATP13A3 is the major polyamine import protein in CHO cells (Hamouda et al., 2021) via the study of CHO-MG cells (Mandel & Flintoff, 1978).

1.2 Maximum tolerated concentration of the PTI

As several of the assays are measured via changes in relative cell growth, it is important to first determine the maximum tolerated concentration (**MTC**) of the PTI candidate. The MTC is the concentration of compound that provides >90% relative growth compared to an untreated control. This parameter is very important as PTI toxicity can give rise to false negatives in Assay 1 and false positives in Assay 2. We typically determine this value for 72 h of incubation as we found that many cell lines require 72 h to show a significant DFMO growth inhibition effect. Rather than running the separate assays at different time periods (48 h vs 72 h for example), we typically run all these assays over 72 h. Note: the time of incubation in Assay 2 can be changed to shorter times (e.g., 48 h), if the cells respond well to DFMO and are able to be rescued significantly with exogenous spermidine.

1.3 PTI solubility and the use of dimethylsulfoxide (DMSO)

There are two types of PTIs. Type 1 PTIs are polyamine-based and include trimer44NMe (Muth et al., 2014) and AMXT-1501 (Samal et al., 2013). These are water soluble and function as competitive inhibitors of polyamine uptake. Type 2 PTIs like GW5074 do not have polyamines within their structure, are insensitive to spermidine concentration and often have limited water solubility (Dobrovolskaite, Madan, Pandey, Altomare, & Phanstiel, 2021). Since polyamines are protonated at physiological pH, the Type 1 PTIs are typically readily water soluble and can be prepared as stock solutions in PBS. In cell-based assays Type 1 PTIs are preferred.

The Type 2 PTIs, however, are often not water soluble and require DMSO as a co-solvent. For this reason, it is important to determine the sensitivity of the selected cell line to DMSO concentration. We found that most cells tolerate 0.5% DMSO as a final concentration in the well over the time course of our assays. In short, DMSO sensitivity is cell line dependent and needs to be determined during the cell line selection process via a plot of DMSO concentration vs relative cell growth. If too high a DMSO concentration is used, then it can inhibit cell growth and create false negatives in Assay 1 and false positives in Assay 2. One can attain 0.5% DMSO levels and lower by conducting the 96-well plate experiments in 200 μL total volume.

We start by preparing the PTI candidate as a 10 mM stock solution in 100% DMSO. For example, if the compound has a molecular weight of 400 g/mol, we weigh out 4 mg and dissolve it in 1 mL of DMSO. We then take 0.1 mL of this vortexed 10 mM stock and attempt to dilute it with 0.9 mL PBS. If successful, this gives a 1 mM (1000 μM) stock of the PTI candidate in 10% DMSO. Adding 1 μL of this vortexed 1 mM stock solution to 199 μL of cells in media gives a 200-fold dilution resulting in 5 μM PTI candidate in 0.05% DMSO in the well, which is 10-fold below the max tolerated DMSO concentration (0.5%). If one needs to drive this DMSO value lower, one can attempt to make the initial 10 mM stock solution in 50% aqueous DMSO and perform serial dilutions with PBS or deionized water. For these reasons, the solubility properties of the PTI agent itself can result in in-well dosing limitations and introduce the DMSO-associated constraints mentioned above.

Once these important parameters are established, then one is able to perform the following assays.

2. Assay 1 and assay 2
2.1 Assay 1 description and results

The Ant44 assay tests the ability of the PTI to inhibit the uptake of a polyamine poison (N^1-anthracen-9-ylmethyl homospermidine, Ant44) into cells over a 72 h incubation period (Phanstiel, Kaur, & Delcros, 2007; Wang, Delcros, Biggerstaff, & Phanstiel, 2003). In this assay, L3.6pl cells were treated with the pre-determined 72 h IC_{50} concentration of Ant44 (3 μM). Ant44 has a homospermidine (4,4-triamine) motif within its structure and prior experiments demonstrated that Ant44 enters cells via the polyamine transport system. A PTI would be expected to block the entry of the Ant44 poison and rescue the cells back to 100% relative growth compared to an untreated control. By titrating the PTI concentration in the presence of a fixed IC_{50} concentration of Ant44, one can determine the **Assay 1 IC_{50}** value which is the concentration of PTI needed to provide 75% relative growth, i.e., halfway between the Ant44 only control (50% relative growth) and the untreated control (100%

relative growth). Changes in relative growth are measured by the MTS reagent. *The lower the Assay 1 IC_{50} value, the more potent the PTI.* This is illustrated in Fig. 2 with the trimer44NMe PTI in L3.6pl cells.

In some cases, an IC_{50} value cannot be obtained. In these cases, an EC_x value can be reported using the scale defined by the controls, where EC_0 means no Ant44 was inhibited (i.e., the Ant44 control, 50% rel growth) and an EC_{100} means 100% of Ant44 was inhibited (untreated control, 100% rel growth). For example, one can list the calculated % Ant44 inhibited from the relative growth observed and the concentration of PTI used (e.g., EC_{25} at 10 μM). This means that a relative growth value was obtained (62.5%) that corresponded to 25% of the Ant44 uptake being inhibited at 10 μM PTI. A typical readout with a PTI is shown in Fig. 3 (Dobrovolskaite, Madan, Pandey, Altomare, & Phanstiel, 2021).

We note that the 'window of response' in Assay 1 can be expanded by using higher concentrations of Ant44. For example, we found that one can inhibit relative cell growth by 75% and still recover the cells with a PTI, but this feature may be cell line dependent (e.g., the larger the beatdown/recovery window, the more robust the assay).

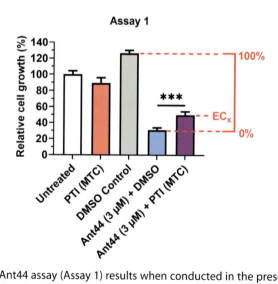

Fig. 3 Typical Ant44 assay (Assay 1) results when conducted in the presence of DMSO. The PTI is used at its MTC and in this case the window ranged from 30% to 130% relative growth. The EC_x was then calculated using the range defined by the two controls (0.5% DMSO and Ant44 in 0.5% DMSO). In this example, the relative growth observed was 50% on the y-axis which corresponds to an EC_{20} at the MTC of the PTI used (100 × (50−30)/(130−30)). In Assay 1, a positive hit results in increased cell growth as the polyamine poison is inhibited from cell entry.

2.2 Assay 2 description and results

The spermidine rescue assay (Assay 2) evaluates the ability of the PTI to inhibit the uptake of a rescuing dose of spermidine (Spd) into DFMO-treated cells over a pre-determined incubation period. The incubation period is established by first determining the sensitivity of the selected cell line to difluoromethylornithine (DFMO).

DFMO is an inhibitor of ornithine decarboxylase (ODC) which is the rate determining step in polyamine biosynthesis (Fig. 2). DFMO often results in a cytostatic effect which is overcome by obligate polyamine transport. We found that DFMO treatment results in a translocation of ATP13A3 from the nucleolus to the cell surface and this plasma membrane localization is dependent upon exogenous spermidine supply (Sekhar, Andl, & Phanstiel, 2022). We note that these cell-based microscopy observations provide a visual mechanism for how DFMO stimulates polyamine uptake.

In terms of the assay, we found that pancreatic cancer cell lines need 72 h of exposure to DFMO to become significantly growth inhibited. To reach 50% growth inhibition compared to an untreated control typically requires the use of 0.5 to 15 mM DFMO and depends upon the cell line. For example, the 72 h DFMO IC_{50} was 4.2 mM in L3.6pl cells, but 11.8 mM in AsPC1 cells (Madan et al., 2016). While these concentrations appear high, they are very typical for in vitro DFMO studies.

High DFMO concentrations above 10 mM are to be avoided if possible and one can typically halve the DFMO IC_{50} value by extending the experiment another 24 h. For example, the DFMO IC_{50} value in L3.6pl cells was 8 mM after 48 h and 4.2 mM after 72 h incubation. The goal here is to have the DFMO IC_{50} value < 5 mM, if possible. Again, extending the assay period to 96 h can help drive the DFMO concentration even lower.

We have noted in several cell lines that DFMO can at times result in a flat dose-response curve over a significant range of DFMO concentrations (1–10 mM for example). Often times the DFMO curve remains flat just above the 50% inhibition mark which is frustrating but again cell line dependent. We found that if the DFMO IC_{50} value is below 5 mM, then the cell line will likely behave well in Assay 2. In sum, one needs a significant growth arrest of cells with DFMO to generate a large enough 'window of response' when one adds exogenous spermidine. The greater the growth inhibition with DFMO, the larger the potential rescue window with spermidine and the better the assay will perform. However, if one pushes the system and uses too high a concentration of DFMO (>10 mM) then one may be unable to rescue those cells back to 100% relative growth.

Once the optimal concentration of DFMO has been determined, one needs to also determine the minimum level of spermidine that can rescue the cells from the 72 h IC_{50} concentration of DFMO. Typically, this is 1 μM spermidine. The goal here is to use the minimum amount of spermidine because Type 1 PTIs are competitive inhibitors of polyamine uptake and are sensitive to spermidine concentration. As such, they (Type 1 PTIs) will not perform well if the spermidine level is too high (Dobrovolskaite, Madan, Pandey, Altomare, & Phanstiel, 2021). We note that Type 2 PTIs can be unaffected by high spermidine concentrations (Dobrovolskaite, Madan, Pandey, Altomare, & Phanstiel, 2021).

A typical Assay 2 experiment will be run in triplicate and have the controls run on each plate each time. Because multiple agents are added to the wells, the cells are initially plated at 170 μL in media containing 10% FBS, 1% Antibiotic cocktail (Pen-Strep) and 250 μM aminoguanidine (AG). AG is needed to suppress the action of amine oxidases present in the culture media. AG is an important additive in all assays containing free polyamines or polyamine structures as its activity can lead to toxic byproducts such as hydrogen peroxide and acrolein. The cells are incubated overnight to allow them time to adhere to the plate.

The following day (drugging day or Day 0) the relevant DFMO and Spd additions are added via 10 μL additions. In certain instances, PBS is added to bring the total volume up to the desired final concentration of the experiment. For example, in the DFMO-only experiment, DFMO is added via 10 μL of an 84 mM DFMO stock in PBS and then PBS (20 μL) is added to bring the total to 200 μL (and the final DFMO concentration to 4.2 mM). In the case of the DFMO+Spd+PTI experiments, each additive is added as a 10 μL aliquot to bring the final volume to 200 μL up from the initial 170 μL. Note this method is for Type 1 PTIs and results in a 20-fold dilution step of the stock solutions used.

For water insoluble Type 2 PTIs like GW5074 (Fig. 1) that require DMSO, we typically plate the cells in 180 μL volume and add the DFMO and Spd as separate 10 μL aliquots and then add the PTI as a 1 μL addition giving a total volume of 201 μL. This allows one to perform essentially a 1:200 dilution in an effort to drive the DMSO concentration below 0.5% DMSO in each well. The controls include: untreated cells (e.g., L3.6pl cells), DFMO-only (dosed at its IC_{50} concentration of 4.2 mM), spermidine only (e.g., Spd at 1 μM), DFMO+Spd (4.2 mM and 1 μM, respectively), and the

PTI at its 72 h MTC (e.g., trimer44NMe, 3 μM). The other wells include fixed concentrations of DFMO (4.2 mM) + spermidine (Spd, 1 μM) and increasing concentrations of PTI up to its MTC value (typically across the range of 0.05 to 3 μM for the trimer44NMe). Changes in relative growth are measured by the MTS reagent. The controls establish the range of the assay. Typically, by design, DFMO treatment gives 50% relative growth, and the DFMO + spermidine control gives 100% relative growth. This approach sets the range of the expected readouts and allows one to determine the Assay 2 EC_{50} value, which is the concentration of PTI needed to give 75% relative growth in the presence of DFMO+Spd, i.e., a reading halfway between the DFMO-only control and the DFMO+Spd control. A typical DFMO-spermidine rescue experiment (Assay 2) using a Type 2 PTI in L3.6pl cells is illustrated in Fig. 4 (Dobrovolskaite, Madan, Pandey, Altomare, & Phanstiel, 2021).

Since the presence of the PTI in this assay results in lower relative cell growth, it is critical that the PTI is not dosed above its MTC value. If the

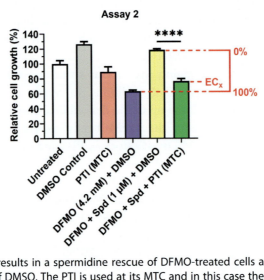

Fig. 4 Typical results in a spermidine rescue of DFMO-treated cells assay (Assay 2) in the presence of DMSO. The PTI is used at its MTC and in this case the window ranged from 60% (DFMO) to 120% relative growth (DFMO+Spd). The EC_x was then calculated using the range defined by the two controls (DFMO and DFMO+Spd). In this example, the relative growth observed for DFMO+Spd+PTI was 75% on the y-axis, which corresponds to an EC_{75} at the MTC of the PTI used (100 x (120−75)/(120−60)). Note: a positive hit in Assay 2 gives decreased relative growth as the rescuing dose of Spd is inhibited from cell entry. Therefore, the EC_x calculation here is different than Assay 1 in Fig. 3.

PTI is dosed at a toxic concentration, then one will get a false positive in this assay as the lower relative growth will come from PTI toxicity and not from its ability to block spermidine uptake. Conceptually, the EC_{50} value reflects the concentration of PTI needed to block 50% of the rescuing dose of spermidine entering into DFMO treated cells. *The lower the EC_{50} value in Assay 2, the more potent the PTI in its ability to inhibit obligate polyamine transport (i.e., DFMO-stimulated import).*

In practice, DFMO and the PTI are co-dosed in animal models to provide a full court press on polyamine metabolism as part of polyamine blocking therapy (PBT) (Alexander, Mariner, Donnelly, Phanstiel, & Gilmour, 2020; Alexander, Minton, Peters, Phanstiel, & Gilmour, 2017; Nakkina et al., 2021; Phanstiel, 2018; Alexander et al., 2023). Therefore, the Assay 2 EC_{50} value is the most clinically relevant parameter when one is designing PTIs for PBT applications.

Please note that Assay 1 and Assay 2 can be run in parallel as they require the same equipment, reagents, and procedure as noted below with differences italicized. While ~200 μL total assay volumes are preferred, we note that these assays can be run at 100 μL total volumes, if PTI solubility and DMSO toxicity are not a problem.

2.3 Equipment
1. 96-well plates (Corning #3599).
2. Vortex Mixer (VWR #945302).
3. Centrifuge (Eppendorf #5810 R).
4. Incubator (Nuaire #NU-8700).
5. Plate Reader (BioTek Synergy MX Multi-Mode Microplate Reader).

2.4 Reagents
1. RPMI 1640 Media (Gibco #11835030) with 1% Antibiotic-Antimycotic (Gibco # 15240062, contains penicillin, streptomycin and Amphotericin B) and 10% FBS (R&D Systems #S11150).
2. Aminoguanidine (AG) (Sigma-Aldrich #A8835).
3. *Ant44* (Synthesized by its published method (Wang et al., 2003)).
4. *DFMO* (Tocris #276110).
5. *Spermidine* (Acros Organics #13274).
6. Tested PTI Compounds.
7. 1X DPBS (Gibco #14190144).
8. MTS Reagent (Abcam #ab197010).

2.5 Procedure

1. Day −1 (Fig. 5) (e.g., Monday).
   a) L3.6pl cells are seeded in a 96-well test plate at 500 cells/well using 170 μL volume of media containing 250 μM AG (see Notes 1 and 2).
   b) We also prepare a separate control plate containing six wells with 500 cells/well using 170 μL volume of media containing 250 μM AG. This plate also contains three additional wells with media only. This control plate allows one to measure how much cell growth has occurred after the overnight incubation (using the MTS reagent). This information will be important later on to calculate the relative new growth that has occurred over the incubation time.
   c) The cells are allowed to adhere overnight at 37°C in a humidified incubator with 5% CO_2.
2. Day 0 (Fig. 5) (e.g., Tuesday) (see Note 3).

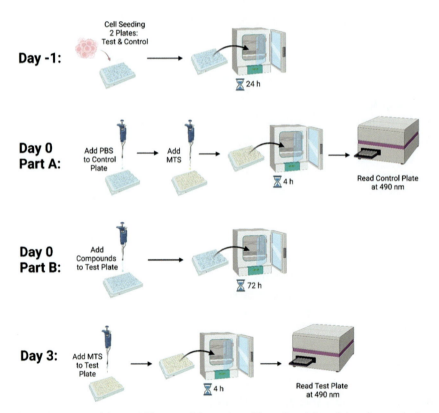

Fig. 5 Assay 1 and Assay 2 Illustrated Procedure. Note: Day 0 Part A involves only the control plate.

a) Prepare all compound stocks, depending on the solubility of the compound, this may be in 1X PBS, DMSO, or a combination of both.
   i. Ant44, DFMO, and Spermidine are all soluble in 1X PBS.
   ii. Stock concentration will depend on tested concentration in the well. For example, adding 10 µL of stock into 200 µL final volume (i.e., using a 60 µM stock of Ant44 for 3 µM in the well) requires a 20X stock solution.
b) Add 30 µL of PBS and then 20 µL MTS reagent (i.e., 3-(4,5-dimethylthiazol-2-yl)-5-(3- carboxymethoxyphenyl)-2-(4-sulfophenyl)-2H-tetrazolium, inner salt) to each well of the control plate only (to generate the Day 0 control readout). These volumes will be added to each of the 3 media-only wells and the six control wells (media with cells). This should be completed at the time of drug/compound addition to the test plate (on Day 0). Note the total volume during the MTS analysis step will now be 170 µL + 30 µL of PBS + 20 µL MTS (or 220 µL total).
   i. Take note of the time of the MTS addition to the control plate (e.g., 11:00 am) as this is the time when the MTS reagent will need to be added for the 72 h reading in the test plate (on Day 3).
   ii. Incubate the Day 0 control plate with PBS and MTS added at 37°C in a humidified incubator with 5% CO_2 for 4 h and read the absorbance readouts at 490 nm on a plate reader and save this control data for later analysis.
   iii. If testing multiple cell lines, make sure to always collect the relevant Day 0 information, i.e., the 490 nm absorbance of the media-only wells and the control wells containing media and cells per cell type (and per media type, if applicable). This information will be used to calculate the *relative new growth* that has occurred over the 72 h incubation period and allows one to calculate (and later subtract out) the amount of cell growth that has occurred up to the time point of drug addition.
c) To the test plate, the appropriate stock solutions of compounds (Ant44, DFMO, Spd and PTI) are added in respective 10 µL volumes. Final volume in each well is made equal to 200 µL by adding 1X PBS (if necessary, e.g., Day 3 blank and control wells).
   i. *Assay 1:*
      i. Use the appropriate stock solution to achieve the desired in-well concentration, i.e., the 72 h IC50 of Ant44 in L3.6pl cells is 3 µM.
   ii. *Assay 2:*

i. Use the appropriate stock solution to achieve the desired in-well concentration, i.e., the 72 h IC_{50} of DFMO in L3.6pl cells is 4.2 mM.
   ii. Use 1 µM of Spd, this can be increased to 2 µM if a larger rescue window is needed to see an effect.
  d) Each plate should have the following controls:
   i. Blank (media only)
   ii. Untreated control (media + cells)
   iii. DMSO control (if using compound solutions containing DMSO)
   iv. Each PTI alone at its MTC
   v. *Assay 1:*
    i. Ant44 alone
   vi. *Assay 2:*
    ii. DFMO alone
    iii. Spd alone
    iv. DFMO + Spd
  e) After addition of each compound, to ensure each component is mixed, firmly hold the plate on a flat surface and move in a 'figure 8' motion for 10-15 s.
  f) Incubate plate at 37°C in a humidified incubator with 5% CO_2 for the required incubation time (72 h). Do not forget to read the 'Day 0' control plate (490 nm) after 4 h incubation with MTS reagent.
3. Day 3 (Fig. 5) (e.g., Friday) (see Note 4).
  a) Add 20 µL MTS reagent to all wells of the test plate for 72 h reading.
  b) Incubate test plate at 37°C in a humidified incubator with 5% CO_2 for 4 h.
  c) Read plate (490 nm) after 4 h incubation with MTS reagent.
  d) Calculate relative % growth for each triplicate, see Section 2.7 below.

2.6 Notes

1. Seeding volume on 'Day −1' may differ depending on the DMSO concentration of the tested compounds. See Section 1.3 for details.
2. Seeding concentration/density on 'Day −1' will differ based on the cell line being used. Choose a cell density that will result in a Day 3 control absorbance reading between 1.0–2.0 (for example L3.6pl cells require 500–1000 cells/well on Day −1).
3. 'Day 0' of the experiment is when compounds are added to the cells on the test plate and a 'Day 0' control plate is read to provide an absorbance reading for cells and media for later calculations.

4. 'Day 3' of the experiment is the day when the plates are read for the 72 h incubation timepoint. Note this time is cell line dependent and can be 48 h or 96 h, if desired.

2.7 Calculations
1. The relative % growth calculation will be the same for both of these assays.
2. Day 0 Control = Day 0 absorbance for the untreated control cells − Day 0 media only background absorbance
3. Day 3 Control = Day 3 absorbance for the untreated control cells − Day 3 media only background absorbance
4. Relative % Growth = $100\% \times \frac{[(\text{Day 3 experimental reading} - \text{Day 3 background}) - (\text{Day 0 control})]}{(\text{Day 3 control} - \text{Day 0 control})}$

3. Assay 3
3.1 Description and results

The radiolabeled polyamine uptake assay is the gold standard for the field and involves determining the concentration of PTI needed to inhibit 50% of the uptake of the respective radiolabeled native polyamine: ^{3}H-putrescine, ^{3}H-spermidine or ^{14}C-spermine into cells (e.g., L3.6pl) over a 15-minute incubation period.

This assay is performed in triplicate by using a fixed concentration of radiolabeled polyamine and incubating cells for a fixed short period of time (e.g., 15 min) at 37 °C in a 5% CO_2 atmosphere. Note, the concentration used will be determined by the radiolabeled polyamine used, see Assay 3 Calculations in Section 3.6 below for these details. The cells are then placed on ice (to stop active transport) and ice-cold PBS is used to wash the cells three times. The resulting cells are then lysed using 0.1% sodium dodecyl sulfate (SDS) and a sample of this lysate (200 μL) is added to scintillation fluid (2 mL) and the sample is read on a liquid scintillation counter which reports counts per minute (cpm). A blank control experiment is run in parallel where no radiolabeled polyamines are added. The blank background reading is typically very low (e.g., ~10 cpm). Note this background value is subtracted out of each radiolabeled output reading to give the corrected final value where the radioactive background has been subtracted out. The corrected data from the untreated control provides the cpm value associated with 100% uptake (e.g., >1000 cpm) and this value is lowered in the presence of increasing PTI concentration. An Assay 3 IC_{50} ^{3}H-Spd uptake can then be obtained by determining the concentration of the PTI needed to inhibit 50% of ^{3}H-spermidine uptake, i.e., needed to give half of the untreated control cpm value.

Typically, the time of incubation is 15 min or 30 min but can be as long as 1 h depending upon the cells. The key here is that this needs to be a kinetic experiment with limited equilibration time and therefore a short time of incubation is preferred. The time period is defined by the polyamine uptake rate of the cell line used as well as the level of radioactivity used. The goal is to create a readout signal >100 fold higher than background to give statistically significant results. A typical readout using a Type 2 PTI in L3.6pl cells is shown in Fig. 6 (Dobrovolskaite, Madan, Pandey, Altomare, & Phanstiel, 2021). This assay yields the best understanding of polyamine transport due to the sensitivity of detecting radiolabeled polyamines intracellularly, allowing for interpretation of the PTI's specific effect on each polyamine substrate.

There are a few factors to take note of regarding cell selection for Assay 3, in addition to the previously stated guidelines above. It is important to use adherent cell lines as this assay requires several washing steps that a loosely adherent cell line may not be able to tolerate. As a result, other precautions may need to be taken for this assay to be successful. Our group has not yet investigated an alternative option for loosely adherent cell lines but some options to consider include (a) centrifuging the 24-well plate between wash steps to ensure cells remain behind or (b) using disposable syringe filters which capture the cells and allow washing steps followed by cell lysis. The use of adherent cell lines, such as L3.6pl, are preferred.

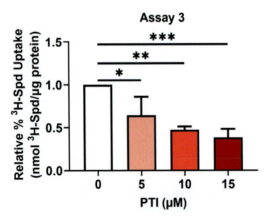

Fig. 6 Typical results in a radiolabeled uptake assay (Assay 3).

When planning this experiment, it is important to investigate the concentration of cells necessary for 60% confluency in the well for a 24-well plate for a 24-hour growth period. In the case of L3.6pl cells, this concentration is 100,000 cells per well.

This is a two-day protocol where Day 1 is the initial seeding of the cells, and Day 2 is the polyamine uptake experiment, preparation of each cell lysate and detection of the radiolabeled polyamines.

3.2 Equipment
1. 24-well plates (Corning #3524).
2. Vortex Mixer (VWR #945302).
3. Microcentrifuge (Eppendorf #5424).
4. Incubator (Nuaire #NU-8700).
5. Water bath (Buchi #B-491).
6. 1.5 mL microcentrifuge tubes (Fisher #02682002).
7. Scintillation vials (Fisher #0333720).
8. Liquid scintillation counter (LSC) (PerkinElmer Tri-Carb 4910TR #A491000).
9. Cell Scrapers (Alkali Scientific #TC7023).
10. Plate Reader (BioTek Synergy MX Multi-Mode Microplate Reader).

3.3 Reagents
1. RPMI 1640 Media (Gibco #11835030) with 1% Antibiotic-Antimycotic (Gibco # 15240062, contains penicillin, streptomycin and Amphotericin B) and 10% FBS (R&D Systems #S11150).
2. Aminoguanidine (AG) (Sigma-Aldrich #A8835).
3. Radiolabeled Polyamines.
   - ^{3}H-putrescine (PerkinElmer #NET1185001MC)
   - ^{3}H-spermidine (PerkinElmer #NET522001MC)
   - ^{14}C-spermine (American Radiolabeled Chemicals, Inc. #ARC3139-50)
4. Hanks' Balanced Salt Solution (HBSS) media (containing Ca^{2+} and Mg^{2+}) (Gibco # 24020117).
5. HBSS media (without Ca^{2+} and Mg^{2+}) (Gibco #14170112).
6. Tested PTI Compounds.
7. 1X DPBS (Gibco #14190144).
8. 0.1% SDS solution made in deionized water (Fisher #BP166–100).
9. Scintillation fluid (Fisher #SX18-4).
10. Pierce BCA Protein Assay Kit (Fisher #PI23225).

3.4 Procedure

CAUTION: The use, handling, and disposal of radioactive materials can be hazardous. Investigators are responsible for abiding by their institution's regulations for working with radioactive materials. Additional radiation safety training is required to perform these assays and involves the safe handling, tracking and disposal of radioactive materials.

1. Day 1 (Fig. 7): L3.6pl cells are seeded in a 24-well plate at 100,000 cells/well using 300 μL volume of media with 250 μM AG. Cells are incubated in a 5% CO_2 humidified incubator at 37°C for 24 h (see Note 1).
2. Day 2 (Fig. 7): Prepare the following before continuing the protocol (see Note 2):

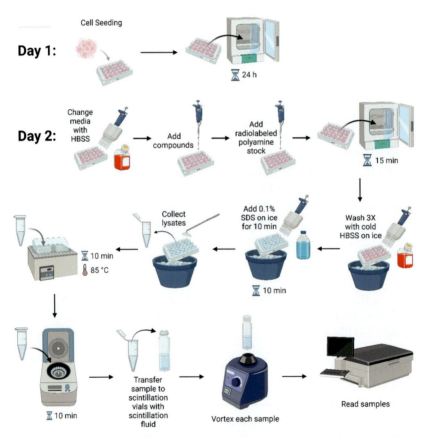

Fig. 7 Assay 3 Illustrated Procedure. Note: the radioactivity data is then normalized by dividing by the protein level of the respective lysate as determined by the BCA method to give nmol of polyamine per μg protein.

a) Preheat HBSS media (containing Ca^{2+} and Mg^{2+}) to 37°C.
b) Have an ice bucket with cold HBSS media (without Ca^{2+} and Mg^{2+}) set aside.
c) Prepare a 0.1% SDS solution made in deionized water.
d) Prepare 10 μM ***hot** polyamine stock for the experiment (see Section 3.6 below for calculations). This is done by combining, for example, 1 μL of 30.8 μM ^{3}H-spermidine with 99 μL of 30.8 μM unlabeled spermidine to give a 100 μL of 30.8 μM spermidine solution wherein the radioactivity is 1% of the original purchased stock. The idea here is to save reagent and use enough radioactivity for robust detection (e.g., >100 fold higher than background) but not too much which might overwhelm the LSC detector.
e) Label all microcentrifuge tubes necessary for lysate collection – one per well. For example, control #1-3, PTI #1-3, DMSO #1-3.
f) Label all scintillation vials necessary for reading each lysate – one vial is used for each well (e.g., control #1-3, PTI #1-3, DMSO #1-3). A hot polyamine control vial is run in parallel to help understand the 100% range on the LSC for that run (e.g., ^{3}H-Spd control).
g) Prepare all compound stocks (see Notes 3 and 4).
h) Prepare BCA standards in 0.1% SDS solution.
i) Calibrate the scintillation counter (Note: some models require calibration every 24 h when used regularly, be sure to check the user manual).

3. Remove the media from the 24-well plate and add 240 μL of preheated 37°C HBSS (containing Ca^{2+} and Mg^{2+}) to each well.
4. Add 30 μL of each compound stock to their respective wells (see Note 3).
   a) Note, for the controls, one adds 30 μL of 1X PBS or DMSO stock (see Note 4).
   b) Each well should be brought up to 270 μL (HBSS + tested condition).
5. Add 30 μL of the 10 μM *****hot** polyamine stock for the experiment (which gives 300 μL total volume in each well) to give 1 μM of *****hot** polyamines in the well (see Note 5).
6. Incubate the plate(s) for 15 min at 37°C in a 5% CO_2 humidified incubator.
   a) Make note of the order in which the *****hot** polyamine stock was added (e.g., well #1 is first, then well #2 is second, etc.), as this will be the same order in which the lysates are collected.
7. After addition of each compound and radioactive polyamine stock, to ensure each component is mixed, firmly hold the plate on a flat surface and

move in a 'figure 8' motion for 10–15 s. Note, there will be a color change from pink to yellow when mixed properly as the HBSS media contains phenol red and the radioactive polyamine stock is acidic.
8. After the 15-minute incubation, transfer the plate(s) onto ice immediately to stop active transport.
9. Wash each well 3 times with 300 μL ice-cold HBSS (without Ca^{2+} and Mg^{2+}) (see Note 6).
10. Add 300 μL of 0.1% SDS to each well while the plate(s) remain on ice, for 10 min.
    a) The purpose of the SDS is to lyse the cells open, so the intracellular radiolabeled polyamines can be measured.
11. Using a cell scraper, scrape the lysate from each well and collect it into the previously labeled microcentrifuge tubes (e.g., control #1–3, PTI #1–3, DMSO #1–3).
    a) Be sure that each well has its own respective microcentrifuge tube, do not combine wells for the triplicates.
12. Incubate the microcentrifuge tubes at 85°C for 10 min.
13. Centrifuge the microcentrifuge tubes at 15,000 rpm for 10 min.
    a) Note: in step 15 below we will take 200 μL of the supernatant.
14. As the samples are spinning down, prepare the following:
    a) Add 2 mL of scintillation fluid into each of the previously-labeled scintillation vials (e.g., control #1-3, PTI #1-3, DMSO #1-3).
    b) Prepare the radioactive polyamine standard(s) by adding 20 μL of the 10 μM radioactive polyamine stock for the experiment and 180 μL of 0.1% SDS into an empty scintillation vial that will serve as the standard, then add the 2 mL of scintillation fluid on top, then vortex to mix.
        i. Respective radioactive polyamine standards are needed for each polyamine being tested in the experiment (e.g., ^{3}H-Spd control).
    c) Have one blank scintillation vial which will contain scintillation fluid only, this will act as the background reading which will later be subtracted during the data calculations.
15. Transfer 200 μL of lysate from each microcentrifuge tube (see step 13) and add it to its respective, labeled scintillation vial which should already have 2 mL of scintillation fluid. There should be 1 scintillation vial per well/tube, do not combine triplicates.
16. Once each vial has received its respective lysate and the standards and blank have been prepared, close each vial and vortex each individual vial for 10–15 s. *This step is very important,* do not skip the vortex step as the

samples must be mixed well before reading, otherwise the scintillation counts may be too low to analyze.
17. Place the vials in the LSC to be read (see Note 7).
18. While the samples are being read in the LSC, prepare the BCA protein detection kit. Use the leftover lysates to test for protein analysis in addition to the BCA standards prepared in 0.1% SDS and follow the kit manufacturer's instructions. It is important to complete a BCA analysis for each experiment to normalize the intracellular radiolabeled polyamine levels vs cellular protein levels, as each sample amount may differ slightly.
19. Calculate concentration of polyamine normalized to protein for each triplicate, see Section 3.6 below for calculations. The data are reported in nmol polyamine/μg protein.

3.5 Notes

1. Seed enough wells on Day 1 so there are 3 wells for each experimental condition tested so that each measurement is tested in triplicate.
2. Preparation is key for this protocol as the time frame between each step is rapid, so it is best to have everything ready in advance.
3. For the tested compounds, stocks should be 10 times higher (10X) than the well concentration (e.g., using a 100 μM stock for 10 μM in the well, or 10% DMSO stock for 1% DMSO in the well). If using DMSO-based compounds, stocks should match the highest final DMSO concentration in the well (1%).
   - If the compound has solubility issues in PBS: add 30 μL of 1X PBS and 3 μL of compound stock in 100% DMSO (e.g., using a 1 mM compound stock in 100% DMSO for 10 μM compound in 1% DMSO in the well).
   - If comparing water-soluble compounds to DMSO-based compounds: DMSO will have to be added to ensure comparison across the entire experiment. This may also mean optimizing the water-soluble compound in the presence of DMSO as this may change its PTI potency.
4. For the untreated condition: for water soluble PTI candidates use 30 μL of 1X PBS; if using a vehicle such as DMSO for poorly water soluble PTI candidates, make sure to include a vehicle control such as 30 μL of a DMSO-only stock. In our experiments, we do not exceed 1% DMSO in the well. For future experiments, if it is established that the vehicle has no effect on the results, the DMSO/vehicle only control can be used in place of the untreated condition wells to save on radioactive reagents moving forward.

5. The best way to dispense radiolabeled polyamine stock in step 5 is via an electronic repeater pipette to ensure each well receives it at relatively the same time for this time sensitive section of the experiment protocol.
6. For the cold wash in step 9, be sure to use a separate aspirator and waste container to dispose of the radioactive liquid waste properly. Also, using a multichannel pipette for the wash steps can help make this step quick so the extracellular radiolabeled polyamines are removed effectively.
7. In step 17, make sure to use the appropriate protocol and LSC flag/tags on the loading rack for the radiolabeled polyamine being used (i.e., ^{3}H-spermidine samples will have a different protocol and flag/tag compared to ^{14}C-spermine samples as the tritium and carbon-14 isotopes will be detected differently). In the case that multiple radiolabeled polyamines are being tested in the same experiment, be sure to match the samples and standards with their appropriate protocol and tags and perform back-to-back readings on the LSC.

3.6 Calculations

1. Hot and Cold Polyamine Stocks.

    Regardless of which polyamine is being used (^{3}H-putrescine, ^{3}H-spermidine or ^{14}C-spermine), make a 10 µM stock which will result in 1 µM in the well. However, depending on the polyamine being used, the ratio of radiolabeled (hot) to unlabeled (cold) polyamines will differ when making the respective stocks. Several examples are given for clarity below.

    Calculate the required concentration of the purchased radiolabeled polyamine, which will differ based on the manufacturer and lot number. Then, make an unlabeled polyamine stock of the same concentration in 0.01 M HCl. Next, these two will be combined in a ratio specific to the polyamine being tested, provided below. Use $M_1V_1 = M_2V_2$ to make the 10 µM mixed ratio *__hot__ stock that will be used in the experiment. The purpose of creating this ratio stock is to not overwhelm the LSC as it is very sensitive to detecting the signal.

    The idea here is to first dilute the radioactivity level into the desired range (0.01 mCi/mL for ^{3}H-Spd, 0.04 mCi/mL for ^{3}H-Put, 0.2 mCi/mL for ^{14}C-Spm) for the **diluted radioactivity stock**, while keeping the polyamine concentration the same. Next, one dilutes the polyamine (e.g., spermidine) to the desired polyamine concentration for the uptake assay (Assay 3). In general, using these radioactivity ranges, we see readouts after incubation with L3.6pl cells for 15 min of >1000 cpm for the untreated controls after workup, which is well above the background of ~10 cpm.

Polyamine transport inhibitors

To treat one full 24-well plate would require 720 µL of 10 µM *hot spermidine stock (30 µL per well). However, additional *hot standard is needed for the LSC reading too, which is an additional 20 µL, bringing the minimum volume needed to 740 µL. Even if using low-retention pipette tips, there will still be some loss to account for, so it is best to make excess stock solution, giving a final volume of 800 µL.

The following descriptions allow the user to generate 800 µL of hot stock for the experiment of each of the native polyamines for one full 24-well plate. These can be scaled accordingly.

2. ^{3}H-spermidine.

3H-spermidine was purchased as: 1 mCi/mL in 0.01 M HCl with a concentration of 32.4 Ci/mmol, resulting in a 30.8 µM solution. The **diluted radioactive** spermidine stock was made by combining 1 part of the purchased ^{3}H-labeled spermidine (30.8 µM, 3 µL) with 99 parts of unlabeled 30.8 µM spermidine solution in 0.01 M HCl by volume (297 µL). This dilutes the purchased radioactivity to 0.01 mCi/mL, while maintaining the concentration of spermidine at 30.8 µM in 300 µL volume. To make the 10 µM **hot stock for the experiment**, one simply needs to perform a ~1:2 dilution of this diluted radioactive spermidine stock. For example, one mixes 260 µL of the diluted radioactive spermidine stock with 540 µL of the 0.01 M HCl to make 800 µL of 10 µM hot spermidine stock for the experiment. The mixture should be vortexed to ensure good mixing before use.

3. ^{3}H-putrescine.

^{3}H-putrescine was purchased as 1 mCi/mL in 0.01 M HCl with a concentration of 28.0 Ci/mmol, resulting in a 35.7 µM solution. A similar approach was used here to prepare the diluted radioactive putrescine stock with 4 parts of the purchased radioactive putrescine stock (12 µL) and 96 parts of the unlabeled 35.7 µM putrescine solution in 0.01 M HCl by volume (288 µL). This dilutes the purchased radioactivity to 0.04 mCi/mL, while maintaining the concentration of putrescine at 35.7 µM in 300 µL volume. The appropriate dilution was then performed with 0.01 M HCl to generate the desired 10 µM hot putrescine stock for the experiment. For example, one mixes 224 µL of the diluted radioactive putrescine stock with 576 µL of the 0.01 M HCl to make 800 µL of 10 µM hot putrescine stock for the experiment.

4. ^{14}C-spermine.

^{14}C-spermine was obtained as 1 mCi/mL in 0.01 M HCl with a concentration of 2.258 Ci/mmol, resulting in a 442.9 µM solution. A similar approach was used here to prepare the diluted radioactive spermine stock with 1 part of the purchased radioactive spermine stock (5 μL) and 4 parts of the unlabeled 442.9 μM spermine solution in 0.01 M HCl by volume (20 μL). This effectively dilutes the purchased radioactivity to 0.20 mCi/mL, while maintaining the concentration of spermine at 442.9 μM in 25 μL volume. The appropriate dilution was then performed with 0.01 M HCl to generate the desired 10 μM hot spermine stock for the experiment. For example, one mixes 18 μL of the diluted radioactive spermine stock with 782 μL of the 0.01 M HCl to make 800 μL of 10 μM hot spermine stock for the experiment.

4. Assay 3 data analysis calculations

1. Create BCA Standard Curve based on Pierce kit instructions to obtain the slope-intercept form of the line (y = mx+b).
2. Determine the protein concentration (μg/μL) for each well by setting the absorbance for 'y' and solving for 'x'.
3. Multiple this value by 200 to obtain the total protein (μg) in the scintillation vial (because there was 200 μL of each lysate per vial analyzed).
4. In a separate column, list the cpm value for each sample and subtract the background cpm reading.
5. Divide these corrected cpm readings by the ***hot** polyamine standard cpm reading to determine nmol of polyamine in the vial.
6. Divide those values from step 5 by the total protein values calculated in step 3, this will result in nmol of polyamine per μg protein in the 200 μL sample.
7. Average the triplicates in step 6 and calculate standard deviation and % error.

5. Results and discussion

Together these assays provide important tools for PTI screening and validation. Assays 1 and 2 avoid the use of radioactive materials and represent complementary assays wherein a PTI gives increased growth in Assay 1 (Ant44 assay) and decreased growth in DFMO+Spd assay (Assay 2). These orthogonal readouts provide a powerful screening tool to identify false negatives and positives. Both Assays 1 and 2 are routinely run in 96-well format but have also been scaled down to 1536-well format and used in a high throughput PTI screen conducted using CHO-K1 cells and a library of over 300,000 compounds at the Sanford Burnham Orlando facility. Note: Assay 1 had z' scores ranging from 0.56–0.62 in 1536-well format and identified the c-Raf inhibitor GW5074 as a Type 2 PTI (Dobrovolskaite, Madan, Pandey, Altomare, & Phanstiel, 2021) and another construct as a FUBP1 inhibitor (Dobrovolskaite et al., 2022). Assay 2 had a z' score of 0.43 in 1536-well format (Dobrovolskaite et al., 2022). True PTIs are expected to be positive hits in both Assays 1 and 2.

Assay 3 is the gold standard to demonstrate inhibition of polyamine uptake. To avoid the use of radioactivity we note that other labs have used fluorescently labeled (e.g., BODIPY-Spm) polyamines to monitor uptake with promising results consistent with radiolabeled polyamine uptake experiments performed in parallel (Dobrovolskaite, Madan, Pandey, Altomare, & Phanstiel, 2021; Houdou et al., 2023).

6. Conclusions

In summary, several assays exist to assist in the identification and development of PTIs for clinical use in combination with DFMO. The availability of orthogonal assays which can be used in a high throughput fashion is an advantage and provides improved scientific rigor in the drug discovery process. Final validation using radiolabeled or fluorescently labeled polyamines is warranted as in some cases compounds which were positive hits in Assay 2 operated via an unexpected alternative mechanism (Dobrovolskaite et al., 2022).

Acknowledgments

The authors wish to thank Drs. Jean Guy Delcros and Jacques Moulinoux at the University of Rennes, France and Drs. Holly Moots and Aiste Dobrovolskaite at the University of Central Florida for their help in developing and optimizing these assays. Note: Figs. 5 and 7 were created in https://BioRender.com.

References

Alexander, E. T., Fahey, E., Phanstiel, IV, O., & Gilmour, S. K. (2023). Loss of anti-tumor efficacy by polyamine blocking therapy in GCN2 null mice. *Biomedicines, 11*(10), 2703. doi:10.3390/biomedicines11102703.

Alexander, E. T., Mariner, K., Donnelly, J., Phanstiel, O., & Gilmour, S. K. (2020). Polyamine blocking therapy decreases survival of tumor-infiltrating immunosuppressive myeloid cells and enhances the antitumor efficacy of PD-1 blockade. *Molecular Cancer Therapeutics, 19*(10), 2012–2022. https://doi.org/10.1158/1535-7163.MCT-19-1116.

Alexander, E. T., Minton, A., Peters, M. C., Phanstiel, O., & Gilmour, S. K. (2017). A novel polyamine blockade therapy activates an anti-tumor immune response. *Oncotarget, 8*(48), 84140–84152. https://doi.org/10.18632/oncotarget.20493.

Bruns, C. J., Harbison, M. T., Kuniyasu, H., Eue, I., & Fidler, I. J. (1999). In vivo selection and characterization of metastatic variants from human pancreatic adenocarcinoma by using orthotopic implantation in nude mice. *Neoplasia (New York, N. Y.), 1*(1), 50–62. https://doi.org/10.1038/sj.neo.7900005.

Dobrovolskaite, A., Madan, M., Pandey, V., Altomare, D. A., & Phanstiel, O., IV (2021). The discovery of indolone GW5074 during a comprehensive search for non-polyamine-based polyamine transport inhibitors. *The International Journal of Biochemistry & Cell Biology, 138*, 106038. https://doi.org/10.1016/j.biocel.2021.106038.

Dobrovolskaite, A., Moots, H., Tantak, M. P., Shah, K., Thomas, J., Dinara, S., ... Phanstiel, O., IV (2022). Discovery of anthranilic acid derivatives as difluoromethylornithine adjunct agents that inhibit far upstream element binding protein 1 (FUBP1) function. *Journal of Medicinal Chemistry, 65*(22), 15391–15415. https://doi.org/10.1021/acs.jmedchem.2c01350.

Gitto, S. B., Pandey, V., Oyer, J. L., Copik, A. J., Hogan, F. C., Phanstiel, O., & Altomare, D. A. (2018). Difluoromethylornithine combined with a polyamine transport inhibitor is effective against gemcitabine resistant pancreatic cancer. *Molecular Pharmaceutics, 15*(2), 369–376. https://doi.org/10.1021/acs.molpharmaceut.7b00718.

Hamouda, N. N., Van den Haute, C., Vanhoutte, R., Sannerud, R., Azfar, M., Mayer, R., ... Vangheluwe, P. (2021). ATP13A3 is a major component of the enigmatic mammalian polyamine transport system. *The Journal of Biological Chemistry, 296*, 100182. https://doi.org/10.1074/jbc.RA120.013908.

Hayes, C. S., Burns, M. R., & Gilmour, S. K. (2014). Polyamine blockade promotes antitumor immunity. *Oncoimmunology, 3*(1), e27360. https://doi.org/10.4161/onci.27360.

Hayes, C. S., Shicora, A. C., Keough, M. P., Snook, A. E., Burns, M. R., & Gilmour, S. K. (2014). Polyamine-blocking therapy reverses immunosuppression in the tumor microenvironment. *Cancer Immunology Research, 2*(3), 274–285. https://doi.org/10.1158/2326-6066.CIR-13-0120-T.

Houdou, M., Jacobs, N., Coene, J., Azfar, M., Vanhoutte, R., Van den Haute, C., ... Vangheluwe, P. (2023). Novel green fluorescent polyamines to analyze ATP13A2 and ATP13A3 activity in the mammalian polyamine transport system. *Biomolecules, 13*(2), https://doi.org/10.3390/biom13020337.

Madan, M., Patel, A., Skruber, K., Geerts, D., Altomare, D. A., & Phanstiel, O. (2016). ATP13A3 and caveolin-1 as potential biomarkers for difluoromethylornithine-based therapies in pancreatic cancers. *American Journal of Cancer Research, 6*(6), 1231–1252. https://e-century.us/files/ajcr/6/6/ajcr0027826.pdf.

Mandel, J. L., & Flintoff, W. F. (1978). Isolation of mutant mammalian cells altered in polyamine transport. *Journal of Cellular Physiology, 97*(3 Pt 1), 335–343. https://doi.org/10.1002/jcp.1040970308.

Muth, A., Madan, M., Archer, J. J., Ocampo, N., Rodriguez, L., & Phanstiel, O. (2014). Polyamine transport inhibitors: Design, synthesis, and combination therapies with difluoromethylornithine. *Journal of Medicinal Chemistry, 57*(2), 348–363. https://doi.org/10.1021/jm401174a.

Nakkina, S. P., Gitto, S. B., Beardsley, J. M., Pandey, V., Rohr, M. W., Parikh, J. G., ... Altomare, D. A. (2021). DFMO improves survival and increases immune cell infiltration in association with MYC downregulation in the pancreatic tumor microenvironment. *International Journal of Molecular Sciences, 22*(24), https://doi.org/10.3390/ijms222413175.

Nakkina, S. P., Gitto, S. B., Pandey, V., Parikh, J. G., Geerts, D., Maurer, H. C., ... Altomare, D. A. (2021). Differential expression of polyamine pathways in human pancreatic tumor progression and effects of polyamine blockade on tumor microenvironment. *Cancers (Basel), 13*(24), https://doi.org/10.3390/cancers13246391.

Phanstiel, O., IV (2018). An overview of polyamine metabolism in pancreatic ductal adenocarcinoma. *International Journal of Cancer. Journal International du Cancer, 142*(10), 1968–1976. https://doi.org/10.1002/ijc.31155.

Phanstiel, O., Kaur, N., & Delcros, J. G. (2007). Structure-activity investigations of polyamine-anthracene conjugates and their uptake via the polyamine transporter. *Amino Acids, 33*(2), 305–313. https://doi.org/10.1007/s00726-007-0527-y.

Poulin, R., Casero, R. A., & Soulet, D. (2012). Recent advances in the molecular biology of metazoan polyamine transport. *Amino Acids, 42*(2–3), 711–723. https://doi.org/10.1007/s00726-011-0987-y.

Samal, K., Zhao, P., Kendzicky, A., Yco, L. P., McClung, H., Gerner, E., ... Sholler, G. (2013). AMXT-1501, a novel polyamine transport inhibitor, synergizes with DFMO in inhibiting neuroblastoma cell proliferation by targeting both ornithine decarboxylase and polyamine transport. *International Journal of Cancer. Journal International du Cancer, 133*(6), 1323–1333. https://doi.org/10.1002/ijc.28139.

Sekhar, V., Andl, T., & Phanstiel, O., IV (2022). ATP13A3 facilitates polyamine transport in human pancreatic cancer cells. *Scientific Reports, 12*(1), 4045. https://doi.org/10.1038/s41598-022-07712-4.

Wang, C., Delcros, J. G., Biggerstaff, J., & Phanstiel, O. (2003). Synthesis and biological evaluation of N1-(anthracen-9-ylmethyl)triamines as molecular recognition elements for the polyamine transporter. *Journal of Medicinal Chemistry, 46*(13), 2663–2671. https://doi.org/10.1021/jm030028w.

Yang, J. (2024). Approval of DFMO for high-risk neuroblastoma patients demonstrates a step of success to target MYC pathway. *British Journal of Cancer, 130*(4), 513–516. https://doi.org/10.1038/s41416-024-02599-6.

Phanstiel, IV, O., & Archer, J. J. (2016). Polyamine transport inhibitors as novel therapeutics. US Patent 2016/0151312 A1.

CHAPTER FIVE

Evaluation of platinum drug toxicity resulting from polyamine catabolism

Kamyar Zahedi[a,b,]*, Sharon Barone[a,b], and Manoocher Soleimani[a,b]
[a]Division of Nephrology, Department of Internal Medicine, University of New Mexico Health Sciences Center Albuquerque, NM, United States
[b]Research Services, New Mexico Veterans Health Care System, Albuquerque, NM, United States
*Corresponding author. e-mail address: kzahedi@salud.unm.edu

Contents

1. Introduction	94
2. Rationale	96
3. Cisplatin induced renal injury	96
3.1 Animals, reagents and equipment	96
3.2 Procedures	97
3.3 Notes	98
4. Harvesting of biological samples	99
4.1 Reagents and equipment	99
4.2 Procedure	100
5. Assessment of renal function	100
5.1 Reagents and equipment	100
5.2 Procedures	100
6. RNA isolation, real-time quantitative RT-PCR (qRT-PCR) and northern blot analysis	101
6.1 Reagents and equipment	101
6.2 Procedures	102
6.3 Notes	106
7. Preparation of kidney protein extracts and western blot analysis	106
7.1 Reagents and equipment	106
7.2 Procedures	107
7.3 Notes	109
8. Histological staining of harvested kidneys for the evaluation of tubular injury and fibrosis	109
8.1 Reagents and equipment	109
8.2 Procedure	110
9. Immunofluoresence microscopic examination of kidney sections	111
9.1 Reagents and equipment	111
9.2 Procedure	111
9.3 Notes	112

Methods in Enzymology, Volume 715
ISSN 0076-6879, https://doi.org/10.1016/bs.mie.2025.01.065
Copyright © 2025 Elsevier Inc. All rights reserved, including those for text and data mining, AI training, and similar technologies.

10. Measurement of tissue polyamine levels	112
10.1 Reagents and equipment	112
10.2 Procedure	112
10.3 Notes	114
11. General equipment	114
Acknowledgments	114
References	114

Abstract

Polyamines, spermidine (Spd) and Spermine (Spm), are polycations that serve a number of important biological functions. The tissue contents of polyamines are tightly regulated through their cellular import and export, as well as their metabolism (anabolism and catabolism). Polyamine catabolism in mediated via the spermidine/spermine N1-acetyltransferase (SAT1)/acetylpolyamine oxidase (APOX) cascade and oxidation of Spm by spermine oxidase (SMOX). The expression of SAT1 and SMOX increases in injured organs in response to trauma, ischemia/reperfusion, sepsis, and exposure to toxic compounds. Cisplatin is a highly effective chemotherapeutic agent that is used for the treatment of a variety of solid tumors. Its anti-tumor activity is mediated via its ability to form stable DNA adducts that inhibit the growth of actively proliferating cells. However, cisplatin also can lead to severe off-target deleterious effects (e.g., nephrotoxicity and ototoxicity), and because of such adverse effects the use of cisplatin has to be discontinued in many patients. Understanding and decoupling the therapeutic and toxic effects of cisplatin will lead to more effective use of this and other platinum-derived compounds in the treatment of cancer patients. Acute and chronic exposure to cisplatin in mice leads to severe renal tubular injuries and an increase in the expression of SAT1 and SMOX while the ablation of their genes in mice reduces the severity of nephrotoxic injuries caused by cisplatin. Furthermore, neutralization of the toxic by-products of polyamine degradation reduce the severity if cisplatin nephrotoxicity. These observations suggest that interventions targeting the adverse effects of enhanced polyamine catabolism may provide effective therapies by reducing the toxic effects of cisplatin without affecting its anti-neoplastic activity.

1. Introduction

Cisplatin, a platinum-based chemotherapeutic agent, is used to treat a variety of solid tumors (Ozkok & Edelstein, 2014). Its antitumor activity is mediated through the formation of stable DNA adducts that result in the induction of an anomalous cell repair process, cell cycle arrest, and death (Zhu, Pabla, Tang, He, & Dong, 2015). Cisplatin also exhibits off-target effects and can induce nephrotoxic and ototoxic injuries (Miller, Tadagavadi, Ramesh, & Reeves, 2010; Santos, Ferreira, & Santos, 2020; Santoso, Lucci et al. 2003). In more than 25 % of patients, the off-target effects of cisplatin can lead to discontinuation of treatment (Miller, Tadagavadi et al. 2010, Pabla & Dong, 2008).

Polyamines, spermine (Spm), and spermidine (Spd), are naturally occurring polycationic alkylamines (Murray Stewart, Dunston, Woster, & Casero, Jr., 2018; Sagar, Tarafdar, Agarwal, Tarafdar, & Sharma, 2021). They are regulators of DNA structure, DNA-protein, and protein–protein interactions, as well as free radical scavengers (Hasan, Alam, & Ali, 1995; Igarashi & Kashiwagi, 2000, Igarashi & Kashiwagi, 2019, Landau, Bercovich, Park, & Kahana, 2010). As such, polyamines are important in the maintenance of genomic integrity and the regulation of cell growth and viability (Igarashi & Kashiwagi, 2019, Janne, Alhonen, & Leinonen, 1991; Landau, Bercovich et al. 2010, Marton & Pegg, 1995). Cellular levels of polyamine are regulated through their export, import, synthesis, and degradation (Pegg & McCann, 1982, Soda, 2022, Zahedi, Barone, & Soleimani, 2022). Polyamine synthesis is initiated by the decarboxylation of ornithine to form putrescine (Put), a reaction that is mediated by ornithine decarboxylase1 (ODC1). This is followed by the sequential addition of aminopropyl groups to Put and Spd by spermidine synthase and spermine synthase, respectively, to generate Spd and Spm (Sagar, Tarafdar et al. 2021). Polyamine catabolism can result in decreased intracellular polyamines through multiple mechanisms (Casero & Pegg, 2009, Seiler, 2004). Spm and Spd are acetylated by Spermidine/spermine N^1-acetyl transferase1 (SSAT or SAT1) and can either be excreted from the cell or serve as substrates for acetylpolyamine oxidase (PAOX) (Nakanishi & Cleveland, 2021, Seiler, 2004, Zahedi, Barone et al. 2022). Additionally, Spm can be directly converted to Spd by spermine oxidase (SMOX), a highly inducible enzyme (Pledgie, Huang et al. 2005, Wang, Devereux et al. 2001). Oxidation of SAT1-generated N^1-acetyl-Spm and -Spd generated by PAOX and Spm by SMOX leads to the generation of cytotoxic compounds (e.g., H_2O_2 and reactive aldehydes) (Pledgie, Huang et al. 2005). Because of the high cellular content of polyamines, which is in the mM range, substantial quantities of H_2O_2 and reactive aldehydes such as 3-aminopropanal, 3-acetoaminopropanal, and acrolein can be produced as a result of their catabolism (Cohen, 1998). These cytotoxic products can lead to nucleic acid and protein damage, as well as a disruption in the integrity of lysosomal and mitochondrial membranes (Boya & Kroemer, 2008, Boya, Gonzalez-Polo et al. 2003, Brunk, Zhang, Dalen, & Ollinger, 1995; Brunk, Zhang, Roberg, & Ollinger, 1995). The adverse effect of enhanced polyamine catabolism has also been demonstrated *in vitro*, where overexpression of SAT1 in cultured cells leads to cellular dysfunction, growth arrest, and apoptosis (Mandal, Mandal, & Park, 2015; Wang, Zahedi et al. 2004, Zahedi, Barone et al. 2017, Zahedi, Bissler et al. 2007).

The expression of polyamine catabolic enzymes increases in response to treatment with cisplatin (Zahedi, Barone et al. 2017, Zahedi, Barone et al. 2024, Zahedi, Wang et al. 2003). In addition, the ablation of genes coding for the polyamine catabolic enzymes, *Sat1* and *Smox*, and neutralization of their toxic by-products reduce the severity of both acute and chronic tissue injury caused by cisplatin treatment (Zahedi, Barone et al. 2017, Zahedi, Barone et al. 2024). In this chapter, we provide a detailed methodology of studies aimed at understanding the role of polyamine catabolism in the mediation of kidney injury in acute high dose and chronic repeated low dose cisplatin model of tissue injury.

2. Rationale

Platinum-derived compounds are important chemotherapeutic agents for the treatment of a variety of solid tumors; however, the off-target toxicity of these compounds can lead to the discontinuation of treatment. The antitumor activity of platinum derived compounds is mediated through the formation of stable DNA adducts that result in the induction of an anomalous cell repair process, cell cycle arrest, and death while their general toxicity is due to the induction of non-specific damage to the vasculature and the normal parenchyma of various organs. We propose that through the identification of molecules and pathways that mediate the off target injuries caused by platinum compounds we can devise novel approaches that decouple the anti-neoplastic and non-specific deleterious effects of these compounds and enhance their therapeutic efficacy. Our studies have identified the polyamine catabolic pathway and its toxic products as important mediators of the non-specific cisplatin injury; and therefore, potential targets for the prevention or reduction of the adverse effects of these compounds without affecting their anti-tumor activity.

3. Cisplatin induced renal injury

3.1 Animals, reagents and equipment

1. Male C57BL/6 J mice (Jackson Laboratories; Bar Harbor, ME), or genetically modified (*Sat1*-KO and *Smox*-KO) mice on C57BL/6J background.
2. Cisplatin (*cis*-Diamineplatinum(II)dichloride; 479306, Sigma-Aldrich, St. Louis, MO).

3. Sterile saline (0.9 % NaCl; Z1376, Intermountain Life Sciences, West Jordan, UT).
4. Heparin sodium salt (Acros, Fair Lawn, NJ).
5. Single use 1 ml Syringes (309659, Becton, Dickenson, Franklin Lakes, NJ).
6. Single use 25Gx5/8 needles (305122, Becton, Dickenson, Franklin Lakes, NJ).
7. Sterile gauze pads.
8. Disposable paper towels.
9. Ethanol (70 % and 100 %).
10. Balance (Scientech, model SL1000 or equivalent).
11. Disposable cages and bedding.
12. Hazardous/toxic waste material disposal containers for sharp objects, contaminated bedding, and contaminated cages.

3.2 Procedures

Cisplatin injury was induced using either a single high dose of cisplatin (20 mg/kg) that leads to acute renal injury or repeated low dose cisplatin (7 mg/kg/week for 4 weeks) that leads to chronic kidney damage (Schematic 1).

3.2.1 Acute cisplatin induced kidney injury (Fig. 1A)

1. Cisplatin (1 mg/ml) was dissolved in sterile saline.
2. Mice were transferred to the toxic material room and housed in disposable cages with disposable bedding.
3. Mice were weighed, their abdomen was cleaned using 70 % ethanol-soaked gauze pads, and then given an intraperitoneal injection of cisplatin (30 μg/g) or saline (15 μl/g).
4. At timed intervals (24, 48, 72 and 96 h post injection) animals were euthanized by an overdose of Euthasol (150 ul) given as an intraperitoneal injection.
5. Cages and bedding were disposed of as toxic compound contaminated waste.

3.2.2 Repeated low dose cisplatin (RLDC) induced kidney injury (Fig. 1B)

1. Cisplatin (1 mg/ml) was dissolved in sterile saline.
2. Mice were transferred to the toxic material room and housed in disposable cages containing disposable bedding.

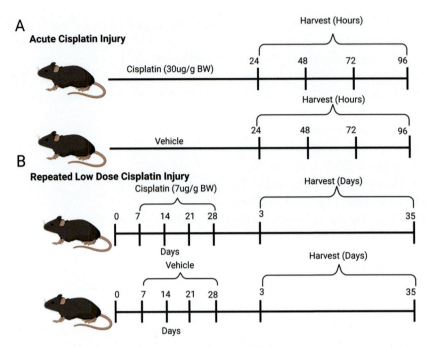

Fig. 1 Visual depiction of the timelines for acute (A) and repeated low dose cisplatin (B) injuries. *Image was generated using BioRender application (biorender.com).*

3. Mice were weighed and then given an intraperitoneal injection of cisplatin (7 μg/g) or saline (7 μl/g). This treatment was repeated weekly for 4 consecutive weeks.
4. At timed intervals (3 and 35 days after the 4th and final cisplatin injection) animals were euthanized by an intraperitoneal injection of 150 μl of Euthasol.
5. Cages and bedding were disposed of as toxic compound contaminated waste.

3.3 Notes

1. Because of toxicity of cisplatin and its presence in feces and urine, investigators should follow the standard procedures outlined for cisplatin use in animals by the animal care facility and Institutional Animal Care and Use Committee at their institution.
2. Male C57BL/6 mice or animals completely bred on C57BL/6 background are used in our studies. The C57BL/6 strain has been chosen because (a) of its susceptibility to renal injury (Sagar, Tarafdar et al.); and

(b) most knock-out strains are bred on to this background. However, if the development of renal fibrosis as a result of RLDC is the primary end point that is being studied the FVB/n mice are the preferred strain since C57BL/6 are more resistant to the development of severe renal fibrosis (Walkin, Herrick et al. 2013).
3. Only male animals are used in our studies since they are more susceptible to cisplatin injury than female mice (Eshraghi-Jazi & Nematbakhsh, 2022).
4. The use of diminazene aceturate and N1,N4-bis(2,3-butadienyl)-1,4-butanediamine (MDL72527), which inhibit SAT1 and polyamine oxidases respectively, as well as available including *Sat1*, *Sat1$^{f/f}$* and *Smox* knockout mice can specifically address the role of polyamine catabolism pathway enzymes in the mediation of cisplatin induced kidney injuries.
5. The neutralization of toxic products of polyamine catabolism (e.g., H_2O_2 and reactive aldehydes) by compounds such as cell permeable polyethylene glycol conjugated catalase (PEG-Catalase), N-2-mercaptopropionyl glycine (N-2-MPG) and phenelzine can be used to assess the role of products of polyamine catabolism in the mediation of cisplatin induced renal injuries.

4. Harvesting of biological samples
4.1 Reagents and equipment
1. Euthasol (390 mg pentobarbital sodium and 50 mg phenytoin sodium/ml; Virbac, Kansas City, MO).
2. Heparin sodium (AC41121-0010, Thermo Scientific Chemicals, Fair Lawn, NJ).
3. Dissection instruments. Dissection tray, scissors, forceps.
4. 70% Ethanol.
5. Phosphate buffered saline pH 7.4 (PBS; P3813-10PAK; Sigma-Aldrich, St. Louis, MO).
6. Heparinized 1 ml syringes.
7. 1.5 ml micro centrifuge tubes (Corning Inc., Reynosa, TX).
8. Microcentrifuge (Labnet International Inc., Iselin, NJ).
9. Liquid Nitrogen.
10. 4% paraformaldehyde fixative solution in PBS (J61984. AP, Thermo Fisher Scientific, Waltham, MA).

4.2 Procedure

1. Animals were euthanized by an overdose of Euthasol (150ul) given as an intraperitoneal injection.
2. At the time of euthanasia, animals were affixed to the dissection tray and a medial incision was made exposing the pleural and peritoneal cavities.
3. Blood was collected by direct heart puncture using a 1 ml Heparinized syringes. The collected blood was centrifuged at 13,000 rpm in a microcentrifuge for 30 min at 4 °C. Serum fraction was collected and frozen at −20°C.
4. Kidneys were harvested and snap frozen in liquid nitrogen and stored at −80°C or fixed in 4 % paraformaldehyde and preserved in 70 % Ethanol.
5. Following euthanasia and tissue harvest, animal remains should be disposed of in designated cisplatin-contaminated biohazard waste.

5. Assessment of renal function
5.1 Reagents and equipment

1. Creatinine Assay Kit (DICT-500; BioAssay Systems, Hayward, CA).
2. Blood Urea Nitrogen (BUN) Assay Kit (DIUR-100; BioAssay Systems, Hayward, CA).
3. Microplate photometer (accuSkan FC, Fisher Scientific, Pittsburgh, PA).

5.2 Procedures

To confirm the onset of cisplatin renal injury, kidney function needs to be assessed by measuring the blood urea nitrogen (BUN) and serum creatinine levels of sera harvested from cisplatin and vehicle treated mice.

1. The creatinine levels were measured using the QuantiChrome™ Creatinine Assay Kit (BioAssay Systems, Hayward, CA) following the manufacturers protocol. The optical density of samples at 510 nm were determined at 0 and 5 min. The differences in OD_{510} were used to calculate the serum creatinine levels.
2. The BUN levels were determined using the QuantiChrome™ Urea Assay Kit (BioAssay Systems, Hayward, CA) following the manufacturers protocol. The OD_{530} of samples was measured and used to determine the BUN levels.

6. RNA isolation, real-time quantitative RT-PCR (qRT-PCR) and northern blot analysis

6.1 Reagents and equipment

1. Tri Reagent RNA/DNA/Protein Isolation Reagent (Molecular Research Center, INC., Cincinnati, OH).
2. BCP Phase separation reagent (Molecular Research Center, INC., Cincinnati, OH).
3. Probe homogenizer.
4. High speed centrifuge (Sorvall RC6 Plus, F13S-14×50cy Rotor).
5. Ultra-Pure distilled water (10977-015, Invitrogen, Grand Island, NY).
6. FORMAzol (Molecular Research Center, INC., Cincinnati, OH)
7. Spectrophotometer or Nanodrop instrument.
8. Fast SYBR Green Master Mix (4385612, Applied Biosystems, Foster City, CA).
9. Appropriate qrt-PCR oligonucleotide pairs (*Sat1*-MP216112, *Smox*-MP216116, *Odc1*-MP210450, *Gapdh*-MP205604; OriGene, Rockville, MD).
10. Quantstudio 5 Real-Time PCR instrument (Applied Biosystems, East Lyme, CT)
11. RNA Master Mix: 6 ml 10 × RNA gel buffer (see below), 10 ml formaldehyde, 12 ml H$_2$O, and 2 ml of 1 mg/ml solution ethidium bromide.
12. 6XLoading buffer: 0.25% bromophenol blue, 0.25% Xylene cyanol, and 30% glycerol in H$_2$O.
13. 37% Formaldehyde (VWR International LLC, Radnor, CA).
14. 10XRNA gel buffer: 200 mM MOPS, 50 mM sodium acetate (CH$_3$CO$_2$Na$_3$.3H$_2$O), and 10 mM EDTA (20 ml of 0.5 M EDTA solution) dissolve in 500 ml of H$_2$O, adjust the pH to 7.0 with 10 N NaOH, add H$_2$O to 1000 ml, and filter.
15. Seakem LE agarose (Lonza, Rockland, ME).
16. 20X SSPE: Dissolve 175.3 g of NaCl, 27.6 g of NaH$_2$PO$_4$, and 9.4 g of EDTA in 800 ml of H$_2$O. Adjust the pH of the solution to 7.4 with 10 N NaOH and add H$_2$O to 1000 ml (final concentrations of components are 150 mM NaCl, 10 mM NaH$_2$PO$_4$ and 1 mM EDTA).
17. Sodium phosphate buffer (0.5 M PiB): Dissolve 134 g of Na$_2$HPO$_4$ in 800 ml of H$_2$O, adjust the pH to 7.2 with H$_3$PO$_4$, and add H$_2$O to 1000 ml.
18. Nytran-N membrane (Whatman, Sanford, ME).
19. 3MM filter paper (Whatman, Sanford, ME).

20. UV crosslinker (Stratalinker, Stratagene).
21. Pre-hybridization buffer: 0.1XSSPE and 1 % SDS.
22. Hybridization Buffer: 1 % crystalline bovine serum albumin (BSA), 7 % sodium dodecyl sulfate (SDS), 0.25 M PiB, and 1 mM EDTA.
23. PCR generated cDNA probe for Spermidine/spermine-N1-acetyltransferase, spermine oxidase, and ornithine decarboxylase: All cDNAs were generated by RT-PCR using RNA isolated from kidneys of mice subjected to renal IRI and Super Script III First Strand Synthesis Kit for RT-PCR (Invitrogen, Carlsbad, CA). The amplified cDNAs were gel purified on 1.5 % Ultra-Pure LMP low melting point agarose (Invitrogen, Carlsbad, CA) TAE (0.04 M Tris–acetate, 0.001 M EDTA) gel.
24. High Prime (Roch, Indianapolis, IN) or any Klenow polymerase-based kit is be used for generation of radiolabeled DNA probes.
25. Easytides [α^{32}P]-dCTP (BLU513H250UC, Revvity, Hopkinton, MA).
26. Quick Nucleotide Removal Kit (Qiagen, Valencia, CA).
27. Hybridization oven HB-1D (Bio-Techne, Minneapolis, MN).
28. Wash Buffer A. 0.5 % BSA, 0.5 % SDS, 20 mM PiB, and 1 mM EDTA.
29. Wash Buffer B: 0.1 % SDS, 20 mM PiB, 1 mM EDTA.
30. Phosphor Screen cassette and Typhoon Phosphor Imager Scanner (GE Life Sciences, Piscataway, NJ).
31. Stripping buffer: 1 %SDS and 0.1 ×, M.SSPE.

6.2 Procedures

6.2.1 RNA isolation

1. Kidneys were homogenized in 1 ml/g of tissue of TRI Reagent (MRC, Cincinnati, OH) for 1 min at maximum setting. Store the homogenates for 5 min on ice.
2. Add 100 µl of BCP/1 ml of TRI Reagent, mix using a vortex mixer, store the resulting mixture at room temperature for 10 min, and centrifuge at 12,000 × g for 15 min at 4 °C.
3. Collect the aqueous (top) phase, transfer to a new tube, and add 0.5 ml of isopropanol/1 ml of TRI Reagent. Store the samples at room temperature for 10 min, and centrifuge at 12,000 × g for 8 min at 4 °C (save the organic phase for DNA or protein extraction).
4. Wash the RNA pellet with 75 % ethanol, centrifuge at 12,000 × g for 6 min at 4 °C, discard the supernatant, let the pellet dry for 5 min, and resuspend the RNA pellet in 100 µl of Formazol (MRC, Cincinnati, OH).
5. Determine the RNA concentration by measuring the absorbance of 1:10 dilution of each sample at 260 nm.

6.2.2 First strand synthesis and quantitative rt-PCR

1. First strand synthesis was performed using the SuperScript III rt-PCR first strand synthesis system (ThermoFisher, Waltham, MA) and 1 μg of total RNA. For these studies, we followed the manufacturer's protocol using the random hexamer primer that is included in the kit.
2. The qrt-PCR quantitation of target transcripts of interest (*Sat1*, *Smox*, *Odc1* and *Gapdh*) was performed using Power SYBR Green PCR Master Mix and oligonucleotide pairs (OriGene, Rockville, MD).
3. The composition of the PCR reaction mix is as follows: Power SYBR Green PCR Master Mix (Appliedbiosystems, Warrington, UK) 20 μl; H$_2$O 15 μl; Oligonucleotides (OriGene, Rockville, MD) 1 μl each, template or TE (control) 1 μl.
4. A QuantStudio 5 instrument was used to perform qrt-PCR using the following protocol: Stage 1-Hold: 50 °C for 2 min and 95 °C for 10 min; Stage 2-PCR: 95 °C for 15 s and 60 °C for 1 min (X50); Stage 3-Melt Curve: 95 °C for 15 s, 60 °C for 1 min and 95 °C for 1 s

6.2.3 Northern blot analysis

6.2.3.1 RNA size fractionation by agarose gel electrophoresis

These instructions are for IBI 1050 Multi-Purpose horizontal gel Apparatus (IBI Scientific, Peosta, IA).

1. Prepare 1.2% agarose/formaldehyde denaturing gel by microwaving 1.2 g of SeaKem LE agarose in 74 ml of H$_2$O for 2 min. Let the agarose solution cool to 60 °C then add 10 ml of 10XRNA gel buffer and 16.4 ml of formaldehyde (this recipe is for a 100-ml gel). Place the gel tray in a chemical hood, place the appropriate comb in place, pour the gel, and allow time for it to solidify.
2. Prepare 500 ml of 1XRNA gel buffer by mixing 50 ml of 10XRNA gel buffer with 450 ml of H$_2$O.
3. Prepare the RNA mixture (10–30 μg of RNA in 15 μl of Formazol and 15 μl of RNA Master Mix). Denature the RNA mixture at 65 °C for 10 min, add 2ul of loading dye and place on ice.
4. Place the solidified agarose gel and the tray in the electrophoresis apparatus, pour in the running buffer to fill the chamber, and cover the gel. Remove the comb and load 32 μl of each RNA sample. Run the gel at 200 V for 10 min then reduce to 100 V. Stop the electrophoretic separation of RNA when the bromophenol blue dye front is approximately 12 cm into the gel.
5. Examine the integrity of isolated RNA and loading accuracy through visualizing the RNA by UV-light trans-illumination and take a picture of the gel.

6.2.3.2 Transfer protocol

1. Set up the transfer apparatus by placing a raised, flat plexiglas support surface in a buffer reservoir (Pyrex baking dish). Add the appropriate amount of 10XSSPE to the buffer reservoir (approximately 0.5 cm lower than the support surface). Place a piece of 3MM paper prewetted with 10XSSPE on the support surface with two ends of the paper immersed in the buffer. Place two additional layers of 3MM paper soaked in 10XSSPE over the first layer (the latter set up should be at least 1 cm larger than the gel in all direction). Smooth out the air bubbles between the filters using a glass rod.
2. Wash the gel in H_2O for 30 min (3 changes, 10 min each) to remove the formaldehyde. Cut the gel to appropriate dimensions, flip the gel over, and place it on the transfer setup. Be certain to squeeze out the air bubbles between the gel and filter papers. Place a piece of all-purpose laboratory wrap around the gel and buffer reservoir. Cut the top portion of the wrap away from the gel to provide access to the gel. Cut a piece of Nytran-N membrane (the dimensions of this membrane should be equal or about 1 mm larger than the gel on all sides) and prewet in deionized H_2O and then 10XSSPE.
3. Place the membrane on the gel and smooth out the air bubbles between the gel and Nytran-N membrane with a glass rod. Situate three layers of 10XSSPE soaked 3MM paper on the Nytran-N membrane and add additional 5–8 cm stack of paper towels on top of the 3MM paper (the aforementioned layers need to be approximately 1 mm smaller than the gel on all sides).
4. Position a glass plate on top of the stack of paper towels and place a 500 g weight atop the glass plate. Allow the transfer of size-fractionated RNA to take place (approximately 12–18 h).
5. When the transfer is complete, remove the paper towels and 3MM paper on top of the Nytran-N membrane, label, and peel off the membrane. Confirm the complete transfer of RNA by examining the gel under UV light.
6. Wash the membrane in 5XSSPE for 5 min, take a picture of the membrane, and crosslink the RNA to the Nytran-N membrane using the Auto Crosslink Function of the Stratagene UV Stratalinker™1800.

6.2.3.3 Generation of radiolabeled DNA probes

Gel-purified cDNAs of interest (generated by RT-PCR) can be used as templates to generate radiolabeled probes. Alternatively, non-radioactive cDNA probes can be used.

1. Denature 25 ng of appropriate DNA template in 11 μl of H$_2$O by heating in 95 °C heating block for 5 min
2. Transfer the DNA to an ice ethanol bath for 1 min. Add 4 μl of High Prime and 5 μl of [α^{32}P]-dCTP to the cooled DNA mixture, allow the labeling reaction to proceed for 10 min at 37 °C and then stop the reaction by adding 2 μl of 0.2 M EDTA pH8.0.
3. Remove the unincorporated dNTPs using Quick Nucleotide Removal Kit (QIAGEN; Germantown, MD) following the manufacturer's protocol.

6.2.3.4 Hybridization and visualization of mRNA bands

Carry out all hybridization steps at 65 °C.
1. Wet the hybridization tube with 10 ml of pre-hybridization buffer preheated to 65 °C.
2. Preheat the hybridization tube by adding 10 ml of 65 °C pre-hybridization buffer and placing it in the hybridization oven for 5 min
3. Place the membrane (RNA side facing toward the center of the tube) in the hybridization tube containing pre-hybridization buffer and incubate at 65 °C for 1 h.
4. Discard the pre-hybridization buffer and add 10 ml of pre- heated (65 °C) hybridization buffer and incubate at 65 °C for 30 min
5. Denature the radiolabeled probe by boiling in a water bath for 5 min and then transfer the tube to an ethanol/ice bath for 1 min
6. Replace the hybridization buffer in the hybridization tube with fresh buffer, add the denatured probe to the side of the hybridization tube, return the vessel to the hybridization oven, and allow hybridization to proceed for 12–16 h at 65 °C.
7. Discard the hybridization buffer containing the labeled probe in a designated radioactive waste receptacle.
8. Wash the membrane in 15 ml of wash buffer A for 10 min at 65 °C (repeat 2 times).
9. Wash the membrane in 15 ml of wash buffer B for 10 min at 65 °C (repeat 4 times).
10. Rinse the membrane in 25 °C 10 mM phosphate buffer (do not allow the membrane to dry if it is going to be re-probed).
11. Place the membrane in a sealable bag, smooth out the air bubbles, and seal the bag.
12. Place the membrane in an exposure cassette containing a phosphor screen and visualize the signal using a Phospho-Imager.

6.2.3.5 Stripping and re-probing the northern blot membranes

To re-probe the northern blot it has to be stripped and the dissociation of labeled probe has to be confirmed.

1. Preheat 500 ml of the stripping buffer to 95 °C. At the same time, preheat a Pyrex dish containing 200 ml of deionized H$_2$O to 95 °C.
2. Discard the deionized H$_2$O and pour 200 ml of stripping buffer into the preheated Pyrex dish.
3. Place the blot in preheated stripping buffer in the Pyrex dish for 10 min. Maintain the temperature of the buffer at 95 °C.
4. Place the blot in 10 mM phosphate buffer for 5 min at 25 °C.
5. Place the membrane in a sealable bag if it needs to be stored or expose it as described in section 6.2.6 to confirm the complete stripping of the probe.
6. For re-probing follow the steps outlined in section 6.2.6 using the appropriate radiolabeled probe.

6.3 Notes

For the analysis mRNA expression levels, we utilize both qrt-PCR and northern blot analyses. While the former is more quantitative the presentation of northern blot images is visually more impactful.

7. Preparation of kidney protein extracts and western blot analysis

7.1 Reagents and equipment

1. RIPA Buffer (Thermo Scientific, Rockford, IL).
2. Halt Protease and Phosphatase Inhibitor (Thermo Scientific, Rockford, IL).
3. Polytron P2100 probe homogenizer (VWR, Batavia, IL).
4. High speed centrifuge (Sorvall RC6 Plus, F13S-14×50cy Rotor).
5. BCA protein assay (Thermo Scientific, Rockford, IL).
6. 2X Laemmli Sample Buffer: 62.5 mM Tris–HCl, pH 6.8, 25 % glycerol, 2 % SDS, and 0.01 % Bromophenol Blue. Add 50 ml of Dithiothreitol per 950 ml of Laemmli buffer before use (Bio-Rad Laboratories, Hercules, CA).
7. Dithiothreitol (Sigma Life Science, St. Louis, MO).
8. X-Cell II PAGE/Transfer apparatus (Invitrogen, Carlsbad, CA).
9. BenchMark pre-stained protein ladder (Invitrogen, Carlsbad, CA).
10. Tris/Glycine/SDS Running Buffer: 25 mM Tris pH 8.6, 192 mM glycine, and 0.1 % SDS (dilute 1:10 in H$_2$O) (Bio-Rad Laboratories, Hercules, CA).

11. Precast SDS 4-12% Tris–Glycine polyacrylamide gels (Invitrogen, Carlsbad, CA).
12. Polyvinylidene difluoride membrane (Invitrogen, Carlsbad, CA).
13. Tris/glycine Transfer Buffer: 25 mM Tris, 192 mM glycine, 20% methanol, and 0.1% SDS (pH 8.3) (Bio-Rad Laboratories, Hercules, CA).
14. Tris buffered saline with Tween-20 (T-TBS): 137 mM NaCl, 20 mM Tris–HCl pH 8.0%, and 0.1% Tween-20.
15. Blotting Grade Blocker (nonfat dry milk, Bio-Rad Laboratories, Hercules, CA).
16. Blocking Buffer: 5% Blotting Grade Blocker in T-TBS.
17. Primary antibodies (Rabbit anti-SAT1 antibody, Rabbit anti-SMOX antibody, Protein Tech. Rosemount, IL; Mouse anti-β-actin, SCBT, Dallas, TX).
18. Wash Buffer: 2.5% dry milk in T-TBS.
19. Horseradish peroxidase-conjugate donkey antirabbit IgG or goat antimouse IgG (Pierce, Rockford, IL).
20. Super Signal West Pico Chemiluminescent Substrate (Thermo Scientific, Rockford, IL).
21. Autoradiography film BX (Midsci, St. Louis, MO).

7.2 Procedures

7.2.1 Preparation of kidney protein extracts

Snap frozen kidneys harvested from treated and control mice will be used for the preparation of protein extracts.

1. Place the frozen kidneys one at a time in 4 ml of ice-cold RIPA buffer supplemented with Halt protease and phosphatase inhibitor cocktail.
2. Homogenize the kidneys with three 30 s bursts using a probe homogenizer.
3. Centrifuge the homogenate at $1000 \times g$ for 10 min at 4 °C. The nuclei and cell debris will be in the pellet.
4. Collect the supernatant and determine the protein content using a BCA protein assay following the manufacturer's protocol.

7.2.2 Size-fractionation of kidney protein extracts

The protocol in the following section is specific for Novex X-Cell II PAGE and Transfer apparatus.

1. Dilute 50 μg of protein from each sample with an identical volume of Laemmli buffer containing 50 μl/ml of dithiothrietol (1:1).
2. Heat the samples 10 min at 95 °C. Do not heat the molecular weight standard (BenchMark Pre-stained Protein Ladder).

3. Load the sample onto an 4-12% TRIS/Glycine/SDS precast gel and perform the electrophoretic separation of proteins at 110 V until the bromophenol blue dye front is within 3 mm of the bottom of the gel.

7.2.3 Transfer of size-fractionated kidney protein extracts to nitrocellulose membrane

For these transfers we utilized Nitrocellulose Membrane Filter Paper Sandwiches (LC2001; Life Technologies Corp., Carlsbad, CA) and Novex X-Cell II Transfer apparatus.

1. Prepare 1X transfer buffer by diluting 100 ml of 10X tris/glycine buffer in 700 ml ddH$_2$O and 200 ml of methanol.
2. Pre-wet nitrocellulose membranes in ddH$_2$O and then 1X transfer buffer.
3. Pre-wet sheets of 3MM paper in 1X transfer buffer.
4. Pre-wet 4 foam pads in distilled water and then transfer buffer.
5. In a shallow container filled with transfer buffer, assemble the transfer cassette. Place two pre-wetted foam pads on the blot module. Roll out the air bubbles using a glass rod. Place two sheets of pre-wetted 3MM paper on the foam pads and roll out the air bubbles. Disassemble the gel unit and extract the gel. Place the gel on 3MM paper and roll out the bubbles. Place the pre-wetted polyvinylidene difluoride membrane on the gel and roll out the bubbles. Place two sheets of prewetted 3MM paper on the polyvinylidene difluoride membrane and roll out the bubbles. Place two pre-wetted foam pads on the filter papers and roll out the bubbles.
6. Place the transfer sandwich in the blot module such that the nitrocellulose membrane is between the gel and the anode. Transfer takes place by electrophoresis, place the module in the buffer reservoir, lock in place, and add transfer buffer to fill the module. Fill the buffer reservoir with cold water ddH$_2$O.
7. Transfer the proteins onto nitrocellulose at a constant power of 25 m Amps for 90 min.
8. Upon completion of the transfer, disassemble the transfer module by removing the top sponge and 3MM paper layers, followed by removing the gel. Mark the top left corner of the membrane (the pre-stained protein standards should be visible).

7.2.4 Immunodetection of specific proteins

1. Prepare 10X Tris buffered saline (10XTBS; 15.76 g Trizma HCl, 90.9 g NaCl dissolve in 700 ml of H$_2$O, adjust pH to 8.0 and add H$_2$O to 1 liter).

2. Prepare 1X TBS-Tween-20 (T-TBS, 100 ml 10XTBS, 1 ml Tween-20, 899 ml H$_2$O).
3. Prepare the blocking buffer by dissolving 0.5 g of Blotting-grade Blocker (Bio-Rad Laboratories, Hercules, CA) in T-TBS.
4. Block the membrane in 15 ml of blocking buffer for 30 min at 25 °C on a rocking platform.
5. Discard the buffer and add fresh blocking buffer containing pre-determined dilution of the appropriate primary antibody. Allow the binding to proceed at 4 °C on a rocking platform for 12–18 h.
6. Remove the primary antibody and wash the membrane 5 times for 5-min each with 15 ml of T-TBS.
7. Prepare a fresh 1:1000 dilution of horseradish peroxidase conjugated antibody in blocking buffer. Incubate the membrane in secondary antibody for 1 h at 25 °C on a rocking platform.
8. Discard the secondary antibody and wash the membrane 5 times for 5-min each with T-TBS.
9. Mix 12 ml each of solution A and B of the Super Signal West Pico Chemiluminescent Substrate.
10. Incubate the blot in the substrate for 2 min
11. Place the blot in a plastic sleeve.
    The remaining steps should be performed in a dark room.
12. Place the sleeve in an X-ray cassette.
13. Expose the membrane to X-ray film for 1 min and then based on the strength of the signal adjust the exposure time as needed.

7.3 Notes

For the documentation of western blot analysis results, we utilize both digital imaging (LICOR Odyssey DLX; LICOR Bioscience, Lincoln NE) and bioluminescence-autoradiography documentation. For final publication images, we use autoradiographic film images because of their resolution and clarity.

8. Histological staining of harvested kidneys for the evaluation of tubular injury and fibrosis

8.1 Reagents and equipment

1. Microscope slides.
2. Glass staining jars or canisters.

3. Xylene.
4. Ethanol (100%, 95%, and 70% diluted in distilled/deionized water).
5. VectaMount permanent mounting medium (H-5000, Vector Laboratories, Burlington, CA).
6. Cover slips.
7. H&E Stain Kit (KTHNEPT; StatLab Medical Products, McKinny, TX).
8. Masson's Trichrome Stain Kit (KTMTR; StatLab Medical Products, McKinny, TX).
9. Heat blocks (adjusted to 80, 85 and 90 °C).
10. Microscope with high resolution camera for light and immunofluorescence microscopy (Zeiss LSM800, Zeiss, Oberkochen, Germany).

8.2 Procedure

The extent of renal injury and fibrosis was determined by examination of H&E and Masson's Trichrome stained kidney sections.

1. Harvested kidneys were fixed in a solution of 4% Paraformaldehyde in PBS for 18 h.
2. Paraformaldehyde fixed tissues were transferred to 70% ethanol for preservation.
3. Samples were then embedded in paraffin.
4. Paraffin embedded sections were cut into 5 μm thin sections and affixed to microscope slides for further studies.
5. Kidney sections were stained with H&E following the manufacturers protocol (KTHNEPT; StatLab Medical Products, McKinny, TX).
6. The histology of cisplatin-treated kidneys was compared by light microscopic examination of H&E-stained sections. The extent of renal injury was assessed by examining the cortical and corticomedullary regions of the kidney for tubular dilatation, interstitial edema, cast formation, and leukocyte infiltration. Kidneys were assigned an injury score of 0 to 3 (0– no injury; 1– mild injury; 2– moderate injury; 3– severe injury) to compare the severity of tissue damage.
7. Kidney sections were stained with Masson's Trichrome (KTMTR; StatLab Medical Products, McKinny, TX) as instructed by the manufacturer.
8. The sections of cisplatin-treated and control kidneys were stained with Masson's Trichrome and examined microscopically for the severity of renal fibrosis. The extent of renal fibrosis was quantitated using the Image J software (The National Institutes of Health and the Laboratory for Optical and Computational Instrumentation release 1.53 K).

9. Immunofluoresence microscopic examination of kidney sections

9.1 Reagents and equipment

1. Microscope slides.
2. Glass staining jars or canisters.
3. Xylene.
4. Ethanol (100%, 95%, and 70% diluted in distilled/deionized water).
5. R-Universal Epitope Recovery Buffer (Electron Microscopy Sciences, Hatfield, PA).
6. 2100 Retriever (Electron Microscopy Sciences, Hatfield, PA).
7. Primary antibodies (Rabbit anti-SAT1 antibody, Rabbit anti-SMOX antibody, Protein Tech. Rosemount, IL).
8. Alexa Fluor 594 Secondary antibody (Goat anti-Rabbit, Invitrogen, Eugene, OR).
9. VectaMount permanent mounting medium (H-5000, Vector Laboratories, Burlington, CA).
10. Cover slips.

9.2 Procedure

1. Paraffin embedded sections were cut into 5 μm thin sections and affixed to microscope slides for further studies and baked at 65 °C for 60 min.
2. Slides were deparaffinized in xylene and then dehydrated in a series of ethanol dilutions.
3. Antigen retrieval was accomplished by pressure cooking of slides in R-Universal Epitipe Recovery Buffer (Electron Microscopy Sciences; Hatfield, PA) using a 2100 Retriever unit (Electron Microscopy Sciences; Hatfield, PA).
4. Slides were incubated in Image-iT™ FX signal enhancer for 30 min.
5. Slides were incubated with the appropriate primary antibodies diluted in PBS pH7.4, 10% normal goat serum and 0.05% Triton X-100 at 4 °C for 18 h.
6. Slides were washed 3 times for 10 min each in PBS.
7. Slides were incubated in the appropriate secondary antibody (Alexa Fluor IgG, Invitrogen, Carlsbad, CA) diluted (1:200) in PBS plus 10% normal goat serum for 2 h at 25 °C.
8. Slides were washed 3 times for 10 min each in PBS and allowed to air dry.

9. Vectashield Hard Set (Vector Labs, Burlington, CA) was added and cover slips were applied.
10. Slides were examined and immunofluorescence microscopic images were obtained using a Zeiss LSM800 fluorescence microscope and analyzed utilizing Zen software (version 3.4).

9.3 Notes

In addition to immunostaining for SAT1 and SMOX, the kidney sections can be stained for activated caspase 3 and leukocytes to assess the onset of apoptotic and inflammatory response respectively.

10. Measurement of tissue polyamine levels

10.1 Reagents and equipment

1. Sterile saline, 0.9 %.
2. Perchloric acid (1.2 N).
3. Tissue homogenizer.
4. Microcentrifuge.
5. 2.5 mM 1,7-diaminoheptane in 2 M Na_2CO_3.
6. Dansyl Chloride (Thin Layer chromatography Grade).
7. HPLC grade acetone.
8. Heating Block set to 70 °C.
9. 100 mg Bakerbond™ spe light load Octadecyl (C18) disposable extraction column (Krackeler Scientific Inc., Albany, NY).
10. Vacuum filtration apparatus (Krackeler Scientific Inc., Albany, NY).
11. Methanol (HPLC grade).
12. Distilled deionized H_2O (HPLC grade, Fisher Scientific).
13. HPLC system [column: Econosphere (C18) 5 m 4.6 mm × 250 mm HPLC Column (Grace, Deerfield, IL)]; Waters 616 LC solvent gradient system (Waters Assoc., Milford, MA); Waters Model 2475 fluorescence detector (Waters, Milford, MA); Waters WISP 710B autosampler; and Millennium 32 version 3.2 data collection system and software (Waters, Milford, MA).
14. Solvent A: 70 % 10 mM ammonium acetate – pH 4.4, 30 % acetonitrile.
15. Solvent B: 100 % acetonitrile.

10.2 Procedure

1. Homogenize 100–150 mg of each sample in 1 ml of saline (samples were flash frozen and stored at −70 °C upon harvest).

2. Mix 400 µl of each homogenate with 400 ml of 1.2 N perchloric acid (PCA) in a 1.5-ml microcentrifuge tube.
3. Centrifuge the samples for 20 min at 12,000 × g and transfer the supernatants to new microcentrifuge tubes (save the pellets for determination of protein content needed for final normalization of polyamine content to the protein levels).
4. Preheat a heating block to 70 °C. Dispense 50 µl of PCA extracted samples or 25 µl of 20 µM polyamine standard into 1.5 ml microcentrifuge tubes.
5. Add 200 µl of 2 M Na_2CO_3 containing 2.5 µM 1,7-diamino- heptane to each sample as internal standard.
6. Prepare the dansyl chloride solution by adding 800 µl of 0.1 g/ml stock solution to 4.5 ml HPLC grade acetone.
7. Add 200 µl of the dansyl chloride/acetone to each tube and mix for 15 s using a vortex mixer. Place the tubes in a heating block set at 70 °C for 3 min, release pressure by uncapping the tube briefly, and mix the samples for 15 s using a vortex mixer (repeat 3 times).
8. Store the dansylation reactions in the dark and let them cool to room temperature.
9. Set up the appropriate number of 100 mg BakerbondTM spe light load C_{18} disposable extraction column in a vacuum apparatus. Condition columns with two column volumes of HPLC grade methanol followed by two column volumes of HPLC grade water. Transfer the contents of each tube to a column. Rinse each tube with 300 µl HPLC grade water and add to rinse solution to the appropriate column. Using vacuum, draw sample through the column and discard the eluate. Wash with two column volumes of HPLC grade water and discard the eluate. Insert collection tubes into the vacuum apparatus. Add 1.0 ml of methanol draw through and collect the eluate for HPLC analysis.
10. Inject a 50 µl aliquot of each dansylated sample onto a 250 × 4.6 mm, 5 µ particle size Econosphere C18 column with a column temperature of 50 °C.
11. Elute using a two solvent gradient with the flow rate of 0.79 ml per min. Begin the gradient at 100 % solvent A and progress linearly to 82 % solvent B over 30 min with followed by an additional 15 min hold at 82 % solvent B (elution times: Putrescine, 20–22 min; Spermidine, 25–27 min; Spermine, 28–30 min).

12. Detect the presence of polyamines by assessing the fluorescence of eluted peaks with excitation wavelength of 340 nm and emission wavelength of 520 nM.
13. Collect and analyze the results using Waters Millennium 32 chromatography software version 3.2 (Milford, MA).

10.3 Notes

1. Equilibrate the column with 100 % solvent A for 8 min between sample injections.
2. Express the polyamine levels as pmol of polyamine/mg of protein.

11. General equipment

1. 4 °C refrigerator.
2. −20°C freezer.
3. −80°C freezer.
4. Water bath.
5. Heat blocks.
6. Stir plate.
7. Orbital shaker.
8. Micro-pipets.
9. Pipets and pipet pump.
10. Vortex mixer.

Acknowledgments

This work was supported by Dialysis Clinic Inc. Research Funds C-4149, Merit Review Award 2 I01 BX001000-10 from the Department of Research Services, Veterans Health Administration, Biomedical Research Institute of New Mexico (BRINM) Catherine Do Award 372DO. Manoocher Soleimani is a Senior Clinician Scientist Investigator with the Department of Veterans Health Administration.

References

Boya, P., Gonzalez-Polo, R. A., Poncet, D., Andreau, K., Vieira, H. L., Roumier, T., … Kroemer, G. (2003). Mitochondrial membrane permeabilization is a critical step of lysosome-initiated apoptosis induced by hydroxychloroquine. *Oncogene, 22*(25), 3927–3936.

Boya, P., & Kroemer, G. (2008). Lysosomal membrane permeabilization in cell death. *Oncogene, 27*(50), 6434–6451.

Brunk, U. T., Zhang, H., Dalen, H., & Ollinger, K. (1995). Exposure of cells to nonlethal concentrations of hydrogen peroxide induces degeneration-repair mechanisms involving lysosomal destabilization. *Free Radical Biology & Medicine, 19*(6), 813–822.

Brunk, U. T., Zhang, H., Roberg, K., & Ollinger, K. (1995). Lethal hydrogen peroxide toxicity involves lysosomal iron-catalyzed reactions with membrane damage. *Redox Report: Communications in Free Radical Research, 1*(4), 267–277.

Casero, R. A., & Pegg, A. E. (2009). Polyamine catabolism and disease. *The Biochemical Journal, 421*(3), 323–338.

Cohen, S. S. (1998). *A guide to the polyamines*. New York: Oxford University Press.

Eshraghi-Jazi, F., & Nematbakhsh, M. (2022). Sex difference in cisplatin-induced nephrotoxicity: Laboratory and clinical findings. *Journal of Toxicology*, 3507721 2022.

Hasan, R., Alam, M. K., & Ali, R. (1995). Polyamine induced Z-conformation of native calf thymus DNA. *FEBS Letters, 368*(1), 27–30.

Igarashi, K., & Kashiwagi, K. (2000). Polyamines: Mysterious modulators of cellular functions. *Biochemical and Biophysical Research Communications, 271*(3), 559–564.

Igarashi, K., & Kashiwagi, K. (2019). The functional role of polyamines in eukaryotic cells. *The International Journal of Biochemistry & Cell Biology, 107*, 104–115.

Janne, J., Alhonen, L., & Leinonen, P. (1991). Polyamines: From molecular biology to clinical applications. *Annals of Medicine, 23*(3), 241–259.

Landau, G., Bercovich, Z., Park, M. H., & Kahana, C. (2010). The role of polyamines in supporting growth of mammalian cells is mediated through their requirement for translation initiation and elongation. *The Journal of Biological Chemistry, 285*(17), 12474–12481.

Mandal, S., Mandal, A., & Park, M. H. (2015). Depletion of the polyamines spermidine and spermine by overexpression of spermidine/spermine N(1)-acetyltransferase 1 (SAT1) leads to mitochondria-mediated apoptosis in mammalian cells. *The Biochemical Journal, 468*(3), 435–447.

Marton, L. J., & Pegg, A. E. (1995). Polyamines as targets for therapeutic intervention. *Annual Review of Pharmacology and Toxicology, 35*, 55–91.

Miller, R. P., Tadagavadi, R. K., Ramesh, G., & Reeves, W. B. (2010). Mechanisms of Cisplatin nephrotoxicity. *Toxins (Basel), 2*(11), 2490–2518.

Murray Stewart, T., Dunston, T. T., Woster, P. M., & Casero, Jr, R. A. (2018). Polyamine catabolism and oxidative damage. *The Journal of Biological Chemistry, 293*(48), 18736–18745.

Nakanishi, S., & Cleveland, J. L. (2021). Polyamine homeostasis in development and disease. *Medical Sciences (Basel), 9*(2).

Ozkok, A., & Edelstein, C. L. (2014). Pathophysiology of cisplatin-induced acute kidney injury. *BioMed Research International, 2014* 967826 967821-967817.

Pabla, N., & Dong, Z. (2008). Cisplatin nephrotoxicity: Mechanisms and renoprotective strategies. *Kidney International, 73*(9), 994–1007.

Pegg, A. E., & McCann, P. P. (1982). Polyamine metabolism and function. *The American Journal of Physiology, 243*(5), C212–C221.

Pledgie, A., Huang, Y., Hacker, A., Zhang, Z., Woster, P. M., Davidson, N. E., & Casero, R. A., Jr (2005). Spermine oxidase SMO(PAOh1), Not N1-acetylpolyamine oxidase PAO, is the primary source of cytotoxic H2O2 in polyamine analogue-treated human breast cancer cell lines. *The Journal of Biological Chemistry, 280*(48), 39843–39851.

Sagar, N. A., Tarafdar, S., Agarwal, S., Tarafdar, A., & Sharma, S. (2021). Polyamines: Functions, metabolism, and role in human disease management. *Medical Sciences (Basel), 9*(2).

Santos, N., Ferreira, R. S., & Santos, A. C. D. (2020). Overview of cisplatin-induced neurotoxicity and ototoxicity, and the protective agents. *Food and Chemical Toxicology: An International Journal Published for the British Industrial Biological Research Association, 136*, 111079.

Santoso, J. T., Lucci, J. A., 3rd, Coleman, R. L., Schafer, I., & Hannigan, E. V. (2003). Saline, mannitol, and furosemide hydration in acute cisplatin nephrotoxicity: A randomized trial. *Cancer Chemotherapy and Pharmacology, 52*(1), 13–18.

Seiler, N. (2004). Catabolism of polyamines. *Amino Acids, 26*(3), 217–233.

Soda, K. (2022). Overview of polyamines as nutrients for human healthy long life and effect of increased polyamine intake on DNA methylation. *Cells, 11*(1).

Walkin, L., Herrick, S. E., Summers, A., Brenchley, P. E., Hoff, C. M., Korstanje, R., & Margetts, P. J. (2013). The role of mouse strain differences in the susceptibility to fibrosis: A systematic review. *Fibrogenesis Tissue Repair, 6*(1), 18.

Wang, Y., Devereux, W., Woster, P. M., Stewart, T. M., Hacker, A., & Casero, R. A., Jr. (2001). Cloning and characterization of a human polyamine oxidase that is inducible by polyamine analogue exposure. *Cancer Research, 61*(14), 5370–5373.

Wang, Z., Zahedi, K., Barone, S., Tehrani, K., Rabb, H., Matlin, K., ... Soleimani, M. (2004). Overexpression of SSAT in kidney cells recapitulates various phenotypic aspects of kidney ischemia-reperfusion injury. *Journal of the American Society of Nephrology: JASN, 15*(7), 1844–1852.

Zahedi, K., Barone, S., Brooks, M., Stewart, T. M., Foley, J. R., Nwafor, A., ... Soleimani, M. (2024). Polyamine catabolism and its role in renal injury and fibrosis in mice subjected to repeated low-dose cisplatin treatment. *Biomedicines, 12*(3).

Zahedi, K., Barone, S., Destefano-Shields, C., Brooks, M., Murray-Stewart, T., Dunworth, M., ... Soleimani, M. (2017). Activation of endoplasmic reticulum stress response by enhanced polyamine catabolism is important in the mediation of cisplatin-induced acute kidney injury. *PLoS One, 12*(9), e0184570.

Zahedi, K., Barone, S., & Soleimani, M. (2022). Polyamines and their metabolism: From the maintenance of physiological homeostasis to the mediation of disease. *Medical Sciences (Basel), 10*(3).

Zahedi, K., Bissler, J. J., Wang, Z., Josyula, A., Lu, L., Diegelman, P., ... Soleimani, M. (2007). Spermidine/spermine N1-acetyltransferase overexpression in kidney epithelial cells disrupts polyamine homeostasis, leads to DNA damage, and causes G2 arrest. *American Journal of Physiology. Cell Physiology, 292*(3), C1204–C1215.

Zahedi, K., Wang, Z., Barone, S., Prada, A. E., Kelly, C. N., Casero, R. A., ... Soleimani, M. (2003). Expression of SSAT, a novel biomarker of tubular cell damage, increases in kidney ischemia-reperfusion injury. *American Journal of Physiology. Renal Physiology, 284*(5), F1046–F1055.

Zhu, S., Pabla, N., Tang, C., He, L., & Dong, Z. (2015). DNA damage response in cisplatin-induced nephrotoxicity. *Archives of Toxicology, 89*(12), 2197–2205.

CHAPTER SIX

Measurements of acrolein adducts resulting from polyamine catabolism

Keiko Kashiwagi* and Kazuei Igarashi
Amine Pharma Research Institute, Innovation Plaza at Chiba University, Chiba, Japan
*Corresponding author: e-mail address: keiko1235@gmail.com; kkashiwagi@cis.ac.jp

Contents

1. Introduction	118
2. Acrolein production from spermine	119
2.1 Measurement of acrolein	119
2.2 Acrolein is produced from spermine by spermine oxidase	119
3. Increase in polyamine oxidases and protein-conjugated acrolein in plasma of stroke patients	120
3.1 Measurement of polyamine oxidases (AcPAO and SMOX) and protein-conjugated acrolein (PC-Acro) in plasma	120
3.2 Increase in polyamine oxidases (AcPAO and SMOX) and protein-conjugated acrolein (PC-Acro) in plasma of stroke patients	121
4. Plasma levels of PC-Acro, IL-6 and CRP as markers of silent brain infarction (SBI)	122
4.1 Measurement of PC-Acro, IL-6 and CRP in plasma	122
4.2 Imaging	122
4.3 Statistics	123
4.4 Relationship between biochemical markers (PC-Acro, IL-6 and CRP) and SBI, CA and WMH determined by imaging	123
5. Urinary amino acid-conjugated acrolein and taurine as new biomarkers for detection of dementia	124
5.1 Collection of urine and plasma samples	124
5.2 Measurement of urinary amino acid-conjugated acrolein and taurine contents, and plasma Aβ_{40} and Aβ_{42}	125
5.3 Statistics	125
5.4 Correlation between MMSE and age, and the decrease in urinary AC-Acro and taurine	125
6. Inactivation of GAPDH by acrolein conjugation	127
6.1 Identification of glyceraldehyde-3-phosphate dehydrogenase (GAPDH) as an acrolein-conjugated protein	127
6.2 Determination of amino acid residues of GAPDH conjugated with acrolein	129
6.3 Translocation of acrolein-conjugated GAPDH to nuclei resulting in apoptosis	130

7. Inhibition of dendritic spine extension by acrolein conjugation with cytoskeleton proteins during brain infarction — 130
  7.1 Acrolein conjugation with cytoskeleton proteins — 130
  7.2 Induction of brain infarction in mice — 131
  7.3 Time-dependent decrease of dendritic spine extension in infarct brain — 131
8. Summary and conclusions — 132
Acknowledgments — 132
References — 132

Abstract

Acrolein (CH_2=CH-CHO) is an unsaturated aldehyde, produced primarily from spermine [$NH_2(CH_2)_3NH(CH_2)_4NH(CH_2)_3NH_2$] by spermine oxidase (SMOX), and shows high reactivity with protein cysteine- and lysine-residues to produce protein-conjugated acrolein (PC-Acro) or amino acid-conjugated acrolein (AC-Acro). Increase in PC-Acro in plasma has been identified as a biomarker for stroke, silent brain infarction (SBI) and dementia such as Alzheimer's disease. In contrast, decrease in AC-Acro and taurine in urine was observed in dementia. In addition, properties of proteins such as glyceraldehyde 3-phosphate dehydrogenase (GAPDH) and cytoskeleton proteins are altered by conjugation with acrolein.

Abbreviations

AcPAO   acetylpolyamine oxidase.
CA   carotid atherosclerosis.
PC-Acro protein-conjugated acrolein.
PIT   photochemically induced thrombosis.
ROC   receiver operating characteristic.
RRV   relative risk value.
SBI   silent brain infarction.
SMOX   spermine oxidase.
SSAT   spermidine/spermine N^1-acetyltransferase.
WMH   white matter hyperintensity.

1. Introduction

Cell damage is thought to be caused mainly by reactive oxygen species (ROS) such as superoxide anion radical ($O_2^{-\bullet}$), hydrogen peroxide (H_2O_2), and hydroxyl radical (•OH) (Giorgio, Trinei, Migliaccio, & Pelicci, 2007). However, when the toxicity of acrolein and ROS was compared, it was found that acrolein was more toxic than hydrogen peroxide (Sharmin et al., 2001), and slightly more toxic than hydroxyl radical (Yoshida, Tomitori, Machi, Hagihara et al., 2009a) in cell culture systems. Furthermore, acrolein is thought to be produced by lipid peroxidation

(Uchida et al., 1998), but we found that it was more effectively produced from polyamines, especially from spermine, which are abundant and essential for cell growth in eukaryotic cells (Igarashi & Kashiwagi, 2010).

It has been shown that plasma levels of protein-conjugated acrolein (PC-Acro) increased in patients with stroke (Tomitori et al., 2005), silent brain infarction (SBI) (Yoshida et al., 2010; Yoshida, Tomitori, Machi, Katagiri et al., 2009b) and dementia (Waragai et al., 2012), and urinal level of AC-Acro decreased in patients with dementia (Yoshida et al., 2023). These results indicate that PC-Acro and AC-Acro are good biomarkers for these diseases.

2. Acrolein production from spermine

2.1 Measurement of acrolein

Acrolein was measured according to the method of Alarcon (Alarcon, 1968).

1) The reaction mixture (0.4 mL) containing 30 mM potassium phosphate buffer, pH 7.0, 0.2 mM substrate (spermine, spermidine, putrescine, N^1-acetylspermidine, N^8-acetylspermidine, or 3-aminopropanal) and 0.05 mL fetal bovine serum (FBS) was incubated at 37 °C for 2, 4, or 6 h.
2) The reaction was terminated by the addition of 2.6 mL mixed reagent containing 18 mM *m*-aminophenol, 33 mM hydroxylamine hydrochloride, and 1.15 N HCl.
3) After boiling for 10 min, the precipitate was removed by centrifugation.
4) Fluorescence was measured at an excitation wavelength of 350 nm and an emission wavelength of 510 nm.

2.2 Acrolein is produced from spermine by spermine oxidase

Spermine is oxidized by spermine oxidase (SMOX) in the presence of flavin adenine dinucleotide (FAD) to produce equal molecules of hydrogen peroxide and 3-aminopropanal. Acrolein is spontaneously produced from 3-aminopropanal (Fig. 1) (Wang et al., 2001, 2003). When spermine is converted to N^1-acetylspermine by spermidine/spermine N^1-acetyltransferase (SSAT), N^1-acetylspermine is oxidized by acetylpolyamine oxidase (AcPAO) to produce equal molecules of hydrogen peroxide and 3-acetamidopropanal. Acrolein is less-effectively produced from 3-acetamidopropanal (Fig. 1).

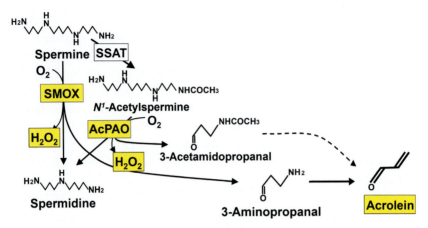

Fig. 1 Pathway of spermine catabolism. SMOX, spermine oxidase; AcPAO, acetylpolyamine oxidase; SSAT, spermidine/spermine N^1-acetyltransferase.

3. Increase in polyamine oxidases and protein-conjugated acrolein in plasma of stroke patients

3.1 Measurement of polyamine oxidases (AcPAO and SMOX) and protein-conjugated acrolein (PC-Acro) in plasma

1) Plasma samples were collected from 35 control subjects without stroke (20 males, 15 females; 67.8 ± 2.3 years of age) and 62 patients with stroke (33 males, 29 females; 70.7 ± 1.7 years of age).
2) Blood containing 3 U/mL heparin was centrifuged at 1500 x g for 10 min at 4 °C. The supernatant (plasma) was carefully collected to avoid contamination by erythrocytes.
3) The reaction mixture (0.075 mL) containing 10 mM Tris-HCl, pH 7.5, 0.2 mM spermine for spermine oxidase (SMOX) or N^1-acetylspermine for acetylpolyamine oxidase (AcPAO), and 0.065 mL plasma was incubated at 37 °C for 48 hr.
4) To 0.02 mL of the reaction mixture, 0.55 mL of 5 % trichloroacetic acid was added and centrifuged at 12,000 x g for 10 min.
5) A 10 μL aliquot of the supernatant was used for the polyamine measurement by HPLC (Igarashi et al., 1986).
6) The activity of SMOX or AcPAO was expressed as nanomole spermidine increase per milliliter plasma.

7) Protein-conjugated acrolein (PC-Acro) [N-(3-formyl-3,4-dehydropiperidino) lysine (FDP-lysine)] was determined by the method of Uchida et al. (Uchida et al., 1998) using FDP-Lys ELISA kit (Takara Bio Inc.) and 0.05 mL plasma.
8) After the reaction was terminated, absorbance at 450 nm was measured by a microplate reader Bio-Rad Model 550.

3.2 Increase in polyamine oxidases (AcPAO and SMOX) and protein-conjugated acrolein (PC-Acro) in plasma of stroke patients

As shown in Fig. 2, polyamine oxidases (AcPAO and SMOX) and protein-conjugated acrolein (PC-Acro) in plasma of cerebral stroke patients significantly increased compared to no stroke subjects, indicating that these biochemical markers can be used for diagnosis of cerebral stroke (Tomitori et al., 2005).

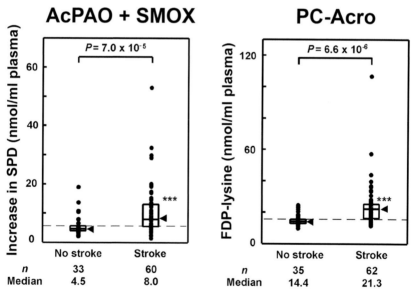

Fig. 2 Levels of AcPAO plus SMOX, and PC-Acro in plasma of no stroke and stroke subjects. AcPAO plus SMOX, and PC-Acro (FDP-lysine, nmol/mL plasma) were shown in median (arrowhead) interquartile deviation shown by box. Dotted line indicates the third quartile of no stroke as cutoff value.

4. Plasma levels of PC-Acro, IL-6 and CRP as markers of silent brain infarction (SBI)

4.1 Measurement of PC-Acro, IL-6 and CRP in plasma

1) Plasma samples with 3 U/mL heparin were collected from 790 elderly volunteers (330 women and 460 men, age 61.0 ± 7.0 years, range 40–88 years).
2) PC-Acro was determined as described in the Section 3.1 using 0.01 mL plasma.
3) IL-6 and CRP were measured using Endogen Human IL-6 ELISA kit (Pierce Biotechnology, Inc.) and human CRP ELISA kit (Alpha Diagnostic), respectively, according to the manufacturer's protocol.
4) After the reaction was terminated, absorbance at 450 nm for IL-6 or CRP was measured by a microplate reader Hitachi MTP-800APC.

4.2 Imaging

All 790 subjects underwent T1- and T2-weighted magnetic resonance imaging (MRI) and fluid-attenuated inversion recovery (FLAIR) in parallel with the collection of blood samples (Yoshida et al., 2010; Yoshida, Tomitori, Machi, Katagiri et al., 2009b). All MRI was performed in 5–10 mm thickness with 1–2 mm slice gap with a 1.5-T MRI unit (Signa; GE Medical Systems). The diagnosis of SBI was made as follows: (1) spotty areas ≥ 3 mm in diameter showing in high intensity in the T2 and FLAIR images and low intensity in the T1 image, (2) lack of neurological signs and/or symptoms that can be explained by the MRI lesions, and (3) no medical history of clinical stroke (Hoshi et al., 2005; Kobayashi, Okada, Koide, Bokura, & Yamaguchi, 1997). The subjects with WMH [deep and subcortical white matter hyperintensity (DSWMH) and periventricular hyper intensity (PVH)] were defined as subjects with spots showing in high intensity in the T2 and FLAIR images without any consistent finding in the T1 image (Shinohara et al., 2007), and subjects who lack neurological signs and/or symptoms. Carotid ultrasound examination was performed using a LOGIQ S6 (GE Healthcare) or a LOGIQ 500 MD (GE Yokogawa Medical Systems) scanner equipped with a 7.5 MHz linear array imaging probe (Makita, Nakamura, & Hiramori, 2005; Yamagami et al., 2005). Intima-medial thickness (IMT) was measured as the distance between the luminal–intimal interface and the medial–adventitial interface. The IMT value was measured in three points for each right and left side (observation possible areas of common carotid artery, carotid bifurcation, and internal carotid artery), and the largest value (max-IMT) in each side was used for analysis. Control subjects were defined as subjects with no spotty areas and carotid atherosclerosis (CA) and no neurological signs and/or symptoms. Through this diagnosis, 790 subjects were classified into 260 control subjects (108 women and 152 men, aged 54.0 ± 6.0 years), 214 SBI subjects (88 women and 126 men, aged 68.0 ± 7.0 years), 263 CA subjects (105 women and 158 men, aged 66.0 ± 6.0 years) and 245 WMH subjects (105 women and 140 men, aged 61.0 ± 5.8 years). There were 187 subjects with two or three pathologies, and of those, 88 subjects had SBI and CA, 9 subjects SBI and WMH, 85 subjects CA and WMH, and 5 subjects SBI, CA and WMH. Accordingly, subjects with SBI and CA, or with WMH and CA consisted of 93 and 90 subjects, respectively.

4.3 Statistics

Statistical calculations were performed with GraphPad Prism® Software (GraphPad Software). Values are indicated as median ± interquartile deviation. Groups were compared using Wilcoxon rank sum test or Kruskal–Wallis test. Relative risk value (RRV) was calculated with artificial neural networks by back propagation method using NEUROSIM/L software version 4 (Fujitsu) (Ellenius, Groth, Lindahl, & Wallentin, 1997). Age and three biochemical markers of 260 control and 214 SBI subjects were used as prediction output values 0 and 1, respectively, and the rules to build RRV were obtained. Then, RRV (0−1) for control, SBI with CA, SBI, CA, WMH with CA and WMH subjects was calculated according to the rules. Sensitivity and specificity for SBI subjects vs. control subjects were evaluated using a receiver operating characteristic (ROC) curve (Hanley & McNeil, 1982). Sensitivity, specificity, positive- and negative-predictive values (PPV and NPV), and positive- and negative-likelihood (LR+ and LR−) were calculated with the standard method (Brenner & Gefeller, 1997). Cutoff value was set up as the closest point on ROC curve from the P point, that is sensitivity = 1 and 1 − specificity = 0.

4.4 Relationship between biochemical markers (PC-Acro, IL-6 and CRP) and SBI, CA and WMH determined by imaging

The ROC curve for the detection of SBI by age, with measurement of PC-Acro plus IL-6 and CRP is shown in Fig. 3A. Sensitivity and specificity were 84.1% and 83.5% respectively. Relative risk value (RRV) was then calculated for SBI, CA and WMH. A value of 1 is the highest value, and 0 is the lowest value as an index of the degree of tissue damage. The median

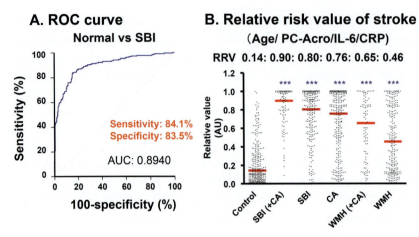

Fig. 3 ROC curve of age/PC-Acro/IL-6/CRP for SBI vs control subjects (A) and relationship between RRV and SBI, CA and WMH (B). (A) ROC curve analysis was performed as described in the Section 4.3. AUC, area under curve. (B) RRV of control, SBI with CA, SBI, CA, CA with WMH and WMH calculated from ROC curve was shown with median (horizontal line). ***$p < 0.001$ compared with control subjects.

RRV for SBI with CA (93 subjects), SBI (214 subjects), CA (263 subjects), WMH with CA (90 subjects), WMH (245 subjects) and control (260 subjects) was 0.90, 0.80, 0.76, 0.65, 0.46 and 0.14, respectively (Fig. 3B). RRV was also correlated with severity in each group of SBI, CA and WMH (Yoshida et al., 2010; Yoshida, Tomitori, Machi, Katagiri et al., 2009b).

5. Urinary amino acid-conjugated acrolein and taurine as new biomarkers for detection of dementia

5.1 Collection of urine and plasma samples

1) Samples of urine and plasma were collected from 57 control subjects without dementia (25 males, 32 females; aged 71.0 ± 8.5 years), 62 mild cognitive impairment (MCI) patients (23 males, 39 females; aged 82.0 ± 5.0 years), and 42 Alzheimer's disease (AD) patients (16 males, 26 females; aged 81.0 ± 4.5 years).

2) Diagnosis of AD and MCI subjects was carried out for AD (McKhann et al., 1984) and for MCI (Petersen et al., 1999). Mini-Mental State Examination (MMSE) (Folstein, Folstein, & McHugh, 1975) was evaluated when subjects were examined by medical doctors.

3) The severity of tissue damage in control subjects and MCI and AD patients was evaluated as Z-score using VSRAD (Voxel-Based Science Regional Analysis System for Alzheimer's Disease) as described previously (Waragai et al., 2012).

5.2 Measurement of urinary amino acid-conjugated acrolein and taurine contents, and plasma Aβ$_{40}$ and Aβ$_{42}$

1) Urinary amino acid (lysine)-conjugated acrolein (AC-Acro) was measured using FDP-Lys ELISA kit (Takara Bio Inc.) (Uchida et al., 1998).
2) Urinary taurine content was measured using a taurine assay kit (Cell Biolabs, Inc.).
3) Values were normalized by creatinine (Cre) content which was measured with the colorimetric assay kit (Cayman Chemical).
4) Plasma Aβ$_{40}$ and Aβ$_{42}$ were measured using Human β Amyloid (1-40) ELISA Kit II (Wako Chemical Co.) and Human β Amyloid (1-42) ELISA Kit (Wako Chemical Co.).
5) After the reaction was terminated, absorbance at 450 nm was measured by a microplate reader (Hitachi MTP-800APC).

5.3 Statistics

Statistical analysis was performed as described in the Section 4.3. RRV was evaluated with artificial neural networks using Neural Works Predict (SET Software Co., Ltd.) (Ellenius et al., 1997). Age and three biochemical markers (AC-Acro, taurine and creatinine) of 57 control subjects and 62 subjects with MCI or 42 with AD were used as prediction output values 0 and 1, respectively, and the rules to build RRV were obtained. Then, RRV (0−1) for control subjects, and subjects with MCI and AD was calculated according to the rules. Correlation between cognitive assessment and biomarkers in control, MCI and AD subjects were compared using Kruskal-Wallis or chi-square test. For examination of the statistical correlation between MMSE or plasma Aβ$_{40/42}$ ratio and urinary biochemical markers, Spearman's rank correlation coefficient was used.

5.4 Correlation between MMSE and age, and the decrease in urinary AC-Acro and taurine

Since urine is relatively easy to collect, sensitive biomarkers for dementia in urine were sought. How MMSE, age, AC-Acro/Cre, and taurine/Cre are changed between control subjects, and subjects with MCI and AD was evaluated (Fig. 4A,B,C and D). Median MMSE values decreased from

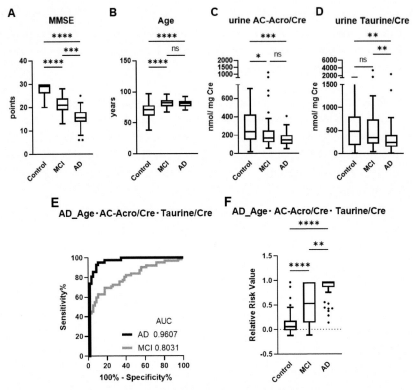

Fig. 4 Comparison of MMSE (A), age (B), AC-Acro/Cre in urine (C), and taurine/Cre in urine (D) in control, MCI and AD groups, and ROC curves (E) and relative risk values (F) evaluated by age, AC-Acro/Cre, and taurine/Cre for AD versus control subjects. (A–D) Distribution of MMSE, age, AC-Acro, and taurine among control, MCI, and AD subjects is shown. (E) ROC curve analysis was performed as described in the Section 4.3. AUC, area under curve. (F) Relative risk values (RRVs) were calculated from artificial neural networks. A horizontal line within the box indicates median, the bottom and the top of the box indicates the 25th and 75th percentiles, and the whiskers (vertical lines) indicate the 5th and 95th percentiles. ****$p < 0.0001$; ***$p < 0.001$; **$p < 0.01$; *$p < 0.05$; ns, not significant.

29.0 ± 2.0 (control) to 21.0 ± 2.6 (MCI) and 15.5 ± 2.1 (AD), and median ages increased from 71.0 ± 8.5 years old (control) to 82.0 ± 5.0 years old (MCI) and 81.0 ± 4.5 years old (AD). Significant differences were not seen in age between MCI and AD, and also MMSE and age in male and female (data not shown). Next, AC-Acro/Cre and taurine/Cre in urine were measured. As shown in Figs. 4C and 4D, the level of AC-Acro/Cre in control, and MCI and AD was 236.0 ± 138.4, 167.2 ± 63.5, and 147.6 ± 48.9 nmol/mg Cre, respectively, and that of taurine was 476.7 ± 310.6, 339.3 ± 257.9, and

235.0 ± 127.1 nmol/mg Cre, respectively. The values of these biomarkers were similar between female and male in control, MCI, and AD (data not shown). It is notable that the level of AC-Acro/Cre did not alter significantly between MCI and AD, and that of taurine/Cre did not alter significantly between control and MCI. Accordingly, measurements of these two markers can distinguish three groups, i.e., control, MCI, and AD. When plasma levels of PC-Acro and $A\beta_{40/42}$ ratio were measured, MCI subjects could not be distinguished from AD subjects (Waragai et al., 2012).

We next examined whether MCI and AD could be distinguished by AC-Acro/Cre and taurine/Cre using a ROC curve. Because age was one of the important factors for evaluation of MCI and AD (Fig. 4B), all values were analyzed including age as a factor in artificial neural networks. The ROC curve for detection of dementia through measurements of AC-Acro/Cre plus taurine/Cre together with age is shown in Fig. 4E. Area under curve (AUC) for AD was 0.9607, and that for MCI was 0.8031, and the median RRV was 0.06, 0.53, and 0.96 for control, MCI, and AD, respectively (Fig. 4F). Sensitivity and specificity were 92.9 % and 91.2 % for AD, and 72.6 % and 73.7 % for MCI, respectively. The results indicate that severity of MCI and AD are evaluated properly through measurement of AC-Acro/Cre and taurine/Cre in urine together with age. When the evaluation was performed with one marker, sensitivity decreased greatly. In the case of AC-Acro/Cre, AUC for AD was 0.7289, and that for MCI was 0.6324, and in case of taurine/Cre, AUC for AD was 0.6637, and that for MCI was 0.5116 (data not shown). These results indicate that two factors, AC-Acro/Cre and taurine/Cre, are essential for the evaluation of the severity of dementia (Yoshida et al., 2023).

6. Inactivation of GAPDH by acrolein conjugation

6.1 Identification of glyceraldehyde-3-phosphate dehydrogenase (GAPDH) as an acrolein-conjugated protein

1) Mouse mammary carcinoma FM3A cells were treated with 0 or 40 µM acrolein for 9 h.
2) S100 proteins were separated by 1D-SDS polyacrylamide gel electrophoresis (PAGE) and 2D-SDS PAGE (Fig. 5A).
3) The level of a protein of approximately 37 kDa decreased in acrolein-treated FM3A cells compared to control cells. The results suggest that

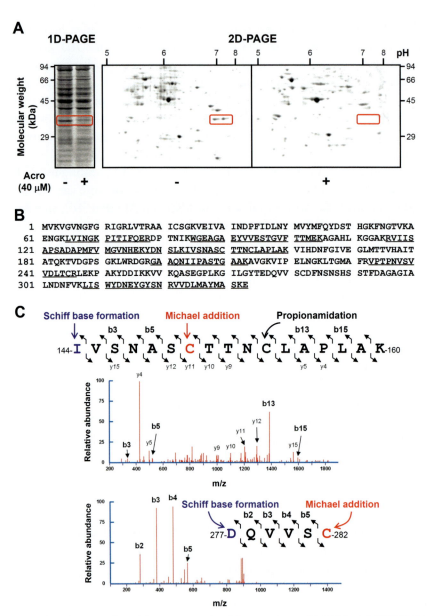

Fig. 5 Identification of acrolein-conjugated 37 kDa proteins as GAPDH by LC-MS/MS. (A) FM3Acells were treated with 0 or 40 μM acrolein for 9 h, and S100 proteins were analyzed by one-dimensional (left) or two-dimensional (right) polyacrylamide gel electrophoresis. Disappearance of 37 kDa proteins by acrolein was observed. Acro, acrolein. (B) The gel containing 37 kDa proteins was analyzed by LC-MS/MS. Amino acid sequence of mouse glyceraldehyde-3-phosphatedehydrogenase (GAPDH) (Swiss-Prot:P16858.2) was

the acrolein conjugated 37 kDa protein became insoluble or shifted from the cytoplasm to mitochondria or nuclei.

4) Next, this 37 kDa protein obtained from control cells was identified by determining the peptide sequences by LC-MS/MS after protease digestion.
5) As shown in Fig. 5B, amino acid sequences of 11 peptides were determined.
6) These sequences were identical to the peptide sequence of glyceraldehyde-3-phosphate dehydrogenase (GAPDH). Among 333 amino acid residues, 137 amino acid residues (41%) were identified. Although two spots were observed in proteins of control cells in 2D PAGE, both spots were identified as GAPDH.
7) The results indicate that GAPDH is one of the major acrolein-conjugated proteins in cells.

6.2 Determination of amino acid residues of GAPDH conjugated with acrolein

Since GAPDH protein could not be immunoprecipitated by antibody against GAPDH after acrolein treatment in FM3A cells, acrolein-conjugated GAPDH was purified using HA-tagged GAPDH and acrolein-conjugated amino acid residues of GAPDH were determined.

1) Mouse neuroblastoma Neuro2a cells transfected with pcDNA-GAPDH (HA) were treated with 40 μM acrolein for 9 h.
2) HA tagged-GAPDH protein was immunoprecipitated with anti-HA antibody, separated by SDS–PAGE and extracted from polyacrylamide gel.
3) Acrolein conjugated amino acid residues of GAPDH were then determined by LC-MS/MS using this protein after trypsin and endoproteinase Asp-N digestion.

When a cysteine residue is conjugated with acrolein by Michael addition, intramolecular Schiff base formation with the N-terminal amino group of the digested peptide would occur (Bradley, Markesbery, & Lovell, 2010). Cysteine residues unreacted with acrolein are observed as the propionamidated forms.

---

shown together with 11 identified peptides (underlined) by LC-MS/MS. (C) Neuro2a cells transfected with pcDNA-GAPDH (HA) were treated with 40 μM acrolein for 9 h. The amino acid residues of HA-tagged GAPDH conjugated with acrolein were determined by LC-MS/MS. MS/MS spectra of the peptides 144–160 and 277–282 containing cysteine residues conjugated with acrolein are shown. Cys-150 and Cys-282 were conjugated with acrolein by Michael addition and Schiff base was formed intramolecularly at the N-terminus of the peptide.

As shown in Fig. 5C, acrolein conjugated with two cysteine residues — Cys-150 and Cys-282. Since Cys-150 is at the active site of GAPDH (Tristan, Shahani, Sedlak, & Sawa, 2011), GAPDH is probably inactivated by acrolein through acrolein conjugation with Cys-150.

6.3 Translocation of acrolein-conjugated GAPDH to nuclei resulting in apoptosis

It has been reported that inactive GAPDH that is conjugated with nitric oxide (NO) at Cys-150 translocates to nuclei together with Siah, an E3 ubiquitin ligase, and activates p300/CBP acetylase resulting in apoptosis (Bae et al., 2006; Sen et al., 2008). If acrolein-conjugated GAPDH also translocates to nuclei, the same phenomena may occur. Thus, the location of inactivated GAPDH by acrolein was compared with that of normal GAPDH. As determined by Western blotting, GAPDH in control cells was mainly located in the cytoplasm, whereas GAPDH in cells treated with 8 μM acrolein for 6 h was located in both cytoplasm and nuclei. The location of GAPDH was confirmed by immunostaining of GAPDH in cells. In control cells there was little overlap in GAPDH staining and DNA staining, but in cells treated with 4 and 8 μM acrolein for 6 h, there was marked overlap of GAPDH and nuclear (DNA) staining. It has been also reported that inactivated GAPDH is acetylated (Bae et al., 2006; Sen et al., 2008). When FM3A cells were treated with 8 μM acrolein for 6 h, the level of acetyllysine in GAPDH localized in nuclei increased. These results confirmed that acrolein treated GAPDH translocates to nuclei like NO-treated GAPDH. The number of TUNEL positive cells increased after acrolein treatment. TUNEL positive cells treated with 0, 4 and 8 μM acrolein for 24 h were 1 %, 39 % and 78 %, respectively. The results indicate that acrolein-conjugated GAPDH is acetylated by p300/CBP and causes apoptosis (Nakamura et al., 2013) similar to nitrosylated GAPDH (Sen et al., 2008).

7. Inhibition of dendritic spine extension by acrolein conjugation with cytoskeleton proteins during brain infarction

7.1 Acrolein conjugation with cytoskeleton proteins

The cytoskeleton consists of 25 nm microtubule (α-and β-tubulin proteins), 8–12 nm intermediate filament (vimentin), and 5–9 nm microfilament (actin) (Pegoraro, Janmey, & Weitz, 2017).

Acrolein conjugated amino acid residues were determined using purified cytoskeleton proteins (Uemura et al., 2019, 2020). They were Cys25, 295, 347 and 376 in α-tubulin, Cys12, 129, 211, 239, 303 and 354 in β-tubulin, Cys328 in vimentin, and Cys207, Cys257, Cys285 and Lys118 in actin.

7.2 Induction of brain infarction in mice

All animal experiments were approved by the Institutional Animal Care and Use Committee of Chiba University and performed according to the Guidelines for Animal Research of Chiba University.

1) Induction of brain infarction was conducted by photochemically induced thrombosis (PIT) technique using 12-month-old male C57BL/6 mice (22–26 g) (Saiki et al., 2009, 2013).
2) Mice were anesthetized with inhalation of 3 % isoflurane (Abbot Japan), and then maintained with 1.5 % isoflurane during operation using a small-animal anesthesia system.
3) For induction of ischemia, immediately after intravenous injection of photosensitizer Rose Bengal (20 mg/kg) through a jugular vein, green light (wavelength, 540 nm) emitted from a xenon lamp (Hamamatsu Photonics, Japan) illuminated the middle cerebral artery for 10 min
4) At 24 h after the induction of PIT infarction, the brains were removed and stored for further use. If necessary, the locus of brain infarction was confirmed by staining with a triphenyltetrazolium chloride solution (Saiki et al., 2009).

7.3 Time-dependent decrease of dendritic spine extension in infarct brain

Neuro2a cells were cultured with 10 μM acrolein, and intermediate filament (vimentin) was stained using anti-vimentin antibody (Fig. 6A). Dendritic spine extension was time-dependently inhibited by acrolein. In infarct brain, dendritic spine extension had almost disappeared at 24 after induction of infarction (Fig. 6B). Among three cytoskeleton proteins, decrease in vimentin was most evident in infarct brain, and acrolein conjugation of vimentin was confirmed by immunoprecipitation of vimentin using antibody against vimentin followed by Western blotting using rabbit polyclonal anti-protein-conjugated acrolein antibody (MoBiTec) (Uemura et al., 2020).

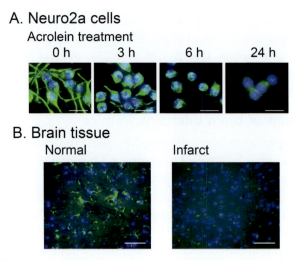

Fig. 6 Inhibition of dendritic spine extension due to the acrolein conjugation with cytoskeleton proteins. (A) Neuro2a cells were treated with 10 μM acrolein, and dendritic spine extension was examined using an antibody against vimentin at 0, 3, 6 and 24 h. Bar = 10 μm. (B) Brain tissue was taken from normal and infarct brain (24 h after the onset of infarction), and dendritic spine extension was examined using an antibody against vimentin. Bar = 100 μm.

8. Summary and conclusions

Acrolein ($CH_2=CH-CHO$) is produced primarily from spermine by spermine oxidase (SMOX), and produces protein-conjugated acrolein (PC-Acro) or amino acid-conjugated acrolein (AC-Acro). Increase in PC-Acro in plasma has been found to be a biomarker for stroke, silent brain infarction (SBI) and dementia such as Alzheimer's disease. Furthermore, decrease in AC-Acro and taurine in urine is a biomarker for dementia. In addition, properties of proteins such as glyceraldehyde 3-phosphate dehydrogenase (GAPDH) and cytoskeleton proteins are altered by conjugation with acrolein.

Acknowledgments

We thank Dr. A. J. Michael for his help in preparing the manuscript.

References

Alarcon, R. A. (1968). Fluorometric determination of acrolein and related compounds with *m*-aminophenol. *Analytical Chemistry*, *40*(11), 1704–1708. https://doi.org/10.1021/ac60267a019.

Bae, B. I., Hara, M. R., Cascio, M. B., Wellington, C. L., Hayden, M. R., Ross, C. A., et al. (2006). Mutant huntingtin: Nuclear translocation and cytotoxicity mediated by GAPDH. *Proceeding of the National Academy of Sciences of the United States of America, 103*(9), 3405–3409. https://doi.org/10.1073/pnas.0511316103.

Bradley, M. A., Markesbery, W. R., & Lovell, M. A. (2010). Increased levels of 4-hydroxynonenal and acrolein in the brain in preclinical Alzheimer disease. *Free Radical Biology & Medicine, 48*(12), 1570–1576. https://doi.org/10.1016/j.freeradbiomed.2010.02.016.

Brenner, H., & Gefeller, O. (1997). Variation of sensitivity, specificity, likelihood ratios and predictive values with disease prevalence. *Statics in Medicine, 16*(9), 981–991. https://doi.org/10.1002/(sici)1097-0258(19970515)16:9<981::aid-sim510>3.0.co;2-n.

Ellenius, J., Groth, T., Lindahl, B., & Wallentin, L. (1997). Early assessment of patients with suspected acute myocardial infarction by biochemical monitoring and neural network analysis. *Clinical Chemistry, 43*(10), 1919–1925.

Folstein, M. F., Folstein, S. E., & McHugh, P. R. (1975). "Mini-mental state". A practical method for grading the cognitive state of patients for the clinician. *Journal of Psychiatric Research, 12*(3), 189–198. https://doi.org/10.1016/0022-3956(75)90026-6.

Giorgio, M., Trinei, M., Migliaccio, E., & Pelicci, P. G. (2007). Hydrogen peroxide: A metabolic by-product or a common mediator of ageing signals? *Nature Reviews. Molecular Cell Biology, 8*(9), 722–728. https://doi.org/10.1038/nrm2240.

Hanley, J. A., & McNeil, B. J. (1982). The meaning and use of the area under a receiver operating characteristic (ROC) curve. *Radiology, 143*(1), 29–36. https://doi.org/10.1148/radiology.143.1.7063747.

Hoshi, T., Kitagawa, K., Yamagami, H., Furukado, S., Hougaku, H., & Hori, M. (2005). Relations of serum high-sensitivity C-reactive protein and interleukin-6 levels with silent brain infarction. *Stroke, 36*(4), 768–772. https://doi.org/10.1161/01.STR.0000158915.28329.51.

Igarashi, K., & Kashiwagi, K. (2010). Modulation of cellular function by polyamines. *The International Journal of Biochemistry and Cell Biology, 42*(1), 39–51. https://doi.org/10.1016/j.biocel.2009.07.009.

Igarashi, K., Kashiwagi, K., Hamasaki, H., Miura, A., Kakegawa, T., Hirose, S., et al. (1986). Formation of a compensatory polyamine by *Escherichia coli* polyamine-requiring mutants during growth in the absence of polyamines. *Journal of Bacteriology, 166*(1), 128–134. https://doi.org/10.1128/jb.166.1.128-134.1986.

Kobayashi, S., Okada, K., Koide, H., Bokura, H., & Yamaguchi, S. (1997). Subcortical silent brain infarction as a risk factor for clinical stroke. *Stroke, 28*(10), 1932–1939. https://doi.org/10.1161/01.str.28.10.1932.

Makita, S., Nakamura, M., & Hiramori, K. (2005). The association of C-reactive protein levels with carotid intima-media complex thickness and plaque formation in the general population. *Stroke, 36*(10), 2138–2142. https://doi.org/10.1161/01.STR.0000181740.74005.ee.

McKhann, G., Drachman, D., Folstein, M., Katzman, R., Price, D., & Stadlan, E. M. (1984). Clinical diagnosis of Alzheimer's disease: Report of the NINCDS-ADRDA Work Group under the auspices of Department of Health and Human Services Task Force on Alzheimer's disease. *Neurology, 34*(7), 939–944. https://doi.org/10.1212/wnl.34.7.939.

Nakamura, M., Tomitori, H., Suzuki, T., Sakamoto, A., Terui, Y., Kashiwagi, K., Igarashi, K., et al. (2013). Inactivation of GAPDH as one mechanism of acrolein toxicity. *Biochemical and Biophysical Research Communications, 430*(4), 1265–1271. https://doi.org/10.1016/j.bbrc.2012.12.057.

Pegoraro, A. F., Janmey, P., & Weitz, D. A. (2017). Mechanical properties of the cytoskeleton and cells. *Cold Spring Harbor Perspectives in Biology, 9*(11), 1–13. https://doi.org/10.1101/cshperspect.a022038.

Petersen, R. C., Smith, G. E., Waring, S. C., Ivnik, R. J., Tangalos, E. G., & Kokmen, E. (1999). Mild cognitive impairment: Clinical characterization and outcome. *Archives of Neurology, 56*(3), 303–308. https://doi.org/10.1001/archneur.56.3.303.

Saiki, R., Hayashi, D., Ikuo, Y., Nishimura, K., Ishii, I., Kashiwagi, K, Igarashi, K., et al. (2013). Acrolein stimulates the synthesis of IL-6 and C-reactive protein (CRP) in thrombosis model mice and cultured cells. *Journal of Neurochemistry, 127*(5), 652–659. https://doi.org/10.1111/jnc.12336.

Saiki, R., Nishimura, K., Ishii, I., Omura, T., Okuyama, S., Kashiwagi, K., Igarashi, K., et al. (2009). Intense correlation between brain infarction and protein-conjugated acrolein. *Stroke, 40*(10), 3356–3361. https://doi.org/10.1161/STROKEAHA.109.553248.

Sen, N., Hara, M. R., Kornberg, M. D., Cascio, M. B., Bae, B. I., Shahani, N., et al. (2008). Nitric oxide-induced nuclear GAPDH activates p300/CBP and mediates apoptosis. *Nature Cell Biology, 10*(7), 866–873. https://doi.org/10.1038/ncb1747.

Sharmin, S., Sakata, K., Kashiwagi, K., Ueda, S., Iwasaki, S., Shirahata, A., & Igarashi, K. (2001). Polyamine cytotoxicity in the presence of bovine serum amine oxidase. *Biochemical and Biophysical Research Communications, 282*(1), 228–235. https://doi.org/10.1006/bbrc.2001.4569.

Shinohara, Y., Tohgi, H., Hirai, S., Terashi, A., Fukuuchi, Y., Yamaguchi, T., et al. (2007). Effect of the Ca antagonist nilvadipine on stroke occurrence or recurrence and extension of asymptomatic cerebral infarction in hypertensive patients with or without history of stroke (PICA Study) 1. Design and results at enrollment. *Cerebrovascular Disease, 24*(2-3), 202–209. https://doi.org/10.1159/000104478.

Tomitori, H., Usui, T., Saeki, N., Ueda, S., Nishimura, K., Kashiwagi, K., Igarashi, K., et al. (2005). Polyamine oxidase and acrolein as novel biochemical markers for diagnosis of cerebral stroke. *Stroke, 36*(12), 2609–2613. https://doi.org/10.1161/01.STR.0000190004.36793.2d.

Tristan, C., Shahani, N., Sedlak, T. W., & Sawa, A. (2011). The diverse functions of GAPDH: Views from different subcellular compartments. *Cellular Signaling, 23*(2), 317–323. https://doi.org/10.1016/j.cellsig.2010.08.003.

Uchida, K., Kanematsu, M., Morimitsu, Y., Osawa, T., Noguchi, N., & Niki, E. (1998). Acrolein is a product of lipid peroxidation reaction. Formation of free acrolein and its conjugate with lysine residues in oxidized low density lipoproteins. *Journal of Biological Chemistry, 273*(26), 16058–16066. https://doi.org/10.1074/jbc.273.26.16058.

Uemura, T., Suzuki, T., Ko, K., Nakamura, M., Dohmae, N., Kashiwagi, K., Igarashi, K., et al. (2020). Structural change and degradation of cytoskeleton due to the acrolein conjugation with vimentin and actin during brain infarction. *Cytoskeleton (Hoboken), 77*(10), 414–421. https://doi.org/10.1002/cm.21638.

Uemura, T., Suzuki, T., Ko, K., Watanabe, M., Dohmae, N., Kashiwagi, K., Igarashi, K., et al. (2019). Inhibition of dendritic spine extension through acrolein conjugation with α-, β-tubulin proteins. *International Journal of Biochemistry and Cell Biology, 113*, 58–66. https://doi.org/10.1016/j.biocel.2019.05.016.

Wang, Y., Devereux, W., Woster, P. M., Stewart, T. M., Hacker, A., & Casero, R. A., Jr. (2001). Cloning and characterization of a human polyamine oxidase that is inducible by polyamine analogue exposure. *Cancer Research, 61*(14), 5370–5373.

Wang, Y., Murray-Stewart, T., Devereux, W., Hacker, A., Frydman, B., Woster, P. M., et al. (2003). Properties of purified recombinant human polyamine oxidase, PAOh1/SMO. *Biochemical and Biophysical Research Communications, 304*(4), 605–611. https://doi.org/10.1016/s0006-291x(03)00636-3.

Waragai, M., Yoshida, M., Mizoi, M., Saiki, R., Kashiwagi, K., Takagi, K., Igarashi, K., et al. (2012). Increased protein-conjugated acrolein and amyloid-$\beta_{40/42}$ ratio in plasma of patients with mild cognitive impairment and Alzheimer's disease. *Journal of Alzheimer's Disease, 32*(1), 33–41. https://doi.org/10.3233/JAD-2012-120253.

Yamagami, H., Kitagawa, K., Hoshi, T., Furukado, S., Hougaku, H., Nagai, Y., et al. (2005). Associations of serum IL-18 levels with carotid intima-media thickness. *Arteriosclerosis, Thrombosis, and Vascular Biology, 25*(7), 1458–1462. https://doi.org/10.1161/01.ATV.0000168417.52486.56.

Yoshida, M., Higashi, K., Kobayashi, E., Saeki, N., Wakui, K., Kashiwagi, K., Igarashi, K., et al. (2010). Correlation between images of silent brain infarction, carotid atherosclerosis and white matter hyperintensity, and plasma levels of acrolein, IL-6 and CRP. *Atherosclerosis, 211*(2), 475–479. https://doi.org/10.1016/j.atherosclerosis.2010.03.031.

Yoshida, M., Tomitori, H., Machi, Y., Hagihara, M., Higashi, K., Goda, H., ... Igarashi, K. (2009a). Acrolein toxicity: Comparison with reactive oxygen species. *Biochemical and Biophysical Research Communications, 378*(2), 313–318. https://doi.org/10.1016/j.bbrc.2008.11.054.

Yoshida, M., Tomitori, H., Machi, Y., Katagiri, D., Ueda, S., Horiguchi, K., ... Igarashi, K. (2009b). Acrolein, IL-6 and CRP as markers of silent brain infarction. *Atherosclerosis, 203*(2), 557–562. https://doi.org/10.1016/j.atherosclerosis.2008.07.022.

Yoshida, M., Uemura, T., Mizoi, M., Waragai, M., Terui, Y, Kashiwagi, K., Igarashi, K., et al. (2023). Urinary amino acid-conjugated acrolein and taurine as new biomarkers for detection of dementia. *Journal of Alzheimer's Disease, 92*(1), 361–369. https://doi.org/10.3233/JAD-220912.

CHAPTER SEVEN

Alterations in polyamine metabolism induced by the pathogen *Helicobacter pylori*: Implications for gastric inflammation and carcinogenesis

Alain P. Gobert[a,b,c,1], Caroline V. Hawkins[a,1], Kara M. McNamara[a,c], and Keith T. Wilson[a,b,c,d,e,*]

[a]Division of Gastroenterology, Hepatology, and Nutrition, Department of Medicine, Vanderbilt University Medical Center, Nashville, TN, United States
[b]Center for Mucosal Inflammation and Cancer, Vanderbilt University Medical Center, Nashville, TN, United States
[c]Program in Cancer Biology, Vanderbilt University Medical Center, Nashville, TN, United States
[d]Department of Pathology, Microbiology, and Immunology, Vanderbilt University Medical Center, Nashville, TN, United States
[e]Veterans Affairs Tennessee Valley Healthcare System, Nashville, TN, United States
*Corresponding author. e-mail address: keith.wilson@vumc.org

Contents

1. Introduction	138
2. Culture of *H. pylori*	140
2.1 Equipment	140
2.2 Reagents	140
2.3 Procedure	141
2.4 Notes	142
3. Animal models of *H. pylori* infection	143
3.1 Equipment	143
3.2 Reagents	143
3.3 Procedures	143
3.4 Notes	144
4. Generation and infection of gastric organoids	145
4.1 Equipment	145
4.2 Reagents	145
4.3 Procedure	146
4.4 Notes	147
5. AcroleinRED assay with organoids	148
5.1 Equipment	148
5.2 Reagents	148

[1] Authors share co-first authorship.

5.3 Procedure	148
5.4 Notes	149
6. Immunodetection of acrolein adducts	149
6.1 Equipment	149
6.2 Reagents	150
6.3 Procedure	150
Statements and Declarations	151
Competing interest	151
Grant Support	151
References	151

Abstract

Helicobacter pylori is a pathogen of the human stomach that infects half of the world's population. The chronicity of the infection is a major risk factor for the development of gastric cancer. Multiple parameters, including the heterogenicity of the virulence factors, the environmental parameters, or the intensity of the host responses, regulate the evolution of the disease. Thus, we have reported that the dysregulation of the polyamine pathway in the infected gastric mucosa represents a major component of the inflammation and the risk for neoplastic transformation. Using different animal models and gastric organoids, we recently highlighted the critical role of spermine oxidase-derived acrolein in the pathogenesis of *H. pylori* infection and the progression to carcinoma. Herein, we describe the experimental procedures that have been used, from the culture of the bacteria to the various models of infection in vivo and in vitro, and the assessment of acrolein generation.

1. Introduction

Helicobacter pylori is a Gram-negative bacterium that infects exclusively the mammalian stomach. This infection represents one of the most exquisite examples of host/bacteria interaction. Half of the world's population is colonized (Hooi et al., 2017) and hundreds of thousands of years of co-evolution have shaped a particular mix of commensalism and pathogenicity (de Sablet et al., 2011; Kodaman et al., 2014). Colonized persons universally develop gastritis, which is inflammation of the stomach. This state becomes chronic, and thus it frequently leads to the development of pre-cancerous lesions, from atrophic gastritis to intestinal metaplasia and dysplasia. This sequence, well-known as the Correa's cascade, increases the risk for developing gastric adenocarcinoma, also termed gastric cancer (GC) (Correa, Piazuelo, & Wilson, 2010; Correa, 1988), which occurs in 1–3 % of *H. pylori*-infected patients. Thus, *H. pylori* is a Type I carcinogen and GC currently represents the fifth most common cancer and the fifth most common cause of cancer-related death worldwide (Bray et al., 2024).

The progression from gastritis to GC is mainly driven by (i) the chronicity of the infection; (ii) the expression of several key *H. pylori* virulence factors including the cancer-promoting protein cytotoxin-associated gene A (CagA) that is directly injected into gastric epithelial cells (GECs) by a type IV secretion system; (iii) environmental factors such as high-salt diet, obesity, and smoking; and (iv) the intensity of the host immune response. In recent years, we have demonstrated that the dysregulation of the metabolism of polyamines by *H. pylori* regulates gastric inflammation and carcinogenesis. Ornithine decarboxylase (ODC), which synthesizes the first polyamine putrescine from L-ornithine is induced in the gastric tissues of *H. pylori*-infected patients (Chaturvedi et al., 2010; Konturek et al., 2003) and mice (Chaturvedi et al., 2010); both gastric epithelial cells (GECs) and myeloid cells show increased ODC expression (Chaturvedi et al., 2010; Xu et al., 2012). Consequently, polyamine levels are enhanced in the infected stomach (Chaturvedi et al., 2010; Linsalata, Russo, Notarnicola, Berloco, & Di Leo, 1998). The enzyme spermine oxidase (SMOX) catalyzes the oxidation of spermine to spermidine, thus generating H_2O_2 and 3-aminopropanal (Wang et al., 2001); this last compound is rapidly and spontaneously converted into the highly reactive monocarbonyl electrophile acrolein. Our group found that SMOX is overexpressed in macrophages (Chaturvedi et al., 2004) and in gastric epithelial cells (Xu et al., 2004) infected in vitro with *H. pylori* and in GECs and macrophages of infected humans, mice, and gerbils (Chaturvedi et al., 2011). The role of these enzymes has been further assessed by genetic and pharmacological approaches. We found that in C57BL/6 mice, *Odc* deletion in myeloid cells enhances the innate immune response of macrophages leading to reduced colonization of the stomach by *H. pylori* (Hardbower et al., 2017), suggesting that macrophage ODC dampens the anti-bacterial immunity. In contrast, we showed that the specific deletion of *Odc* in GECs attenuates gastritis and epithelial proliferation (Latour et al., 2022). Accordingly, when Mongolian gerbils, which develop invasive gastric adenocarcinoma when infected with *H. pylori*, were treated with the ODC inhibitor α-difluoromethylornithine (DFMO) there was a reduction in dysplasia and carcinoma (Chaturvedi et al., 2015). Altogether, these data indicate that polyamines support gastric inflammation and carcinogenesis. Further, we found that *Smox* deletion protects against gastritis in C57BL/6 mice (Sierra et al., 2020). We have also utilized transgenic FVB/N insulin-gastrin (INS-GAS) mice, which express the human gastrin gene regulated by the rat insulin-1 promoter, leading to overexpression of gastrin in the pancreas, and accelerated neoplastic transformation with infection with *H. pylori* (Wang et al., 2000). We generated INS-GAS mice with deletion of *Smox* and demonstrated that in *H. pylori*-infected mice there was reduced GEC proliferation and DNA damage, and less progression from dysplasia to carcinoma (McNamara et al., 2025). Infected gerbils also exhibit less dysplasia and carcinoma when treated with the SMOX inhibitor N(1),N(4)-di(buta-2,3-dien-1-yl)butane-1,4-diamine dihydrochloride (MDL 72527) (Chaturvedi et al., 2015). Mechanistically, we recently attributed the deleterious effect of SMOX activity in GECs to the formation of acrolein (McNamara et al., 2025). This molecule has been shown to cause mutagenesis and disruption of gene expression by reacting with deoxyguanosine in DNA and nucleophilic residues in proteins (Stevens & Maier, 2008).

In this review, we describe the methods to study the effect of *H. pylori* infection in animal models and in gastric organoids, including assessment of acrolein generation.

2. Culture of *H. pylori*

2.1 Equipment
- Amsco Century SV-120 Scientific Prevacuum Sterilizer (Steris).
- 1300 Series Class II, Type A2 Biological Safety Cabinet (ThermoFisher Scientific).
- Heracell™ 240i CO_2 Incubator, 240 L (ThermoFisher Scientific).
- GENESYS™ 30 Visible Spectrophotometer (Thermo Fisher Scientific).
- MaxQ 2000 CO_2 Plus Shaker (Thermo Fisher Scientific).
- Sorvall™ Legend™ Micro 17 R Microcentrifuge (Thermo Fisher Scientific).

2.2 Reagents
- 88 mm × 15 mm Petri Dish (Corning).
- 25-75 cm^2 flask with ventilated cap (Corning).
- Heat inactivated fetal bovine serum (FBS; Corning).
- Tryptic Soy Agar (MP Biomedicals).
- BBL™ Brucella broth (BD Biosciences).
- Sheep blood (HemoStat Laboratories).
- Nalidixic acid (Sigma Aldrich).
- Vancomycin (Thermo Fisher Scientific).
- Amphotericin (Sigma Aldrich).

- Bacitracin (Sigma Aldrich).
- Polymyxin B (Sigma Aldrich).

2.3 Procedure

2.3.1 Preparation of blood-agar plates

1. Dissolve the Tryptic Soy Agar in deionized water at the final concentration of 30 g/l in a glass bottle.
2. Autoclave for 30 min at 120 °C.
3. Leave the bottle at room temperature for 30 min and then in a water bath at 37 °C under agitation.
4. When the temperature of the bottle reaches 37 °C, add 10 % sheep blood pre-warmed at 37 °C and mix gently.
5. Add nalidixic acid (10 µg/ml), vancomycin (100 µg/ml), amphotericin (2 µg/ml), bacitracin (200 µg/ml), polymyxin B (2500 U/l); mix gently.
6. Under sterile conditions using a biological safety cabinet, pour 20 ml of the medium in Petri dishes; avoid the formation of bubbles at the surface.
7. Cover the plates and leave them overnight at room temperature.
8. Keep the plates upside down in a closed plastic bag at 4 °C for up to 1 month.

2.3.2 Preparation of liquid culture medium

1. Dissolve the Brucella broth in deionized water at the final concentration of 28 g/l in a glass bottle.
2. Autoclave for 30 min at 120 °C.
3. This medium, termed BB thereafter, can be maintained at room temperature for up to 1 month.

2.3.3 Preparation of H. pylori for infections

1. Using a disposable inoculating loop, pick up an aliquot of frozen *H. pylori* stock and spread the bacteria by a back-and-forth motion at the surface of the whole agar plate.
2. Put the plate upside down in an incubator at 37 °C and 5 % CO_2.
3. Monitor the growth of the bacteria every day.
4. *H. pylori* colonies on a blood agar plate are small, round, and translucent (Fig. 1). They can gradually grow larger but should be used as small colonies.
5. Under sterile conditions, harvest *H. pylori* colonies from one plate using a sterile cotton tip.

Fig. 1 *H. pylori* colonies on blood-agar plates.

6. Disperse these bacteria in 10 ml BB supplemented with 10 % FBS in a 50 ml tube.
7. Measure the absorbance at 600 nm and adjust the volume of BB-FBS to an $A_{600\ nm}$ of 0.1.
8. Place 10–25 ml of this bacterial suspension in a 25 cm^2 flask with a ventilated cap (Corning) on an orbital shaker platform (170 rpm) at 37 °C and 5 % CO_2 for 18 h. This is performed by using the CO_2-resistant shaker inside the Heracell™ 240i CO_2 Incubator.
9. Measure $A_{600\ nm}$ and dilute this bacterial suspension in fresh BB + FBS medium to an $A_{600\ nm}$ of 0.1; Use a 25–75 cm^2 flask with a ventilated cap (Corning) to grow *H. pylori* on an orbital shaker platform (170 rpm) at 37 °C and 5 % CO_2.
10. Monitor $A_{600\ nm}$ and stop the culture at an optical density between 0.4 and 0.8.
11. Spin at 10,000 x *g* for 10 min
12. Discard the supernatant and resuspend the pellet in the appropriate medium, e.g., BB to infect animals or in cell culture medium for co-culture with eukaryotic cells.

2.4 Notes

1. For H. pylori, A600 nm of 1 corresponds to 5×10^8 bacteria per ml.

3. Animal models of *H. pylori* infection
3.1 Equipment

- T 25 digital Ultra-Turrax (IKA).
- Paraffin Embedding Station EG1150H (Leica).
- Shandon Citadel 2000 (Thermo Fisher Scientific).
- RM2235 Manual Rotary Microtome (Leica).

3.2 Reagents
- Ethanol.
- TechniCloth® TX604 Nonwoven Dry Cleanroom Wipers (Techni-Tool).
- 5 ml polystyrene round-bottom tubes (Falcon).
- PROTOCOL™ Zinc Formalin (Thermo Fisher Scientific).
- CAT Hematoxylin (Biocare).
- Ammonium hydroxide (Sigma).
- Rubens Eosin-Phloxine Solution, Biocare.

3.3 Procedures
3.3.1 Infection
1. Concentrate *H. pylori* in BB to 10^{10} bacteria/ml (see #2.3).
2. Infect mice directly in the stomach with 100 μl of the bacterial suspension by oral gavage using a curved feeding needle. Mongolian gerbils are infected with 200 μl of the same bacterial suspension.
3. Repeat the infection after two days.
4. Sham animals are treated the same volume of BB.

3.3.2 Dissection
1. Inside a biological safety cabinet, sanitize the mouse abdomen surface with 70% ethanol. Use scissors and a forceps to open the abdomen and remove the stomach, place it on an ice-cold aluminum tray.
2. Unfold the stomach on a filter paper and divide it into 4 longitudinal strips. The two lateral strips are snap frozen in liquid nitrogen using cryotubes. The two central pieces are used to determine colonization and to assess histology (see below).

3.3.3 Determination of colonization
1. Add 1 ml of BB containing the antibiotic cocktail (see #2.2.1) in 5 ml tubes and weigh them.
2. Add one strip of the stomach in the tube immediately after dissection and weigh them.

3. The gastric tissue is then homogenized using the Ultra-Turrax.
4. Perform four 1/10-serial dilution of the homogenate in PBS.
5. Plate three 10 μl-spots of each dilution on blood-agar plates and let them dry for 15 min
6. Put the plates up-side down at 37 °C and 5% CO_2 in the incubator. After 5–10 days, count the number of colony forming units (CFUs) for each dilution.
7. Colonization is expressed as CFU per gram of gastric tissue.

3.3.4 Evaluation of histological damage

1. One central strip of the stomach is fixed in 10% zinc formalin and embedded in paraffin.
2. 5 μm sections of embedded tissues are generated using a Microtome and placed on glass microscope slide.
3. Deparaffinize sections with xylene bath, four times for 2 min
4. Re-hydrate sections with 100% alcohol, two times for 2 min, and then once with 95% alcohol for 2 min
5. Wash in distilled water.
6. Incubate in hematoxylin for 2 min
7. Wash in distilled water.
8. Incubate 5 s in a solution of 0.02% ammonium hydroxide in deionized water.
9. Rinse in running tap water several times.
10. Counterstain in eosin-phloxine solution for 1 min
11. Incubate two times in 95% alcohol for 2 min and then two times in 100% alcohol for 2 min
12. Clear in xylene bath, three times for 2 min
13. Mount with a coverslip.
14. The histological score (0–12) of the stomach is determined by a gastrointestinal pathologist by the sum of acute inflammation in the antrum (0–3) and the corpus (0–3) plus chronic inflammation in the antrum (0–3) and the corpus (0–3) (Chaturvedi et al., 2010; Hardbower et al., 2017; Latour et al., 2022; Sierra et al., 2020).

3.4 Notes

1. Mice are infected between 6 to 10 weeks old.
2. Mice on the C57BL/6 background infected with the $cagA^+$ strain PMSS1 develop gastritis (Hardbower et al., 2017; Latour et al., 2022; Sierra et al., 2020).

3. Transgenic INS-GAS mice infected with PMSS1 develop gastritis and neoplastic transformation, from low grade dysplasia (LGD) to intra-mucosal carcinoma (IMC) (Gobert et al., 2021; Gobert et al., 2023).
4. The *cagA*[+] *H. pylori* strain 7.13 is used to infected Mongolian gerbils, which then develop gastric diseases ranging from gastritis to LGD, and invasive adenocarcinoma (Gobert et al., 2021; Sierra et al., 2018; Sierra et al., 2019) (Fig. 2).

4. Generation and infection of gastric organoids

4.1 Equipment
- SG 400 – Class II Type A/B3 Biological Safety Cabinet (Baker Company).
- Heracell™ VIOS 160i CO_2 Incubator, 165L (Thermo Fisher Scientific).
- 5810R Centrifuge (Eppendorf).
- Platform Vari-Mix Platform Rocker (Thermo Fisher Scientific).
- Eclipse Ts2 Inverted Microscope (Nikon).

4.2 Reagents
- Nunc™ 24 Well Plate, TC Surface (Thermo Fisher Scientific).
- 15-ml tubes (Corning)
- Poly-D-Lysine 8-well culture slides (Corning).
- DMEM (Gibco).
- Heat-inactivated-FBS (Corning).
- Penicillin-Streptomycin (P/S; Gibco).
- Hepes (Corning).
- DPBS without Ca^{2+}/Mg^{2+} (Corning).
- 0.5 M EDTA (Corning).
- DPBS without Ca^{2+}/Mg^{2+} (Corning).
- Sucrose (Sigma Aldrich).

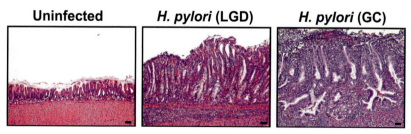

Fig. 2 H&E staining of the stomach of Mongolian gerbils infected or not with *H. pylori* for 8 wks. Scale bar, 50 μm.

- D-sorbitol (Sigma Aldrich).
- 50% W-LRN Conditioned media (see Note 1).
- Y27632 (ToCris).
- SB431542 (ToCris).
- Gentamicin (Gibco).
- Matrigel matrix (Corning).
- 0.25% Trypsin (Gibco).

4.3 Procedure

4.3.1 Creation of gastric organoids

1. Prepare dissociation buffer: 200 ml of DPBS without Ca^{2+}/Mg^{2+} + 2 g D-sorbitol + 3 g sucrose.
2. Cool centrifuge to 4 °C and allow media to warm to room temperature.
3. Collect gastric tissues (see Note 2) in 15 ml-tube containing 5 ml of DMEM plus 10% FBS, 1% P/S, and 1% Hepes.
4. Transfer the tissues into a sterile 15 ml-tube containing 5 ml of 10 mM EDTA in PBS solution.
5. Incubate for 30 min on a shaker in the cold room.
6. After incubating, move tissue pieces into 1 ml of sterile filtered cold dissociation buffer and shake tissues vigorously until you can see glands floating (see Note 3).
7. Move tissue into a new 15 ml-tube with 1 ml cold dissociation buffer and repeat the shaking.
8. Repeat steps 6-7 until an adequate number of glands have been collected.
9. Verify that all the fractions have been collected and choose the ones with the whole glands. Fractions can be combined.
10. Move chosen fraction(s) to 15 ml-tubes and centrifuge at 650 x g for 5 min at 4 °C.
11. Remove the supernatant (see Note 4).
12. Place 15 ml-tube with the pellet on ice.
13. Carefully resuspend the pellet in 300 µl Matrigel Matrix.
14. Plate 50 µl of Matrigel containing the glands per well in pre-warmed (37 °C) 24-well plates.
15. Place the plate back at 37 °C in the incubator to warm.
16. Allow 15–30 min for Matrigel to solidify before adding the media.
17. Add 500 µl of media composed of 50% L-WRN Conditioned media plus 10 µM SB431542, 10 µM Y27632, and 50 µg/ml gentamicin.
18. Change media every 3–4 days and passage every 7–14 days (See Note 5).

4.3.2 Passaging gastric organoids
1. Passage the organoids when they are large and round.
2. Cool centrifuge to 4 °C and warm media to room temperature.
3. Remove the media and add 500 µl of cold DPBS without Ca^{2+}/Mg^{2+}.
4. Scrape the Matrigel and place in a 15 ml tube.
5. Add another 500 µl of cold DPBS to well and place contents in the same tube.
6. Place 24-well plate back at 37 °C.
7. Centrifuge at 650 x g for 5 min at 4 °C and discard the supernatant.
8. Resuspend pellet in 1 ml 0.25% trypsin.
9. Incubate at 37 °C for 10 min and then resuspend vigorously.
10. Add 5 ml DMEM containing 10% FBS.
11. Centrifuge at 650 x g for 5 min at 4 °C.
12. Discard the supernatant, and place on ice.
13. Repeat the procedure 13–18 from Section 4.3.1.

4.3.3 Plating gastric organoid monolayers for infection
Organoids are plated in regular 24-well plates or in Poly-D-Lysine 8-well culture slides for subsequent imaging.

Coat the wells with 1:100 Matrigel in cold DPBS and incubate at 37 °C for 1 h.

Remove the coating solution.

Follow steps 1–10 from Section 4.3.2.

Count the cells.

Centrifuge at 650 x g for 5 min at 4 °C.

Resuspend in appropriate volume of 50% W-LRN media (see Note 6).

Add cells to pre-coated well or chamber slides and place in the incubator overnight at 37 °C.

Infect the cells with *H. pylori* (see Section 2.3.3) with a multiplicity of infection of 10–100.

4.4 Notes
1. 50% L-WRN Conditioned media is generated according to the protocol from Miyoshi and Stappenbeck (Miyoshi & Stappenbeck, 2013).
2. For humans, endoscopic biopsies of gastric tissues obtained for research with patient informed consent under protocols approved by the Institutional Review Board, are separated by antrum and corpus. They are processed as two samples. A strip including both corpus and antrum is used for mouse organoids.

3. After each shaking stage, harvest an aliquot of 50 μl to observe under an inverted microscope to evaluate whether the glands have been isolated. It is important to not over-shake, which can result in single cell isolation, which will not grow efficiently.
4. The pellet can be loose; do not remove the supernatant with a vacuum.
5. Gentamicin is removed after the first passage.
6. 3×10^5 and 10^5 cells are plated in 24-well plates and chamber slides, respectively.

5. AcroleinRED assay with organoids
5.1 Equipment
- Heracell™ VIOS 160i CO2 Incubator, 165L (Thermo Fisher Scientific).
- Cytation C10 Confocal Imaging Reader with Gen 5 + software (Agilent BioTek).

5.2 Reagents
- DPBS without Ca^{2+}/Mg^{2+} (Corning).
- DMEM (Gibco).
- Heat-inactivated FBS (Corning).
- Hepes (Corning).
- AcroleinRED (DiagnoCine).
- VECTASHIELD HardSet Antifade Mounting Medium with DAPI (Thermo Fisher Scientific).

5.3 Procedure
1. After infection of organoids in chamber slides with *H. pylori* (Section 4.3.3), wash each well three times with 200 μl DPBS (See Note 1).
2. Add 200 μl/well of 10 mM AcroleinRED in DMEM and incubate at 37 °C for 30 min.
3. Aspirate AcroleinRED media (See Note 2).
4. Add 200 μl/well DPBS to wash; repeat this step three times.
5. Add 200 μl/well of 3.7% formaldehyde at room temperature for 30 min to fix cells.
6. Add 200 μl/well DPBS to wash; repeat this step three times.
7. Chamber slide walls are removed with tools from the kit.
8. Add 6 drops of VECTASHEILD HardSet Antifade Mounting Medium with DAPI (See Notes 3 and 4).
9. Carefully place a coverslip over the mounting media (See Note 5).

Alterations in polyamine metabolism induced by the pathogen Helicobacter pylori   149

10. Place slide at 4 °C to dry overnight.
11. Image (Fig. 3) and quantify using the Cytation C10 (See Notes 6 and 7).

5.4 Notes

1. Wash the wells carefully. There are numerous wash steps for this process and cells can detach easily. Gently add the DPBS each time to increase the yield of cells for imaging. Organoids are particularly prone to washing off. Proceed with caution.
2. Steps following the AcroleinRED incubation step do not need to be sterile.
3. DAPI is a fluorescent probe that binds strongly to adenine–thymine-rich regions in DNA and therefore stains the nucleus.
4. The drops should be evenly distributed across the slide to help eliminate the chance of bubbles.
5. Try to avoid air bubbles when placing the coverslip as they then are hard to image. Once the coverslip is laid down, it cannot be picked up.
6. Excitation/emission wavelengths are 560 nm/585 nm for AcroleinRED and 405 nm/442 nm for DAPI.
7. Slides cannot be imaged if they are still wet. Slides must be imaged within 2 days of preforming the assay to ensure they do not dry out.

6. Immunodetection of acrolein adducts

6.1 Equipment

- 1310 Gravity Convection Oven (VWR).
- Decloaking chamber (Biocare Medical).
- Eclipse E800 microscope (Nikon).

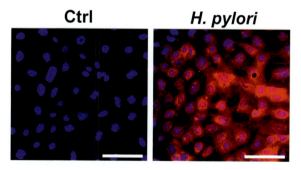

Fig. 3 Human gastric organoids isolated from normal patient biopsies were infected or not with *H. pylori* for 24 h and then stained with AcroleinRED. Acrolein is depicted in red and nuclei are shown in blue. Scale bar, 100 μm.

6.2 Reagents

- Hydrogen peroxide.
- Target Retrieval Solution, pH 9 (Dako).
- Proteinase K, Ready-to-use (Dako).
- Normal horse serum (Jackson ImmunoResearch).
- Protein Block, Serum-Free (Dako).
- Antibody Diluent (Dako).
- Acrolein Monoclonal Antibody (Thermo Fisher Scientific).
- Donkey anti-Mouse IgG (H+L) Highly Cross-Adsorbed Secondary Antibody, Alexa Fluor™ Plus 555 (Thermo Fisher Scientific).
- VECTASHIELD HardSet Antifade Mounting Medium with DAPI (Thermo Fisher Scientific).

6.3 Procedure

1. Paraffin-embedded tissues are sectioned (5 μm) and heated for 1 h at 60 °C in a convection oven.
2. Sections are then deparaffinized in xylene (5 baths, 2 min each) and rehydrated in graded alcohols (100%, 95%, 90%, 80%, 5 min in each).
3. Endogenous peroxidases are blocked with 3% hydrogen peroxide solution for 30 min at room temperature.
4. Incubate the slides in Target Retrieval Solution and incubate 20 min at 110 °C in decloaking chamber.
5. Incubate the sections with Proteinase K (Dako) for 30 s
6. Wash three times in PBS for 5 min
7. Incubate the slides in a solution of 5% normal horse serum in PBS for 1 h at room temperature.
8. Incubate the slides with universal Protein Block (Dako) for 1 h at room temperature.
9. Incubate the slides with the Acrolein Monoclonal Antibody, 1:100 in antibody diluent, overnight at 4 °C.
10. Wash three times in PBS for 5 min
11. Incubate the slides with Donkey anti-Mouse IgG (H+L) Highly Cross-Adsorbed Secondary Antibody, Alexa Fluor™ Plus 555, 1:600 in antibody diluent, for 1 h at room temperature.
12. Mount the slides with VECTASHIELD HardSet Antifade Mounting Medium with DAPI.
13. Images (Fig. 4) are taken on a Nikon Eclipse E800 microscope using SPOT 5.4 Software.

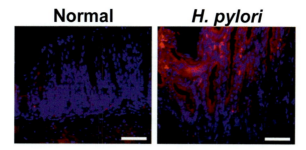

Fig. 4 The antral gastric tissues from normal and *H. pylori*-infected patients were immunostained for acrolein adducts. Acrolein adducts and nuclei are shown in red and blue, respectively. Scale bar, 50 μm.

Statements and Declarations
Competing interest
The authors disclose no conflicts.

Grant Support
This study was funded by NIH grants R01DK128200, P01CA116087, and P01CA028842 (KTW); Veterans Affairs Merit Review grants I01CX002171 and I01CX002473 (KTW); Department of Defense Peer Reviewed Cancer Research Program Impact Award W81XWH-21-1-0617 (KTW); the Tissue Morphology Core of the Vanderbilt Digestive Disease Research Center supported by P30DK058404, the Thomas F. Frist Sr. Endowment (KTW); and the Vanderbilt Center for Mucosal Inflammation and Cancer (KTW).

References
Bray, F., Laversanne, M., Sung, H., Ferlay, J., Siegel, R. L., Soerjomataram, I., & Jemal, A. (2024). Global cancer statistics 2022: GLOBOCAN estimates of incidence and mortality worldwide for 36 cancers in 185 countries. *CA: A Cancer Journal for Clinicians*.

Chaturvedi, R., Asim, M., Hoge, S., Lewis, N. D., Singh, K., Barry, D. P., ... Wilson, K. T. (2010). Polyamines impair immunity to *Helicobacter pylori* by inhibiting L-arginine uptake required for nitric oxide production. *Gastroenterology, 139*, 1686–1698.

Chaturvedi, R., Asim, M., Romero-Gallo, J., Barry, D. P., Hoge, S., de Sablet, T., ... Wilson, K. T. (2011). Spermine oxidase mediates the gastric cancer risk associated with Helicobacter pylori CagA. *Gastroenterology, 141*, 1696–1708.

Chaturvedi, R., Cheng, Y., Asim, M., Bussiere, F. I., Xu, H., Gobert, A. P., ... Wilson, K. T. (2004). Induction of polyamine oxidase 1 by *Helicobacter pylori* causes macrophage apoptosis by hydrogen peroxide release and mitochondrial membrane depolarization. *The Journal of Biological Chemistry, 279*, 40161–40173.

Chaturvedi, R., de Sablet, T., Asim, M., Piazuelo, M. B., Barry, D. P., Verriere, T. G., ... Wilson, K. T. (2015). Increased *Helicobacter pylori*-associated gastric cancer risk in the Andean region of Colombia is mediated by spermine oxidase. *Oncogene, 34*, 3429–3440.

Correa, P. (1988). A human model of gastric carcinogenesis. *Cancer Research, 48*, 3554–3560.

Correa, P., Piazuelo, M. B., & Wilson, K. T. (2010). Pathology of gastric intestinal metaplasia: Clinical implications. *The American Journal of Gastroenterology, 105*, 493–498.

de Sablet, T., Piazuelo, M. B., Shaffer, C. L., Schneider, B. G., Asim, M., Chaturvedi, R., ... Wilson, K. T. (2011). Phylogeographic origin of *Helicobacter pylori* is a determinant of gastric cancer risk. *Gut, 60*, 1189–1195.

Gobert, A. P., Asim, M., Smith, T. M., Williams, K. J., Barry, D. P., Allaman, M. M., ... Wilson, K. T. (2023). The nutraceutical electrophile scavenger 2-hydroxybenzylamine (2-HOBA) attenuates gastric cancer development caused by *Helicobacter pylori*. *Biomedicine & Pharmacotherapy = Biomedecine & Pharmacotherapie, 158*, 114092.

Gobert, A. P., Boutaud, O., Asim, M., Zagol-Ikapitte, I. A., Delgado, A. G., Latour, Y. L., ... Wilson, K. T. (2021). Dicarbonyl electrophiles mediate inflammation-induced gastrointestinal carcinogenesis. *Gastroenterology, 160*, 1256–1268.

Hardbower, D. M., Asim, M., Luis, P. B., Singh, K., Barry, D. P., Yang, C., ... Wilson, K. T. (2017). Ornithine decarboxylase regulates M1 macrophage activation and mucosal inflammation via histone modifications. *Proceedings of the National Academy of Sciences of the United States of America, 114*, E751–E760.

Hooi, J. K. Y., Lai, W. Y., Ng, W. K., Suen, M. M. Y., Underwood, F. E., Tanyingoh, D., ... Ng, S. C. (2017). Global Prevalence of *Helicobacter pylori* Infection: Systematic Review and Meta-Analysis. *Gastroenterology, 153*, 420–429.

Kodaman, N., Pazos, A., Schneider, B. G., Piazuelo, M. B., Mera, R., Sobota, R. S., ... Correa, P. (2014). Human and Helicobacter pylori coevolution shapes the risk of gastric disease. *Proceedings of the National Academy of Sciences of the United States of America, 111*, 1455–1460.

Konturek, P. C., Rembiasz, K., Konturek, S. J., Stachura, J., Bielanski, W., Galuschka, K., ... Hahn, E. G. (2003). Gene expression of ornithine decarboxylase, cyclooxygenase-2, and gastrin in atrophic gastric mucosa infected with *Helicobacter pylori* before and after eradication therapy. *Digestive Diseases and Sciences, 48*, 36–46.

Latour, Y. L., Sierra, J. C., McNamara, K. M., Smith, T. M., Luis, P. B., Schneider, C., ... Wilson, K. T. (2022). Ornithine decarboxylase in gastric epithelial cells promotes the immunopathogenesis of *Helicobacter pylori* infection. *Journal of Immunology, 209*, 796–805.

Linsalata, M., Russo, F., Notarnicola, M., Berloco, P., & Di Leo, A. (1998). Polyamine profile in human gastric mucosa infected by Helicobacter pylori. *Italian Journal of Gastroenterology and Hepatology, 30*, 484–489.

McNamara, K. M., Sierra, J. C., Latour, Y. L., Hawkins, C. V., Asim, M., Williams, K. J., ... Wilson, K. T. (2025). Spermine oxidase promotes *Helicobacter pylori*-mediated gastric carcinogenesis through acrolein production. *Oncogene, 44*, 296–306.

Miyoshi, H., & Stappenbeck, T. S. (2013). In vitro expansion and genetic modification of gastrointestinal stem cells in spheroid culture. *Nature Protocols, 8*, 2471–2482.

Sierra, J. C., Asim, M., Verriere, T. G., Piazuelo, M. B., Suarez, G., Romero-Gallo, J., ... Wilson, K. T. (2018). Epidermal growth factor receptor inhibition downregulates *Helicobacter pylori*-induced epithelial inflammatory responses, DNA damage and gastric carcinogenesis. *Gut, 67*, 1247–1260.

Sierra, J. C., Piazuelo, M. B., Luis, P. B., Barry, D. P., Allaman, M. M., Asim, M., ... Wilson, K. T. (2020). Spermine oxidase mediates *Helicobacter pylori*-induced gastric inflammation, DNA damage, and carcinogenic signaling. *Oncogene, 39*, 4465–4474.

Sierra, J. C., Suarez, G., Piazuelo, M. B., Luis, P. B., Baker, D. R., Romero-Gallo, J., ... Wilson, K. T. (2019). alpha-Difluoromethylornithine reduces gastric carcinogenesis by causing mutations in *Helicobacter pylori cagY*. *Proceedings of the National Academy of Sciences of the United States of America, 116*, 5077–5085.

Stevens, J. F., & Maier, C. S. (2008). Acrolein: Sources, metabolism, and biomolecular interactions relevant to human health and disease. *Molecular Nutrition & Food Research, 52*, 7–25.

Wang, T. C., Dangler, C. A., Chen, D., Goldenring, J. R., Koh, T., Raychowdhury, R., ... Fox, J. G. (2000). Synergistic interaction between hypergastrinemia and *Helicobacter* infection in a mouse model of gastric cancer. *Gastroenterology, 118*, 36–47.

Wang, Y., Devereux, W., Woster, P. M., Stewart, T. M., Hacker, A., & Casero, R. A., Jr (2001). Cloning and characterization of a human polyamine oxidase that is inducible by polyamine analogue exposure. *Cancer Research, 61*, 5370–5373.

Xu, H., Chaturvedi, R., Cheng, Y., Bussiere, F. I., Asim, M., Yao, M. D., ... Wilson, K. T. (2004). Spermine oxidation induced by *Helicobacter pylori* results in apoptosis and DNA damage: implications for gastric carcinogenesis. *Cancer Research, 64*, 8521–8525.

Xu, X., Liu, Z., Fang, M., Yu, H., Liang, X., Li, X., ... Ia, J. (2012). *Helicobacter pylori* CagA induces ornithine decarboxylase upregulation via Src/MEK/ERK/c-Myc pathway: Implication for progression of gastric diseases. *Experimental Biology and Medicine (Maywood), 237*, 435–441.

CHAPTER EIGHT

Yeast reconstituted translation assays for analysis of eIF5A function

Byung-Sik Shin and Thomas E. Dever[*]

Division of Molecular and Cellular Biology, Eunice Kennedy Shriver National Institute of Child Health and Human Development, National Institutes of Health, Bethesda, MD, United States
*Corresponding author. e-mail address: thomas.dever@nih.gov

Contents

1. Introduction	156
2. Overview of in vitro translation assays	158
3. Preparation of initiation factors	160
4. Purification of eIF5A	160
5. Purification of elongation factors	162
5.1 Purification of eEF1A	162
5.2 Purification of eEF2 and eEF3	164
6. Purification of termination factors	165
6.1 Purification of eRF1	165
6.2 Purification of eRF3	166
7. Preparation of mRNA	166
7.1 Preparation of DNA template for T7 transcription	167
7.2 Preparation of mRNA by in vitro T7 transcription	168
8. Preparation of tRNA^{Pro}	169
9. Purification of tRNA synthetases and CCA-adding enzyme	170
9.1 Purification of methionyl-tRNA synthetase (MetRS)	170
9.2 Purification of Prolyl tRNA synthetase (PRS)	171
9.3 Purification of Phenylalanyl and Lysyl tRNA synthetase	171
9.4 Purification of CCA-adding enzyme	172
10. Limited charging of tRNA$_i$ using ^{35}S methionine	173
11. Aminoacylation of tRNA^{Phe}, tRNA^{Pro}, and tRNA^{Lys}	174
12. Purification of 40S and 60S ribosomal subunits	175
13. Peptide formation and release assays	176
13.1 Preparation of initiation complexes	176
13.2 Peptide formation assay	177
13.3 Peptide release (termination) assay	179
Acknowledgments	180
References	180

Abstract

Polyamines are critically important for protein synthesis. Through their positive ionic charge, polyamines readily bind to ribosomes, as well as to mRNAs and tRNAs. Moreover, the polyamine spermidine serves as a substrate for the synthesis of hypusine, an essential post-translational modification on the translation factor eIF5A. Though originally thought to function in translation initiation, eIF5A is now known to generally promote translation elongation and termination. Moreover, translation of certain motifs like polyproline show a greater dependency on eIF5A. In this chapter, we describe the biochemical assays we use to study eIF5A and its regulation. Owing to the complex nature of protein synthesis, these assays require the purification of over 10 translation factors plus ribosomes, tRNAs, and aminoacyl-tRNA synthetases. We describe the methods used to purify these components, to synthesize the mRNA templates for translation, and to resolve the translation products by electrophoretic thin-layer chromatography. With the recent identification of eIF5A as a key target for regulating the synthesis of polyamine synthesis and transport, and the recent identification of mutations in eIF5A causing a neurodevelopmental disorder, the assays described in this chapter will be useful in further elucidating the function and regulation of this enigmatic protein.

1. Introduction

As the only cellular protein containing a polyamine-derived post-translational modification, the translation factor eIF5A is of particular interest to researchers studying polyamines. Translation factors play critical roles in cellular protein synthesis, interacting with the ribosome, mRNA, and aminoacyl-tRNAs to enhancing both the efficiency and fidelity of the process. The translation process is divided into three steps: initiation, elongation, and termination. During initiation, the ribosome binds an mRNA and selects the translation start codon (typically AUG). Translation elongation refers to the codon-dependent addition of amino acids to the growing peptide chain. When the elongating ribosome encounters a stop codon (UAA, UAG, or UGA), translation terminates, and the completed peptide is released from the ribosome.

The eukaryotic translation Initiation Factors (eIFs) function during the initiation phase (reviewed in Dever, Kinzy, & Pavitt, 2016; Hinnebusch & Lorsch, 2012; Kapp & Lorsch, 2004b; Lorsch & Dever, 2010). First, the factor eIF2 binds GTP and initiator methionyl-tRNA$_i^{Met}$ to form a ternary complex and then together with initiation factors eIF1, eIF1A, and eIF3 associates with the small ribosomal subunit to form a preinitiation complex (PIC). While binding of the PIC to an mRNA in cells is facilitated by the factors eIF4A, eIF4B, eIF4E and eIF4G, the PIC can bind unstructured mRNAs in vitro without the addition of these factors (Algire et al., 2002; Lorsch & Herschlag, 1999). Following selection of the start codon by the PIC, the factors eIF5 and eIF5B facilitate release of the other initiation factors and joining of the large ribosomal subunit to form an 80S ribosome on the start codon (Pestova et al., 2000; Shin et al., 2002).

During translation elongation, the eukaryotic elongation factors (eEFs) eEF1A and eEF2 function like their bacterial counterparts EF-Tu and EF-G to bind aminoacyl-tRNAs to the A site of the ribosome and to promote translocation of the tRNAs and ribosome following peptide bond formation, respectively (reviewed in Dever & Green, 2012; Dever, Dinman, & Green, 2018). Yeast have a third essential elongation factor, eEF3, that is proposed to promote late stages of translocation and release of deacylated tRNA from the ribosomal E site. When a stop codon enters the A site of the elongating ribosome, the eukaryotic release factor (eRF) 1 binds in the A site with the assistance of the factor eRF3. The eRF1 then promotes hydrolysis of the acyl bond connecting the peptidyl-tRNA and the nascent peptide to release the completed protein.

While eIF5A was originally identified as a factor promoting translation initiation due to its ability to promote methionyl-puromycin synthesis, a model assay for first peptide bond formation (reviewed in Dever, Gutierrez, & Shin, 2014), this name is a misnomer. The methionyl-puromycin synthesis assay is an elongation assay, and more recent work revealed that the factor promotes translation elongation and termination (Saini, Eyler, Green, & Dever, 2009; Schuller, Wu, Dever, Buskirk, & Green, 2017). While eIF5A generally promotes peptide bond formation during translation elongation and peptide release at termination, stimulating the rate of the reaction at least 1.5–2-fold for all amino acids, certain substrates like polyproline and other proline-rich sequences have a heightened requirement for eIF5A (Gutierrez et al., 2013; Schuller et al., 2017).

Of note, the ability of eIF5A to promote translation elongation and termination is dependent on its post-translational modification. A specific lysine residue in eIF5A (Lys51 in the yeast protein) is post-translationally modified to hypusine (N^6-[4-amino-2-hydroxybutyl]-L-lysine). This modification occurs in two steps: first, the enzyme deoxyhypusine synthase (DPHS; Dys1 in yeast) transfers an n-butylamine moiety from spermidine to the ε-amino group of Lys51; second the enzyme deoxyhypusine hydroxylase (DOHH; Lia1 in yeast) hydroxylates the second carbon of the added moiety (Dever et al., 2014). While the second, hydroxylation, step of the modification is not essential in yeast, deoxyhypusine synthase is essential in yeast, and eIF5A lacking the modification is defective at promoting translation elongation and termination. Recently described patients with mutations in eIF5A (Faundes et al., 2021), DHPS (Ganapathi et al., 2019), or DOHH (Ziegler et al., 2022) exhibit similar neurodevelopmental disorders revealing the importance of eIF5A and its hypusine modification in human health.

Biochemical analysis of eIF5A function in translation elongation and termination requires reconstitution of all steps of translation including translation initiation, elongation, and termination. All these steps are complex and require the preparation of ribosomes, aminoacyl-tRNAs, and many translation factors, including eIF5A with its essential post-translational modification. In vitro reconstituted translation systems for yeast have been developed in the labs of Jon Lorsch (Algire et al., 2002; Lorsch & Herschlag, 1999) and Rachel Green (Eyler & Green, 2011; Saini et al., 2009; Shoemaker, Eyler, & Green, 2010). By programming ribosomes to synthesize peptides with specific sequences, and monitoring peptide synthesis or release by electrophoretic thin-layer chromatography (eTLC) (Eyler & Green, 2011), we exploited these systems for the study of the function and regulation of eIF5A (Gutierrez et al., 2013; Ivanov et al., 2018; Shin et al., 2017; Vindu et al., 2021), revealing eIF5A as the target for autoregulated expression of polyamine synthesis (*AZIN1*) and transport (*HOL1*). In this chapter, we describe in detail the methods developed for reconstituting the yeast translation system to analyze the role of eIF5A in elongation and termination.

2. Overview of in vitro translation assays

Before providing specifics on preparation of the reagents needed to synthesize polypeptides, it will be helpful to understand the basic steps of the process. As summarized in Fig. 1, the assays involve the stepwise assembly of the translation complex, followed by synthesis and differential means to terminate the assay to monitor peptide synthesis versus peptide release. The assays start by assembling the eIF2 ternary complex consisting of eIF2, GTP, and [^{35}S]Met-tRNA$_i^{Met}$. This complex is then mixed with the small (40S) ribosomal subunit, initiation factors eIF1 and eIF1A, and an mRNA encoding the peptide of interest. The eIF2 ternary complex, eIF1

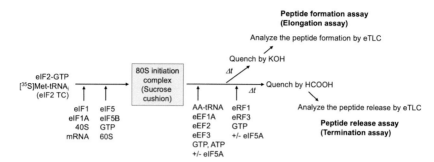

Fig. 1 Schematic of yeast reconstituted in vitro translation elongation or termination assay. eTLC, electrophoretic thin layer chromatography.

and eIF1A readily bind to the 40S subunit to form a 43S preinitiation complex (PIC). While 43S PIC binding to the mRNA typically requires several mRNA interacting factors, appending 15–20 nts of poly(UC) 5' and 3' of the ORF enables efficient binding of the 43S PIC in the absence of additional initiation factors (Algire et al., 2002; Lorsch & Herschlag, 1999). The 43S PIC binds the mRNA at the AUG codon, and addition of eIF5, the GTPase eIF5B, GTP and 60S ribosomal subunits enables subunit joining and conversion of the 43S PIC into an 80S initiation complex with [^{35}S]Met-tRNA$_i^{Met}$ in the P site with its anticodon base-paired with the start codon of the mRNA. The 80S initiation complexes are isolated by sedimentation through a sucrose cushion and then used for translation elongation and termination assays.

To synthesize peptides, the necessary aminoacyl-tRNAs together with elongation factors eEF1A, eEF2, eEF3, and their nucleotide substrates GTP and ATP are added to the 80S initiation complexes. To monitor translation elongation, reactions are stopped by quenching at high pH to cause hydrolysis of the acyl linkages between amino acids and tRNAs and to release the peptides linked to tRNAs on the ribosome. To monitor translation termination, the release factors eRF1 and eRF3 are added with the elongation assay components and reactions are quenched under acid conditions to prevent hydrolysis of the aminoacyl bonds. Accordingly, nascent peptides remain associated with the ribosome, and only peptides released from the ribosome by the action of the release factors are quantified. Peptides produced in the assay are separated by electrophoretic thin-layer chromatography and, owing to the presence of [^{35}S]Met at their N-termini, are visualized and quantified using a phosphorimager.

3. Preparation of initiation factors

In the simplified initiation system outlined in Fig. 1, five initiation factors are required to produce the 80S initiation complexes. Detailed protocols for preparation of the initiation factors have been described previously. As described by Acker et al., initiation factors eIF1, eIF1A and eIF5 are expressed in *E. coli*, while eIF2, consisting of three non-identical subunits, is purified from yeast using an N-terminal polyhistidine-tag on the eIF2γ (Gcd11) subunit (Acker, Kolitz, Mitchell, Nanda, & Lorsch, 2007). An N-terminally truncated form of eIF5B(397–1002) is expressed as a GST fusion protein in yeast and purified as we previously described (Shin & Dever, 2007).

4. Purification of eIF5A

Unmodified, deoxyhypusinated, and hypusinated eIF5A are produced in bacteria by co-expressing eIF5A and its modification enzymes Dys1 (DHPS) and Lia1 (DOHH) using the previously described expression system (Gutierrez et al., 2013). As depicted in Fig. 2A, derivatives of the polycistronic expression system developed by Song Tan (Tan, 2001) are used to produce the three forms of eIF5A in *E. coli*, and purification of the proteins is facilitated by an N-terminal His_6-tag on eIF5A.

1. Transform *E. coli* strain BL21 CodonPlus(DE3)-RIL (Agilent) with pC4183 (pST39-His_6-eIF5A-*DYS1-LIA1*) to produce hypusinated eIF5A or derivatives of the plasmid lacking modification enzyme CDS to produce partially- or un-modified eIF5A. Grow strains in 1 liter LB medium containing 100 μg/ml ampicillin at 37 °C to OD_{600} ~ 0.5.
2. Add IPTG to a final concentration of 0.5 mM and incubate the culture at 25 °C with shaking for 14 h.
3. After harvesting the cells by centrifugation, the cell pellet is suspended in 40 ml eIF5A Lysis Buffer (50 mM Tris-HCl [pH 7.5], 300 mM KCl, 10 mM imidazole, 0.5 mM 4-(2-aminoethyl)benzenesulfonyl fluoride hydrochloride [AEBSF]) and cells are broken in a cold room by sonication using a microtip (5 cycles of 30 s pulse followed by 30 s cooling on ice).
4. Clear the cell lysate by centrifugation in a Sorvall SS-34 rotor at 27,000 x *g* for 30 min, and then mix gently with 1 ml Ni-NTA resin (Qiagen) at 4 °C for 2 h.
5. Transfer the resin to a 5 ml disposable column (Qiagen), wash sequentially with 10 ml eIF5A Lysis Buffer and 10 ml Lysis Buffer

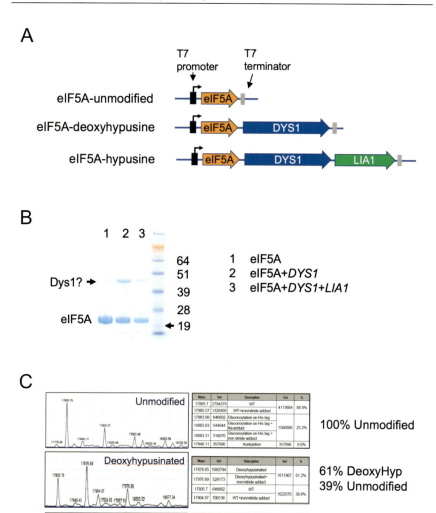

Fig. 2 Purification of eIF5A from *E. coli*. (A) Schematics of vectors used for expression of unmodified, deoxyhypusinated, or hypusinated eIF5A in *E. coli*. (B) SDS-PAGE of eIF5A purified by Ni-NTA affinity resin. Substoichiometric amounts of Dys1 seem to copurify with eIF5A. (C) ESI QTOF MS analysis of the three different forms of purified eIF5A. Note that iron nitride adducts, or gluconaylation or acetylation, alters the mass of a fraction of each form of eIF5A; when calculating the percentages of unmodified, deoxyhypusinated, and hypusinated eIF5A in each purified sample, the proteins +/− adducts and alternatively modified forms are summed together.

containing 20 mM imidazole, and then elute the protein in 10 ml Lysis Buffer containing 0.5 M imidazole.
6. Typically, the eIF5A eluted from Ni-NTA is sufficiently pure (Fig. 2B) to use for in vitro peptide formation assays. However, if further purification is necessary, the Ni-NTA elute is diluted five times with eIF5A Dilution Buffer (20 mM Tris-HCl [pH 7.5], 10% glycerol, 2 mM DTT), and loaded on a HiTrap Q column. The bound eIF5A proteins are eluted in a 100 mM to 1 M KCl gradient in Dilution Buffer.
7. Fractions containing eIF5A (eluting near 0.3 M KCl) are identified by SDS-PAGE, pooled, and then dialyzed against eIF5A Storage Buffer (30 mM Hepes-KOH [pH 7.5], 150 mM KCl, 2 mM DTT, 10% glycerol).
8. Purified eIF5A can be analyzed by ElectroSpray-Ionization Quadrupole-Time-of-Flight Mass Spectrometry (ESI QTOP MS) to determine the extent of modification (Fig. 2C).

5. Purification of elongation factors

Given their high natural abundance, and requirement for dedicated chaperones (Sabbarini et al., 2023) or post-translational modification (Shin et al., 2023), elongation factors eEF1A, eEF2, and yeast-specific eEF3 are purified from yeast.

5.1 Purification of eEF1A

Native elongation factor eEF1A is purified from yeast strain YRP840 (*MATa his4–539 leu2-3 leu2-112 trp1-1 ura3-52 cup1::LEU2/PM*) (*PGK1pG/MFA2pG*) (Olivas & Parker, 2000) using a modified version of a previously described protocol (Eyler & Green, 2011).

1. Cells are grown in 4-liters YPD medium to OD_{600} ~1.0, harvested, and resuspended in 50 ml eEF1A Lysis Buffer (60 mM Tris–HCl [pH 7.5], 50 mM KCl, 5 mM $MgCl_2$, 0.1 mM EDTA [pH 8.0], 10% glycerol, 1 mM DTT, 1× Roche Complete EDTA-free protease inhibitor [PI] tablet, 0.5 mM AEBSF).
2. Cells are pelleted, resuspended in 10 ml eEF1A Lysis Buffer, and broken with 50% volume of glass beads using a Bead Beater (Biospec Products) in a cold room with 5 cycles of 1 min running followed by 1 min cooling on ice.
3. Following removal of the glass beads and unbroken cells by centrifugation in a Beckman JS-4.2 rotor at 1900 x *g* for 10 min, the extract

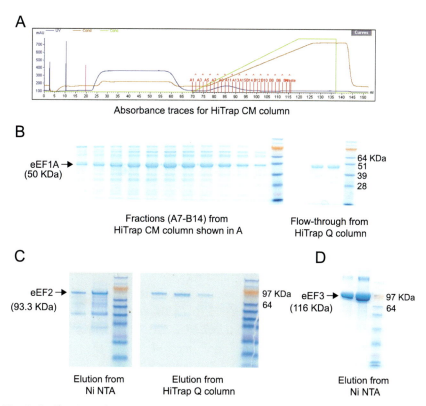

Fig. 3 Purification of native eEF1A (A, B), His$_6$-tagged eEF2 (C), and eEF3 (D) from yeast. (A) Traces of UV absorbance (blue) and KCl gradient (green) from chromatography on a HiTrap CM Sepharose column. (B) Left panel, SDS-PAGE of Fractions A7 to B14 from the HiTrap CM column; right panel, unbound (flow-through) fraction from HiTrap Q column. (C) SDS-PAGE of purified eEF2 eluted from Ni-NTA resin (left panel), and fractions from HiTrap Q column (right panel). (D) SDS-PAGE of eEF3 eluted from Ni-NTA resin. Proteins were stained with GelCode™ Blue Stain Reagent.

is clarified by centrifugation in a Sorvall SS-34 rotor at 27,000 x *g* for 30 min

4. The supernatant is then gently mixed with 10 ml DE52 resin (Whatman, pre-equilibrated with eEF1A Lysis Buffer) for 1 h at 4 °C.
5. The unbound fraction containing eEF1A is isolated by pouring the mixture into a column and collecting the eluate.
6. The unbound fraction is applied to a HiTrap CM Sepharose column (Cytiva), and eEF1A is eluted with a linear gradient to 500 mM KCl in eEF1A Lysis Buffer without protease inhibitors (Fig. 3A).

7. Fractions containing eEF1A are identified by SDS-PAGE and pooled (Fig. 3B, left panel).
8. After the pooled fractions are diluted to 50 mM KCl by adding eEF1A Dilution Buffer (60 mM Tris–HCl [pH 7.5], 5 mM MgCl$_2$, 10 % glycerol, 1 mM DTT), the protein solution is applied to a HiTrap Q column (Cytiva), and the unbound (flow-through) fraction is collected.
9. Fractions containing eEF1A are identified by SDS-PAGE (Fig. 3B, right panel), pooled, and dialyzed against eEF1A Storage Buffer (20 mM Tris–HCl [pH 7.5], 40 mM KCl, 5 mM MgCl$_2$, 10 % glycerol, 1 mM DTT).
10. The protein solution is then concentrated using a Microcon Centrifugal Filter (30 K), and aliquots are stored at −80 °C.

5.2 Purification of eEF2 and eEF3

Six-histidine tagged versions of elongation factors eEF2 and eEF3 are purified from yeast strains TKY675 (*MATa ade2 leu2 ura3 his3 trp1 eft1::HIS3 eft2::TRP1* p[*EFT2–6xHis LEU2 CEN*]) and TKY702 (*MATα ura3–52 leu2–3 leu2–112 trp1–7 lys2 met2–1 his4–713 yef3::LEU2* p [*YEF3–6xHis TRP1 2 μ*]) (Anand, Chakraburtty, Marton, Hinnebusch, & Kinzy, 2003), respectively, using protocols based on previous publications (Andersen et al., 2004; Ortiz, Ulloque, Kihara, Zheng, & Kinzy, 2006), with some modifications.

1. Cells are grown in 2.5-liter YPD to OD$_{600}$ ~1.5, harvested, and then suspended in 20 ml eEF2 Lysis Buffer (50 mM Tris-HCl [pH 8.0], 300 mM KCl, 1 mM 2-mercaptoethanol, 1 mM AEBSF, 10 mM imidazole, and 1X Roche PI tablet).
2. Break cells with glass beads as described above, and then clear the lysate of unbroken cells by centrifugation in a Beckman SS-34 rotor at 17,000 x *g* for 30 min
3. Clarify the lysate by centrifugation in a Beckman Type 70 Ti at 180,000 x *g* for 80 min
4. Gently mix the cleared lysate with 1 ml Ni-NTA resin for 2 h at 4 °C.
5. Wash the resin with 5 vol of eEF2 Wash Buffer (50 mM Tris-HCl [pH 8.0], 300 mM KCl, 1 mM 2-mercaptoethanol, 20 mM imidazole).
6. His-tagged proteins are then eluted in eEF2 Elution Buffer (eEF2 Wash Buffer containing 250 mM imidazole).
7. For eEF2 purification, the eluted proteins are diluted 6-fold with eEF2 Dilution Buffer (20 mM Tris-HCl [pH 7.5], 2 mM DTT, 10 %

glycerol), loaded on a HiTrap Q column, and then eluted by a linear KCl gradient to 1 M. Fractions containing eEF2 are identified by SDS-PAGE. For eEF3 purification, move to step 8.

8. Purified proteins (Fig. 3C and D) are dialyzed against eEF2 Storage Buffer (20 mM Tris-HCl [pH 7.5], 100 mM potassium acetate, 2.5 mM magnesium acetate, 2 mM DTT, 10% glycerol), concentrated using Amicon Ultracel-50K (for eEF2) or Ultracel-100K (for eEF3), and then stored at −80 °C.

6. Purification of termination factors

6.1 Purification of eRF1

Eukaryotic release factor 1 (eRF1) and eRF3 are required for analyzing peptide release from ribosomes poised with a termination codon in the A site. eRF1 is purified from *E. coli* BL21(DE3) CodonPlus-RIL transformed with the plasmid pPROEX-HTb-eRF1 (obtained from Rachel Green, Johns Hopkins University).

1. Cells are grown in 500 ml LB with 100 μg/ml of ampicillin to OD_{600} ∼0.5, and expression of eRF1 is induced by adding 0.2 mM IPTG and incubating at 20 °C for 14 h.
2. Cells are harvested by centrifugation and resuspended in 20 ml eRF1 Lysis Buffer (20 mM HEPES-KOH [pH 7.4], 0.5 M KCl, 10 mM imidazole, 2 mM 2-mercaptoethanol).
3. Cells are lysed by sonication (30 s sonication at 30% output with microtip followed by 30 s cooling on ice for total time of 10 min), and then the lysate is clarified by centrifugation in a Sorvall SS-34 rotor at 27,000 x *g* for 30 min
4. Mix the cleared lysate with 1 ml Ni-NTA resin on a nutator at 4 °C for 1 h.
5. Wash the resin with 10 ml eRF1 Lysis Buffer, and bound proteins are eluted with eRF1 Lysis Buffer containing 250 mM imidazole.
6. Dilute the eluate to 100 mM KCl with eRF1 Dilution Buffer (20 mM HEPES-KOH [pH 7.4], 2 mM DTT), apply to a Hitrap Q column, and elute bound proteins with a linear gradient to 1 M KCl.
7. Fractions containing eRF1 are identified by SDS-PAGE, dialyzed against eRF1 Storage Buffer (20 mM HEPES-KOH [pH 7.4], 100 mM potassium acetate, 2 mM DTT, 10% glycerol) and concentrated with a Microcon centrifugal filter (30 kDa).

6.2 Purification of eRF3

To purify eRF3, a DNA fragment encoding N-terminally truncated eRF3 (amino acids 166–685) was cloned in the vector pGEX-6P-2 to express a GST-eRF3 fusion protein, and the resulting plasmid was transformed into E. *coli* BL21.

1. *E. coli* cells carrying the GST-eRF3 expression vector are grown in 500 ml LB medium containing 100 μg/ml ampicillin to OD_{600} ~ 0.5, and GST-eRF3 expression is induced by adding 0.2 mM IPTG and incubating at 20 °C for 14 h.
2. Cells are harvested, and the cell pellet is suspended in 25 ml ice-cold 1X PBS.
3. Cells are disrupted by sonication as described for purifying eRF1.
4. The expressed proteins are solubilized by adding Triton X-100 (1 % final concentration) to the cell lysate and gently mixing for 30 min at room temperature, and then clarify the lysate by centrifugation in a Sorvall SS-34 rotor at 27,000 x *g* for 30 min at 4 °C.
5. Mix the cleared lysate with 1 ml Glutathione Sepharose 4B resin (Cytiva) and incubate at 4 °C for 1 h.
6. Transfer the resin to a 5 ml disposable column, wash sequentially with 10 column volumes of cold 1X PBS buffer and then 10 column volumes of eRF3 Cleavage Buffer (50 mM Tris-HCl [pH 7.5], 100 mM NaCl, 1 mM EDTA, 1 mM DTT).
7. Load a Protease Mix (mix 160 units of PreScisson Protease [Cytiva] with 920 μl of eRF3 Cleavage Buffer) on the column, seal, and incubate at 4 °C for 14 h.
8. Elute the cleaved eRF3 by gravity and identify fractions containing eRF3 by SDS-PAGE.
9. Dialyze the purified eRF3 against eRF1 Storage Buffer and concentrate with a Microcon centrifugal filter.

7. Preparation of mRNA

Unstructured model mRNAs with poly(UC) sequences that facilitate ribosome binding (Algire et al., 2002; Lorsch & Herschlag, 1999) flanking the coding sequences are prepared by T7 in vitro transcription and used for preparation of 43S-mRNA PICs. Here, we describe the methods to generate the DNA templates for transcription and the protocols for in vitro transcription. The following protocol is for preparation of an mRNA encoding the peptide Met-Phe-Lys-Stop; mRNAs encoding different peptides can be generated using different DNA templates.

7.1 Preparation of DNA template for T7 transcription

Double-stranded DNA templates for transcription reactions are prepared by PCR in two steps. First, two DNA oligonucleotides that partially overlap are annealed and then extended to generate the full-length sequence. Second, the full-length double-stranded DNA is amplified to generate the templates for transcription.

The DNA sequence of the template is (italics, T7 promoter; red bold font, coding sequence for **Met-Phe-Lys-Stop**):

*GGCGTAATACGACTCACTATAG*GAATCTCTCTCTCTCT**ATGTTC AAATAA**CTCTCTCTCTCTCTCTCTCGTTAATAAGCAAAATTCATTAT AACC.

1. To generate the full-length 99-bp template, two oligonucleotides (57 and 62 nt) are designed so that their 3' 22-nt anneal:
   Primer 1
   5'-GGCGTAATACGACTCACTATAGGAATCTCTCTCTCTCT**AT GTTCAAATAA**CTCTC-3'
   Primer 2 3'-AGAGA**TACAAGTTTATT**GAGAGAGAGAGAGAGAG AGAGCAATTATTCGTTTTAAGTAATATTG-5'
2. Prepare 10X PCR buffer (500 mM KCl, 100 mM Tris-HCl [pH 8.6], 1% Triton X100)
3. Prepare 1.2 ml PCR master mix:
   10X PCR buffer 120 μl
   1 M MgCl$_2$ 3 μl
   5 mM dNTPs 60 μl(dNTP mix from Roche)
   Taq polymerase 9 μl(5000 units/ml, New England Biolab)
   Water 1008 μl
4. Prepare annealing mix:
   PCR Master mix 96 μl
   100 μM primer 1, 2.0 μl
   100 μM primer 2, 2.0 μl
5. Anneal and extend primers by PCR to make a double-stranded DNA template
   a. 95 °C for 1 min
   b. Five cycles of 50 °C for 1 min and 72 °C for 1 min
6. Use second PCR to amplify the DNA template

a. Prepare PCR mix:
   PCR Master mix 989 μl
   Extension product 5 μl
   Primer T7F 3 μl Primer specific to T7 promoter
   Primer TP1 3 μlPrimer specific to 3' end of DNA template
   T7F sequence: GGCGTAATACGACTCACTATAG
   TP1 sequence: GGTTATAATGAATTTTGCTTATTAAC
b. Aliquot 50 μl of PCR mix to 20 PCR tubes and perform second PCR
   i. 95 °C for 1 min
   ii. 15 cycles of 95 °C for 40 s, 50 °C for 40 s, and 72 °C for 40 s
c. Pool the 20 PCR reactions and check the PCR product by 2 % agarose gel electrophoresis.
d. Clean up the PCR reaction by phenol/chloroform extraction, and then precipitate the DNA in ethanol.
e. Dissolve the DNA pellet in 100 μl water.

7.2 Preparation of mRNA by in vitro T7 transcription

1. First prepare 10X T7 buffer (400 mM Tris-HCl [pH 8.0], 10 mM spermidine, 0.1 % Triton X100) and then assemble the in vitro transcription reaction using T7 RNA polymerase:
   Water 447 μl
   10X T7 buffer 100 μl
   1 M DTT 10 μl
   1 M MgCl$_2$ 22.5 μl
   25 mM NTP mix 150 μl(NTP mix from Roche)
   2 M KOH 11 μl
   100 mM GMP 50 μl
   DNA template 100 μl
   T7 RNA polymerase* 20 μl
   *Purified from *E. coli* BL21(DE3) harboring p-T7 (Acker et al., 2007), (strain and protocol obtained from Rachel Green, Johns Hopkins University)
2. Incubate the reaction at 37 °C for 3 h. The reaction solution will become cloudy due to Mg^{2+}•PPi precipitates.
3. Add 20 μl 100 mM MnCl$_2$ and 4 μl DNase I (2000 units/ml, New England Biolab), and incubate at 37 °C for 30 min to digest DNA template.
4. Add 75 μl 0.5 M EDTA to dissolve the white Mg^{2+}•PPi precipitates in the solution. If this does not eliminate all precipitates in the solution,

briefly spin the solution in a microcentrifuge and transfer the cleared supernatant to a clean tube.

5. Add 60 μl 5 M NaCl and 1 ml isopropanol to the solution, incubate at room temp for 5 min, and then pellet the precipitated RNA by spinning in a microcentrifuge at maximum speed.
6. Decant the alcohol and dissolve the RNA pellet in 100 μl RNase-free water.
7. Purify the RNA by electrophoresis on a 10% denaturing polyacrylamide gel (7.5 ml 40% Acrylamide/Bis (19:1), 3 ml 10X TBE, 14 g Urea, water to 30 ml) as described by Acker et al. (2007).

8. Preparation of tRNA^{Pro}

The UGG isoacceptor of tRNA^{Pro} is purified from bulk *S. cerevisiae* tRNA (Roche) using the biotinylated oligonucleotide 5'-CCAAAGCG AGAATCATACCACTAGAC-BioTEG-3' (purchased from IDT) and the hybridization method described in Yokogawa, Kitamura, Nakamura, Ohno, & Nishikawa (2010).

1. 400 μl streptavidin beads (Pierce) are washed three times with 400 μl 10 mM Tris-HCl (pH 7.5).
2. Mix the beads with 8 nmol biotinylated oligonucleotide and incubate at 25 °C for 30 min
3. Wash the beads twice with 1 ml 10 mM Tris-HCl (pH 7.5). Pellet the beads after each wash by centrifugation for 30 s at 1000 x g and then remove the supernatant.
4. Mix bulk tRNA (180 nmol in 300 μl water) with an equal volume of 2X TMA buffer (20 mM Tris-HCl [pH 7.5], 1.8 M tetramethylammonium [TMA] chloride, 0.2 mM EDTA).
5. Mix the tRNA solution with the streptavidin beads and incubate at 65 °C for 10 min to denature the tRNA.
6. Slowly cool the mixture to 25 °C over ~10 min to allow annealing of the tRNAs to the oligonucleotides bound to the beads.
7. Wash the beads eight times with 400 μl 10 mM Tris-HCl (pH 7.5).
8. Heat the beads to 65 °C for 5 min, and then elute the tRNAs by centrifugation of the beads in a Ultrafree-MC centrifugal filter unit (0.22 μm, Millipore) preloaded with 2 μl 1 M magnesium acetate.
9. Repeat the melting and elution process, combine the two eluted fractions, and the precipitate the tRNA in ethanol.

10. Resuspend the tRNA in 50 μl RNase-free water and quantify by measuring the A_{260}.

Using different biotinylated oligonucleotides, this protocol can be used to purify other tRNAs from bulk yeast tRNA. Purified yeast tRNA^{phe} can be purchased from Sigma (though, note that a small portion of the commercial tRNA^{phe} may be aminoacylated ["charged"]). While tRNA^{Lys} was previously available from tRNA Probes (College Station, Texas), it is no longer available, and we recommend using biotinylated oligonucleotides to purify tRNA^{Lys} from bulk yeast tRNA.

9. Purification of tRNA synthetases and CCA-adding enzyme

9.1 Purification of methionyl-tRNA synthetase (MetRS)

Bacterial MetRS is able to aminoacylate yeast tRNA$_i^{Met}$. We therefore provide a protocol for overexpression and purification of *E. coli* MetRS. *E. coli* XL1 Blue cells carrying the His$_6$-MetRS expression plasmid pRA101 (Alexander, Nordin, & Schimmel, 1998; obtained from Jagpreet Nanda and Jon Lorsch) was used for purification of MetRS.

1. Cells are grown in 500 ml LB medium containing 100 μg/ml ampicillin at 37 °C to OD_{600} ~ 0.5.
2. Following addition of 0.2 mM IPTG, the culture is incubated with shaking at 20 °C for 14 h.
3. Cells are harvested, the cell pellet is suspended in 20 ml MetRS Lysis Buffer (50 mM Tris–HCl [pH 7.5], 300 mM KCl, 6 mM 2-mercarptoethanol, 10% glycerol), and cells are broken by sonication using a microtip with 50% output (5 cycles of 30 s pulse followed by 30 s cooling on ice).
4. Clear the cell lysate by centrifugation in a Sorvall SS-34 rotor at 27,000 x g for 30 min
5. Gently mix the cleared lysate with 1 ml Ni–NTA resin at 4 °C for 1 h.
6. Transfer the resin to a 1 ml disposable column (Qiagen), and then wash sequentially with 10 ml MetRS Lysis Buffer and 20 ml MetRS Lysis Buffer containing 20 mM imidazole
7. Elute the protein in 4 ml MetRS Lysis Buffer containing 250 mM imidazole.
8. Dialyze the eluted protein overnight against MetRS Storage Buffer (50 mM Tris–HCl [pH 7.5], 25 mM KCl, 10 mM MgCl$_2$, 1 mM DTT, 10% glycerol).

9. Concentrate the protein solution using a Microcon YM-50 centrifugal filter (Amicon), and store frozen aliquots at −80 °C.

9.2 Purification of Prolyl tRNA synthetase (PRS)

The *S. cerevisiae* His$_6$-tagged PRS was purified from *E. coli* XL1 Blue cells transformed with the expression vector pQE30-PRS (obtained from Karin Musier-Forsyth, Ohio State University). Note that yeast PRS expression is toxic in *E. coli*; the XL1-Blue cells should have lacI^Q to suppress PRS expression under noninducing conditions.

1. Grow *E. coli* XL1 Blue transformants in 500 ml LB medium containing 100 µg/ml ampicillin until OD$_{600}$ ~0.5.
2. Induce the expression of PRS by adding 0.1 mM IPTG and incubate with shaking at 22 °C for 22 h.
3. Harvest cells by centrifugation, resuspend in 20 ml ice-cold PRS Lysis Buffer (50 mM KPO$_4$ [pH 7.4], 300 mM KCl), and disrupt the cells by sonication as described above for the purification of eRF1.
4. Clarify the lysate by centrifugation in a Sorvall SS-34 rotor at 27,000 x *g* for 30 min at 4 °C, then mix the lysate with 2 ml Ni-NTA resin on a nutator at 4 °C for 1 h.
5. Wash the resin with 10 ml PRS Lysis Buffer containing 10 mM imidazole, and then elute the bound proteins with PRS Lysis Buffer containing 250 mM imidazole.
6. Dialyze the eluted protein solution overnight at 4 °C against 1-liter PRS Storage Buffer (50 mM KPO$_4$ [pH 7.4], 5 mM magnesium acetate, 2 mM DTT, 10% glycerol).

9.3 Purification of Phenylalanyl and Lysyl tRNA synthetase

Phenylalanine tRNA synthetase (FRS) is composed of two α and two β subunits to make the active tetrameric enzyme. We cloned the N-terminal His$_6$-tagged FRS2 (α subunit) and FRS1 (β subunit) coding sequences in the multimeric expression vector pST39 and co-expressed the two proteins in *E. coli*.

1. Grow *E. coli* BL21(DE3) pLys transformants harboring pST39-FRS2-FRS1 in 0.5-liter LB medium containing 100 µg/ml ampicillin to OD$_{600}$ ~0.5.
2. Add IPTG to 0.2 mM to induce expression of FRS and incubate with shaking at 25 °C for 16 h.
3. Harvest cells by centrifugation, resuspend in 20 ml FRS Lysis Buffer (20 mM Tris-HCl [pH 7.5], 500 mM KCl, 5 mM MgCl$_2$, 1 mM

2-mercaptoethanol, 1 mM AEBSF, 1X Roche PI tablet, 10% glycerol, 10 mM imidazole), and disrupt cells by sonication as described above for purifying eRF1.
4. Clarify the lysate by centrifugation in a Sorvall SS-34 rotor at 27,000 x g for 30 min at 4 °C, and then mix the supernatant with 1 ml Ni-NTA resin on a nutator at 4 °C for 2 h.
5. Wash the resin sequentially with 5 ml FRS Lysis buffer and 10 ml FRS Lysis buffer containing 20 mM imidazole.
6. Elute the protein in 2 ml FRS Lysis buffer containing 250 mM imidazole.
7. Dialyze the eluted FRS against 1 liter FRS Storage Buffer (20 mM Tris-HCl [pH 7.5], 150 mM potassium acetate, 2.5 mM magnesium acetate, 2 mM DTT, 10% glycerol).
8. Clarify the protein solution by centrifugation at 15,000 rpm in a microfuge for 10 min at 4 °C.
9. Concentrate the protein solution to 1 ml using a Microcon centrifugal filter.

To purify yeast Lysyl-tRNA synthetase (KRS), N-terminal His$_6$-tagged *KRS1* coding sequence was cloned into the expression vector pET3a and transformed to *E. coli* strain BL21(DE3) CodonPlus-RIL (Agilent). *E. coli* transformants are grown in 250 ml LB medium containing 100 μg/ml ampicillin at 37 °C to OD$_{600}$ ~ 0.5. Expression of KRS1 is induced by the addition of 0.5 mM IPTG, and the culture is incubated at 20 °C for 16 h. KRS is purified as described above for the purification of FRS.

9.4 Purification of CCA-adding enzyme

The 3' end of all mature tRNAs terminates with the sequence CCA. This sequence is critical for efficient aminoacylation by tRNA synthetases and for proper positioning of the tRNAs on the ribosome. The CCA sequence is not encoded in the tRNA genes but instead is added post-transcriptionally. As its name indicates, the CCA-adding enzyme, also known as tRNA nucleotidyltransferase, adds the essential CCA nucleotides. In our experience, the tRNA^{Pro} purified from bulk yeast tRNA using the hybridization technique described above is inefficiently (generally ~10%) aminoacylated with proline by purified PRS. Though this low fraction of aminoacylated Pro-tRNA^{Pro} is sufficient for peptide formation assays, treatment of the purified tRNA^{Pro} with CCA-adding enzyme increases the aminoacylation efficiency about two-fold. We overexpress and purify the CCA-adding enzyme from an *E. coli* strain harboring the plasmid pET-CCA (obtained from Allen Buskirk, Johns Hopkins University School of Medicine) using the same protocol as used for purification of MetRS described above, except substituting CCA Lysis Buffer (50 mM KPO_4 [pH 7.6], 300 mM KCl, 6 mM 2-mercaptoethanol, 10 mM imidazole) and CCA Storage Buffer (20 mM Tris-HCl [pH 7.5], 100 mM potassium acetate, 0.1 mM magnesium acetate, 2 mM DTT, 10% glycerol) in place of the corresponding MetRS buffers.

9.4.1 CCA-adding reaction

When needed to enhance aminoacylation efficiency, purified tRNAs are treated with purified CCA-adding enzyme as follows:

1. Assemble the CCA-adding reaction (200 μl reaction volume) containing 1X CCA Buffer (50 mM Tris-HCl [pH 7.5], 20 mM $MgCl_2$, 0.5 mM DTT), 6 mM ATP, 12 mM CTP, 16 μM tRNA, and 6 μM CCA-adding enzyme.
2. Incubate the reaction at 37 °C for 30 min
3. Terminate the reaction by adding 0.1 M sodium acetate (pH 5.3).
4. Remove excess NTPs by spin-column chromatography using a ProbeQuant G-50 Micro Column (Cytiva).
5. Add sodium acetate to 0.3 M, clean-up the reaction by phenol/chloroform extraction, and pellet the tRNA by ethanol precipitation.
6. Pellet the precipitated tRNA by centrifugation, dissolve the pellet in 20 μl water, quantitate the tRNA abundance by measuring A_{260}, and store aliquots at −80 °C.

10. Limited charging of tRNA$_i$ using ^{35}S methionine

For in vitro reconstituted translation assays, peptide formation or peptide release is monitored by incorporating [^{35}S]Met-tRNA$_i^{Met}$ into the 43S-mRNA PICs used for these assays. In order to ensure a single-round reaction, the labeled [^{35}S]Met-tRNA$_i^{Met}$ is included at a limiting concentration, usually 2 nM (Algire et al., 2002). While we previously purchased initiator tRNA$_i$ from tRNA Probes (College Station, Texas), this vendor is no longer available. As an alternative, tRNA$_i$ can be purified from bulk yeast tRNA using the biotinylated-oligonucleotide hybridization method described above for purification of tRNA^{Pro} or tRNA$_i$ can be prepared by T7 in vitro transcription as described by Acker et al. (2007). Note, the CCA-3' terminus of the mature tRNA can be incorporated during in vitro transcription to bypass the extra step of adding this sequence motif post-transcriptionally. In addition, though the T7 transcribed tRNA$_i$ lacks tRNA$_i$-specific base modifications, it is readily aminoacylated by MetRS and can be used for peptide formation and release assays (Acker et al., 2007; Kapp & Lorsch, 2004a; Pestova & Hellen, 2001; Shin et al., 2017).

1. Prepare 10X Charging Buffer (1 M Hepes [pH 7.5], 10 mM DTT, 100 mM KCl)
2. Assemble 100 μl aminoacylation reaction containing 1X Charging Buffer, 5 μM tRNA$_i$, 0.3 μM [^{35}S]Met (1175 Ci/mmol, Revvity), 2 mM ATP-Mg^{2+}, 10 mM MgCl$_2$, and 1 μM MetRS.
3. Incubate reactions at 30 °C for 30 min
4. To measure the aminoacylation efficiency, perform control reaction (20 μl) lacking tRNA$_i$ as described by Acker et al. (2007).
5. Following aminoacylation, extract the reactions with phenol/chloroform, add sodium acetate to 0.3 M, and precipitate the tRNA$_i$ with ethanol. Pellet the tRNA$_i$ by centrifugation, and dissolve the pellets in RNase-free water to the desired volume (we typically aim for 60 nM [^{35}S]Met-tRNA$_i^{Met}$).

11. Aminoacylation of tRNA^{Phe}, tRNA^{Pro}, and tRNA^{Lys}

The same protocol is followed for aminoacylation of the different tRNAs used for translation elongation. Of note, aminoacylation efficiencies and deacylation rates vary for the different tRNAs and synthetases. As Pro-tRNA^{Pro} is especially unstable and deacylates easily during storage, we freshly aminoacylate tRNA^{Pro} before the assay. Both Phe-tRNA^{Phe} and Lys-tRNA^{Lys} are more stable than Pro-tRNA^{Pro} and can be stored at −80 °C for several weeks.

1. Prepare 10X Aminoacylation/Charging Buffer (400 mM Tris-HCl [pH 7.6], 100 mM magnesium acetate, 10 mM DTT)
2. Assemble 100 μl aminoacylation reactions containing 1X Aminoacylation/Charging Buffer, 5 μM tRNA, 0.3 mM amino acid, 2 mM ATP-Mg^{2+}, and 1 μM tRNA synthetase.
3. Incubate reactions at 30 °C for 30 min
4. Following aminoacylation, extract the reactions with phenol/chloroform, add sodium acetate to 0.3 M, and precipitate the tRNA (mixture of aminoacylated and free tRNA) with ethanol.
5. Pellet the tRNA by centrifugation, and dissolve pellets in 20 μl RNase-free water.

12. Purification of 40S and 60S ribosomal subunits

Ribosomal subunits are purified from *Saccharomyces cerevisiae* strain YRP840 (*MATa leu2-3 leu2-112 his4-539 trp1 ura3-52 cup1::LEU2/PGK1pG/MFA2pG* (van Hoof, Lennertz, & Parker, 2000)) based on the protocol published by Acker et al. (2007) with some modifications.

1. Grow yeast in 10 liters YPD medium to $OD_{600} \sim 1.0$.
2. Harvest cells and resuspend the cell pellet in 50 ml Ribosome Lysis Buffer (20 mM HEPES–KOH [pH 7.4], 100 mM potassium acetate, 2.5 mM magnesium acetate, 1 mg/ml heparin, 2 mM DTT, 0.4 mM AEBSF).
3. Add 50% volume of glass beads to the cell suspension, and then vigorously agitate on a vortex in the coldroom for 1 min and then incubate on ice for 1 min. Repeat this cycle of breaking and cooling five times.
4. Remove glass beads and unbroken cells by centrifugation in a Beckman JS-4.2 rotor at $\sim 2560 \times g$, then clarify the extract by centrifugation in a Sorvall SS-34 rotor at $\sim 27,000 \times g$ for 30 min
5. Transfer ~ 20 ml of cleared lysate to a centrifuge tube for a Beckman Type 70 Ti rotor, and then using a 5 ml syringe layer 3 ml of a 1 M sucrose solution (20 mM HEPES–KOH [pH 7.4], 100 mM potassium acetate, 2.5 mM magnesium acetate, 500 mM KCl, 1 M sucrose, 2 mM DTT) at the bottom of tube below the lysate.
6. Pellet the ribosomes through the sucrose cushion by centrifugation in a Beckman Type70 Ti rotor at $164,000 \times g$.
7. Resuspend the ribosome pellet from each tube in 1 ml Subunit Separation Buffer (20 mM HEPES–KOH [pH 7.4], 500 mM KCl, 2 mM $MgCl_2$, 2 mM DTT) and then stir on ice for 1 h.
8. To release any nascent chains associated with the ribosomes, add 10 μl 100 mM puromycin to 1 ml ribosome solution, and then sequentially incubate on ice for 15 min, at 37 °C for 10 min, and then on ice for 10 min
9. Gently layer 1 ml of ribosome solution on top of a 5–20% sucrose gradient, and then separate the subunits by centrifugation at $174,000 \times g$ for 6 h at 4 °C in a Beckman SW32 rotor.
10. Fractionate the gradients while monitoring A_{260} and separately pool fractions containing the 40S and 60S subunits (Fig. 4).
11. Pellet the 40S and 60S subunit pools separately by centrifugation at $59,000 \times g$ for 24 h in a Beckman Type 70 Ti rotor.

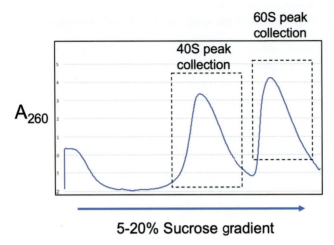

Fig. 4 Ribosomal subunit purification by sucrose gradient fractionation. Gradients were fractionated while monitoring A_{260}, and fractions pooled for 40S and 60S subunits are indicated.

12. Dissolve ribosomal subunit pellets in Ribosome Storage Buffer (20 mM HEPES– KOH [pH 7.4], 100 mM potassium acetate, 2.5 mM magnesium acetate, 250 mM sucrose, 2 mM DTT), and store aliquots at −80 °C.
13. Calculate subunit concentrations by measuring A_{260} and using the conversion 1.0 A_{260} = 50 pmole of 40S subunits and 31 pmole of 60S subunits.

13. Peptide formation and release assays

13.1 Preparation of initiation complexes

Initiation complexes are prepared in 1X Recon Buffer A (30 mM Hepes [pH 7.4], 100 mM potassium acetate, 3 mM magnesium acetate, 1 mM DTT) based on the protocol published by Acker et al. (2007).

1. Prepare [^{35}S]Met-tRNA$_i^{Met}$–GTP–eIF2 ternary complexes by combining 1X Recon Buffer A, 1 mM GTP-Mg^{2+}, 1.2 μM eIF2, and 12 nM [^{35}S]Met-tRNA$_i^{Met}$ in a total volume of 25 μl in a 0.5 ml tube.
2. Incubate the mixture in a 26 °C waterbath for 15 min
3. During incubation of the ternary complex assembly, combine in a separate 0.5 ml tube a 25 μl mixture for 43S PIC formation containing 1X Recon Buffer A, 1.2 μM 40S subunits, 3 μM mRNA, 3 μM eIF1, and 3 μM eIF1A.

4. Add the ternary complex assembly from step 2 to the 43S reagents in step 3 and incubate at 26 °C for 5 min to assemble the 43S complexes.
5. During the 5 min incubation of the 43S complex assembly, combine in a separate 0.5 ml tube a 25 µl mixture for 80S PIC formation containing 1X Recon Buffer A, 1.2 µM 60S subunits, 3 µM eIF5, 1.5 µM eIF5B, and 3 mM GTP-Mg^{2+}.
6. After the 5 min incubation for 43S complex formation, add the 25 µl of 80S components from step 5 to the 50 µl 43S complex mixture from step 4 and incubate for 15 min at 26 °C to generate 80S initiation complexes.
7. Layer the 75 µl 80S initiation complex mixture on 0.8 ml 1 M sucrose in 1X Recon Buffer A and sediment the ribosomal complexes through the sucrose cushion by centrifugation at 90,000 rpm for 1 h in a Beckman TLA 120.2 rotor. As the ribosome pellet is difficult to locate after centrifugation, mark the bottom of tube at the expected pellet position.
8. Dissolve the ribosome pellet in 20 µl 1X Recon Buffer B (30 mM Hepes [pH 7.4], 100 mM potassium acetate, 1 mM magnesium acetate, 1 mM DTT, 1 mM spermidine), and store 5 µl aliquots at −80 °C.

13.2 Peptide formation assay

To assay translation elongation, 80S initiation complexes from Section 11.1 are added to pre-mixed elongation components. The final concentrations for each component in the elongation reaction after mixing in the initiation complex are: 2 µM eEF1A, 1 µM eEF2, 1 µM eEF3, 5 µM eIF5A, 1 µM each aminoacyl-tRNA, 1 mM GTP and 1 mM ATP in 1X Recon Buffer B. To ensure sufficient elongation components for the assays, we typically pre-mix excess elongation components sufficient for one or two additional reactions.

1. Prepare 2X elongation component pre-mix in a 0.5 ml microcentrifuge tube on ice by combining 4 µM eEF1A, 2 µM eEF2, 2 µM eEF3, 10 µM eIF5A, 2 µM each aminoacyl-tRNA, 2 mM GTP and 2 mM ATP in 2X Recon Buffer.
2. Incubate the 2X elongation component pre-mix on ice for 15 min to allow GTP binding to eEF1A and eEF2, and ATP binding to eEF3; eEF1A–GTP–aminoacyl-tRNA ternary complexes will form at the same time.
3. Mix 5 µl initiation complexes with 5 µl 2X elongation component pre-mix and incubate at 26 °C to start the elongation reactions.

4. Monitor the progress of peptide formation by withdrawing 1 μl aliquots from the elongation reactions at different times, and then mix the aliquots with 1 μl 0.2 N KOH to quench the reaction and deacylate tRNAs.
5. With a pencil draw a line across the middle of TLC cellulose plate (20 × 20 cm, EMD Millipore) and then place marks on the line at 1 cm intervals for sample spotting.
6. Spot 0.5 μl of the quenched reactions at the marked positions on the cellulose TLC plate and dry the spots using a heat gun.
7. Place a serological pipet (5 or 10 ml) at the margin of the TLC plate and drop 2 ml pyridine acetate buffer (200 ml glacial acetic acid and 5 ml pyridine in 1 liter, pH 2.8) on the front side of the serological pipet, and then smoothly roll the pipet to the center of TLC plate, stopping just short of the spotted samples. Wait until pyridine acetate buffer diffuses to wet the spotted samples, and then repeat the procedure to wet the other half of the TLC plate. Once the entire surface of the TLC plate is wet with the buffer, place the TLC plate in the electrophoresis tank (Fig. 5A) and run at 1000 V for 30 min in the same pyridine acetate buffer.
8. Following electrophoresis, dry the TLC plate using a heat gun, and detect and quantify the peptide spots by phosphorimage analysis (Fig. 5B).
9. The fractional yield of the synthesized peptides and free [^{35}S]Met in each reaction at different times are quantified and fit using Kaleidagraph (Synergy Software) or similar software to the single exponential equation: $y = Y_{max}(1-\exp(-k_{obs} \times t))$, where Y_{max} is the maximum fraction of peptide formed and k_{obs} is observed rate constant.

The electrophoretic TLC apparatus is not commercially available and must be manually assembled. As depicted in Fig. 5A, the anode and cathode electrodes are separated by placing the anode in a larger rectangular glass jar used as the outside tank (thick blue lines) while the cathode is placed in an inserted acrylic tank (thinner red lines). The TLC plate should be pre-wetted with pyridine-acetate buffer (pH 2.8) and placed in the apparatus such that one end of TLC plate is in the cathode section and the other end is in the anode section, both filled with the pyridine-acetate running buffer. To prevent drying of the TLC plate and to help cool the buffer during electrophoresis at high voltage, the TLC plate is covered by non-miscible Stoddard solvent. Due to the use of organic solvents, the apparatus is stored and run in a chemical fume hood.

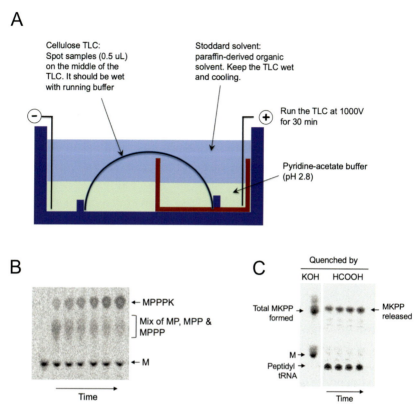

Fig. 5 Electrophoretic TLC detection of translation products. (A) Schematic of electrophoretic TLC apparatus. (B-C) Detection of products from peptide synthesis and release assays. Peptide products from assays programmed with mRNA encoding Met-Pro-Pro-Pro-Lys (MPPPK) (B) and released peptide products from assays programmed with mRNA encoding Met-Lys-Pro-Pro (MKPP) (C) were separated by electrophoretic TLC and then visualized using a phosphorimager. Locations of free Met (M) and peptide species are indicated.

13.3 Peptide release (termination) assay

Reconstituted peptide release assays are an extension of the peptide formation assays described in section 13.2. Rather than terminating assays with KOH, which releases all peptide products, termination factors eRF1 and eRF3 are added to elongation reactions to trigger peptide release from complexes with a stop codon in the A site. The final concentrations of eRF1 and eRF3 in the termination assays are 1 µM.

1. Assemble 3X termination mix in a 0.5 ml tube on ice by combining 3 µM eRF1, 3 µM eRF3, 3 mM GTP-Mg^{2+}, and 15 µM eIF5A in 1X Recon

Buffer A. Note, if assaying the effect of eIF5A on peptide release, either omit eIF5A from the peptide formation portion of the assay or if eIF5A is required for peptide formation (e.g. synthesis of proline-rich peptide) sediment the elongation complex through a sucrose cushion to remove the eIF5A.

2. Preincubate the assembled mix for 10 min on ice to enable GTP binding to eRF3.
3. After the peptide synthesis reactions from section 10.1 reach a plateau for peptide formation (usually after 5 min), mix 10 μl of the elongation complex with 5 μl termination mix and incubate at 26 °C to start the termination reaction.
4. Monitor the progress of peptide release by withdrawing 2 μl aliquots from the termination reactions at different times and mix with 1 μl formic acid to quench the reaction. Note that the acyl bonds linking amino acids and peptides to tRNAs are stable in acidic conditions, so only peptides released from tRNAs during translation termination will be resolved by TLC.
5. To assess total peptide synthesis (sum of released and nascent peptides) in each reaction, mix 1 μl of the final reaction, after the final time point, with 1 μl 0.2 N KOH as described in step 4 of section 13.2.
6. Spot 0.5 μl of quenched reactions on a cellulose TLC plate and analyze released peptides as described in section 13.2 (Fig. 5C).
7. Quantitate the percentage of peptide released relative to total peptide synthesized by comparing the peptide yields from the release assay (formic acid quench) versus from total synthesis (KOH quenched).

Acknowledgments

We thank Jon Lorsch, Rachel Green, Alan Hinnebusch, Allen Buskirk, and members of their labs for sharing protocols and reagents. Research in our laboratory is support by the Intramural Research Program of the National Institutes of Health.

References

Acker, M. G., Kolitz, S. E., Mitchell, S. F., Nanda, J. S., & Lorsch, J. R. (2007). Reconstitution of yeast translation initiation. *Methods in Enzymology, 430,* 111–145.

Alexander, R. W., Nordin, B. E., & Schimmel, P. (1998). Activation of microhelix charging by localized helix destabilization. *Proceedings of the National Academy of Sciences of the United States of America, 95,* 12214–12219.

Algire, M. A., Maag, D., Savio, P., Acker, M. G., Tarun, S. Z., Jr., Sachs, A. B., ... Lorsch, J. R. (2002). Development and characterization of a reconstituted yeast translation initiation system. *RNA (New York, N. Y.), 8,* 382–397.

Anand, M., Chakraburtty, K., Marton, M. J., Hinnebusch, A. G., & Kinzy, T. G. (2003). Functional interactions between yeast translation eukaryotic elongation factor (eEF) 1A and eEF3. *The Journal of Biological Chemistry, 278,* 6985–6991.

Andersen, C. F., Anand, M., Boesen, T., Van, L. B., Kinzy, T. G., & Andersen, G. R. (2004). Purification and crystallization of the yeast translation elongation factor eEF3. *Acta Crystallographica. Section D, Biological Crystallography, 60*, 1304–1307.

Dever, T. E., Dinman, J. D., & Green, R. (2018). Translation elongation and recoding in eukaryotes. *Cold Spring Harbor Perspectives in Biology, 10*, a032649.

Dever, T. E., & Green, R. (2012). The elongation, termination, and recycling phases of translation in eukaryotes. *Cold Spring Harbor Perspectives in Biology, 4*, a013706.

Dever, T. E., Gutierrez, E., & Shin, B. S. (2014). The hypusine-containing translation factor eIF5A. *Critical Reviews in Biochemistry and Molecular Biology, 49*, 413–425.

Dever, T. E., Kinzy, T. G., & Pavitt, G. D. (2016). Mechanism and regulation of protein synthesis in Saccharomyces cerevisiae. *Genetics, 203*, 65–107.

Eyler, D. E., & Green, R. (2011). Distinct response of yeast ribosomes to a miscoding event during translation. *RNA (New York, N. Y.), 17*, 925–932.

Faundes, V., Jennings, M. D., Crilly, S., Legraie, S., Withers, S. E., Cuvertino, S., ... Banka, S. (2021). Impaired eIF5A function causes a Mendelian disorder that is partially rescued in model systems by spermidine. *Nature Communications, 12*, 833.

Ganapathi, M., Padgett, L. R., Yamada, K., Devinsky, O., Willaert, R., Person, R., ... Chung, W. K. (2019). Recessive rare variants in deoxyhypusine synthase, an enzyme involved in the synthesis of hypusine, are associated with a neurodevelopmental disorder. *American Journal of Human Genetics, 104*, 287–298.

Gutierrez, E., Shin, B. S., Woolstenhulme, C. J., Kim, J. R., Saini, P., Buskirk, A. R., & Dever, T. E. (2013). eIF5A promotes translation of polyproline motifs. *Molecular Cell, 51*, 35–45.

Hinnebusch, A. G., & Lorsch, J. R. (2012). The mechanism of eukaryotic translation initiation: New insights and challenges. *Cold Spring Harbor Perspectives in Biology, 4*, a011544.

Ivanov, I. P., Shin, B. S., Loughran, G., Tzani, I., Young-Baird, S. K., Cao, C., ... Dever, T. E. (2018). Polyamine control of translation elongation regulates start site selection on antizyme inhibitor mRNA via ribosome queuing. *Molecular Cell, 70*, 254–264.

Kapp, L. D., & Lorsch, J. R. (2004a). GTP-dependent recognition of the methionine moiety on initiator tRNA by translation factor eIF2. *Journal of Molecular Biology, 335*, 923–936.

Kapp, L. D., & Lorsch, J. R. (2004b). The molecular mechanics of eukaryotic translation. *Annual Review of Biochemistry, 73*, 657–704.

Lorsch, J. R., & Dever, T. E. (2010). Molecular view of 43 S complex formation and start site selection in eukaryotic translation initiation. *The Journal of Biological Chemistry, 285*, 21203–21207.

Lorsch, J. R., & Herschlag, D. (1999). Kinetic dissection of fundamental processes of eukaryotic translation initiation in vitro. *The EMBO Journal, 18*, 6705–6717.

Olivas, W., & Parker, R. (2000). The Puf3 protein is a transcript-specific regulator of mRNA degradation in yeast. *The EMBO Journal, 19*, 6602–6611.

Ortiz, P. A., Ulloque, R., Kihara, G. K., Zheng, H., & Kinzy, T. G. (2006). Translation elongation factor 2 anticodon mimicry domain mutants affect fidelity and diphtheria toxin resistance. *The Journal of Biological Chemistry, 281*, 32639–32648.

Pestova, T. V., & Hellen, C. U. (2001). Preparation and activity of synthetic unmodified mammalian tRNAi(Met) in initiation of translation in vitro. *RNA (New York, N. Y.), 7*, 1496–1505.

Pestova, T. V., Lomakin, I. B., Lee, J. H., Choi, S. K., Dever, T. E., & Hellen, C. U. T. (2000). The joining of ribosomal subunits in eukaryotes requires eIF5B. *Nature, 403*, 332–335.

Sabbarini, I. M., Reif, D., McQuown, A. J., Nelliat, A. R., Prince, J., Membreno, B. S., ... Denic, V. (2023). Zinc-finger protein Zpr1 is a bespoke chaperone essential for eEF1A biogenesis. *Molecular Cell, 83*, 252–265.

Saini, P., Eyler, D. E., Green, R., & Dever, T. E. (2009). Hypusine-containing protein eIF5A promotes translation elongation. *Nature, 459*, 118–121.

Schuller, A. P., Wu, C. C., Dever, T. E., Buskirk, A. R., & Green, R. (2017). eIF5A functions globally in translation elongation and termination. *Molecular Cell, 66*, 194–205.

Shin, B. S., & Dever, T. E. (2007). Molecular genetic structure-function analysis of translation initiation factor eIF5B. *Methods in Enzymology, 429*, 185–201.

Shin, B. S., Ivanov, I. P., Kim, J. R., Cao, C., Kinzy, T. G., & Dever, T. E. (2023). eEF2 diphthamide modification restrains spurious frameshifting to maintain translational fidelity. *Nucleic Acids Research, 51*, 6899–6913.

Shin, B. S., Katoh, T., Gutierrez, E., Kim, J. R., Suga, H., & Dever, T. E. (2017). Amino acid substrates impose polyamine, eIF5A, or hypusine requirement for peptide synthesis. *Nucleic Acids Research, 45*, 8392–8402.

Shin, B. S., Maag, D., Roll-Mecak, A., Arefin, M. S., Burley, S. K., Lorsch, J. R., & Dever, T. E. (2002). Uncoupling of initiation factor eIF5B/IF2 GTPase and translational activities by mutations that lower ribosome affinity. *Cell, 111*, 1015–1025.

Shoemaker, C. J., Eyler, D. E., & Green, R. (2010). Dom34:Hbs1 promotes subunit dissociation and peptidyl-tRNA drop-off to initiate no-go decay. *Science (New York, N. Y.), 330*, 369–372.

Tan, S. (2001). A modular polycistronic expression system for overexpressing protein complexes in Escherichia coli. *Protein Expression and Purification, 21*, 224–234.

van Hoof, A., Lennertz, P., & Parker, R. (2000). Three conserved members of the RNase D family have unique and overlapping functions in the processing of 5S, 5.8S, U4, U5, RNase MRP and RNase P RNAs in yeast. *The EMBO Journal, 19*, 1357–1365.

Vindu, A., Shin, B. S., Choi, K., Christenson, E. T., Ivanov, I. P., Cao, C., ... Dever, T. E. (2021). Translational autoregulation of the *S. cerevisiae* high-affinity polyamine transporter Hol1. *Molecular Cell, 81*, 3904–3918.

Yokogawa, T., Kitamura, Y., Nakamura, D., Ohno, S., & Nishikawa, K. (2010). Optimization of the hybridization-based method for purification of thermostable tRNAs in the presence of tetraalkylammonium salts. *Nucleic Acids Research, 38*, e89.

Ziegler, A., Steindl, K., Hanner, A. S., Kar, R. K., Prouteau, C., Boland, A., ... Park, M. H. (2022). Bi-allelic variants in DOHH, catalyzing the last step of hypusine biosynthesis, are associated with a neurodevelopmental disorder. *American Journal of Human Genetics, 109*, 1549–1558.

CHAPTER NINE

Quantification of spermine oxidase (SMOX) activity in tissues by HPLC

Jackson R. Foley[a] and Cassandra E. Holbert[a,b,*]
[a]Department of Oncology, Johns Hopkins University School of Medicine, Baltimore, MD, United States
[b]Department of Biology, Loyola University Maryland, Baltimore, MD, United States
*Corresponding author. e-mail address: cholber2@jhmi.edu; cholbert@loyola.edu

Contents

1. Introduction	184
2. Rationale	187
3. Tissue sample preparation	188
3.1 Equipment	188
3.2 Reagents	188
3.3 Procedure	190
3.4 Notes	190
4. PAOX and SMOX reactions	190
4.1 Equipment	190
4.2 Reagents	190
4.3 Procedure	191
4.4 Notes	192
5. HPLC analysis of polyamine oxidase activity	192
5.1 Equipment	192
5.2 Reagents	192
5.3 Procedure	193
5.4 Notes	196
Acknowledgments	196
References	196

Abstract

The polyamine oxidases, SMOX and PAOX, are enzymes involved in the normal metabolism of polyamines. Both enzymes are implicated in numerous human diseases including cancer, reperfusion injury, and neurodegenerative diseases. The ability to directly measure the activity of these enzymes is imperative in understanding their role in human health and disease. Most assays currently used to measure both SMOX and PAOX activity use a coupled reaction with horse radish peroxidase (HRP). These assays cannot be used when evaluating certain compounds for potential polyamine oxidase inhibition if these compounds also affect HRP activity. Additionally, since most assays use

H_2O_2 production as an indicator of oxidase activity they cannot be used to evaluate polyamine oxidase activity in the presence of iron or other divalent metals. This prevents the use of these assays to evaluate polyamine oxidase activity in tissue samples. Here we describe the protocols for determining polyamine oxidase activity in an HRP-independent manner via an HPLC-based assay allowing for evaluation of both compounds that may interfere with HRP activity and polyamine oxidase activity in tissue samples.

1. Introduction

Polyamine catabolism is often viewed as a double-edged sword with regard to human health. While induction of polyamine catabolism shows promise as an anticancer therapeutic, elevated polyamine catabolism is linked with the etiologies of numerous human diseases (Wang & Casero, 2006). Polyamine homeostasis is tightly regulated through coordinated biosynthesis, catabolism, and transport. Back-conversion of spermine to spermidine can occur through two individual catabolic pathways (Vujcic, Liang, Diegelman, Kramer, & Porter, 2003). Acetylation by spermidine/spermine N^1-acetyltransferase (SSAT) produces N^1-acetylspermine which is then oxidized by the peroxisomal enzyme, PAOX (FAD-dependent acetylpolyamine oxidase) (Casero & Pegg, 1993; Wu, Yankovskaya, & McIntire, 2003). Alternatively, spermine can be directly oxidized by the FAD-dependent enzyme spermine oxidase (SMOX) to form spermidine without an acetylated intermediary molecule (Wang et al., 2001). Byproducts of SMOX activity include reactive oxygen species (ROS), in the form of H_2O_2, and 3-aminopropanal (3-AP). The 3-AP produced by SMOX activity can spontaneously convert into the highly reactive and toxic aldehyde acrolein (Fig. 1) (Houen, Bock, & Jensen, 1994; Sharmin et al., 2001).

PAOX is a FAD-dependent amine oxidase that is capable of oxidizing N^1-acetlayed polyamines and holds the greatest affinity for N^1-acetylspermine (Hölttä, 1977; Vujcic et. al., 200; Wu et al., 2003). The oxidation of N^1-acetylspermine produces spermidine, 3-acetoamidopropanal, and reactive oxygen species in the form of H_2O_2. PAOX is generally constitutively active in normal tissues as its activity is rate-limited by the production of acetylated substrates by SSAT. SSAT transcription and activity can be impacted by a variety of stimuli including hormones, cytokines, inflammation, stress, and the activity of particular gene products including Ras and nuclear factor κB (NFκB) (Babbar & Casero, 2006; Babbar, Hacker, Huang, & Casero, 2006; Gerner & Meyskens, 2004; Shantz & Levin, 2007; Wang & Casero, 2006).

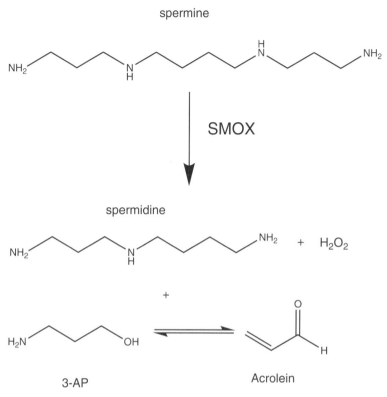

Fig. 1 Metabolism of spermine by spermine oxidase (SMOX). Spermine is oxidized by SMOX to produce spermidine, hydrogen peroxide (H_2O_2), and 3-aminopropanal (3-AP). 3-AP can spontaneously convert into the highly toxic aldehyde acrolein.

An important contrast in regards to SMOX is it's ability to be active in both the nuclear and cytoplasmic compartments of the cell (Murray-Stewart et al., 2008). Inside the nucleus, extensive SMOX activity results in a decrease of the antioxidant spermine as well as ROS production in close proximity to DNA, potentially resulting in oxidative damage and apoptosis. While this is problematic in normal cells as a potential jump-start for inflammation-associated carcinogenesis, polyamine catabolism-induced oxidative damage can be utilized as a tumoricidal tactic in already transformed cells.

Contrary to other polyamine metabolic enzymes, SMOX activity is predominantly regulated at the level of transcription and mRNA stabilization (Wang et al., 2005). It is highly inducible by a variety of stimuli including polyamine analogues and certain pro-inflammatory cytokines, including TNF-α, IL-1β and IL-6 (Babbar, Murray-Stewart, & Casero, 2007;

Devereux et al., 2003; Fan et al., 2019; Hacker, Marton, Sobolewski, & Casero, 2008). The upregulation of SMOX by pro-inflammatory cytokines leads to the production of H_2O_2 and subsequent oxidative DNA damage. As such, elevated SMOX expression has been associated with numerous inflammation-associated conditions.

Elevated expression of SMOX is linked with several inflammation-associated conditions including gastritis, inflammatory bowel disease (IBD), and chronic hepatitis (Chaturvedi et al., 2011; Goodwin et al., 2011; Smirnova et al., 2017) (Fig. 2). Tissues of individuals infected with *Helicobacter pylori* exhibit elevated levels of SMOX mRNA compared to healthy controls. Oxidative damage by SMOX activity can lead to mutational burden in gastric cells, and the apoptotic death of macrophages induced by *H. pylori* is directly associated with the H_2O_2 produced by SMOX and the subsequent fatal DNA damage (Chaturvedi et al., 2004; Xu et al., 2004). Therefore, SMOX activity not only enhances immune evasion by *H. pylori* but can play a direct role in the progression of gastritis to gastric cancer. Expression of SMOX is also induced by enterotoxigenic *Bacteroides fragilis* and the hepatitis C virus (HCV) resulting in a positive correlation with IBD, chronic hepatitis and their associated cancers (Goodwin et al., 2011; Hu et al., 2018; Smirnova et al., 2017; Snezhkina et al., 2016; Toprak et al., 2006) (Table 1).

The catabolism of spermine via SMOX activity is also associated with noncancerous conditions including stroke, epilepsy, neurodegeneration, and ischemic reperfusion injury. The accumulation of toxic polyamine catabolic

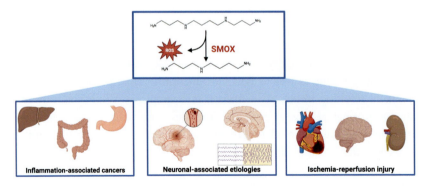

Fig. 2 SMOX expression and its metabolites are associated with numerous disease etiologies. This includes inflammation-associated cancers (hepatic, gastric, and colon), neuronal-associated conditions (stroke, Parkinson's disease, and epilepsy), and ischemia-reperfusion injury of various organs including heart, brain, and kidney.

Table 1 Inflammatory conditions and cancers associated with elevated SMOX expression.

Infectious agent/toxin	Inflammatory condition	Associated cancer
Helicobacter pylori	Gastritis	Gastric cancer
Enterotoxigenic *Bacteroides fragilis* (ETBF)	Inflammatory bowel disease	Colorectal cancer
Hepatitis C virus	Chronic hepatitis	Hepatocellular carcinoma

Adapted from Murray Stewart, Dunston, Woster, and Casero (2018).

metabolites, including ROS, are seen in ischemia reperfusion injury associated with myocardial infarction, stroke, and renal tubular necrosis (Barone et al., 2005; Zahedi et al., 2003; Zahedi, Barone, et al., 2010) (Fig. 2). Increases in acrolein and SMOX levels are seen in renal failure patients, and SMOX ablation reduces tubular injury and preserves kidney function following cisplatin-induced acute kidney injury (Zahedi et al., 2017).

SMOX expression is elevated in the brain following ischemic reperfusion, traumatic, and excitotoxic injuries (Cervelli et al., 2013; Doğan et al., 1999; Ivanova et al., 2002; Zahedi, Huttinger, et al., 2010). SMOX and acrolein levels are elevated in plasma of stroke patients and correlate to stroke size, making them valuable biomarkers in stroke diagnosis (Tomitori et al., 2005). SMOX is upregulated in Parkinson's disease, and overexpression of SMOX in neocortex neurons induces astrogliosis, a hallmark characteristic of Parkinson's disease (Cervelli et al., 2022; Cervetto et al., 2021; Dumitriu et al., 2012). Additionally, mouse models overexpressing SMOX exhibit neuronal loss and increased seizure susceptibility.

Given the importance of SMOX in human diseases, the complexity of in vivo impacts on SMOX activity, and the inability of luminescence-based assays to analyze SMOX activity in tissues, there is a need to estimate the activity of spermine oxidase in tissue samples via alternative methods such as high performance liquid chromatography.

2. Rationale

As indicated above, the polyamine oxidases, PAOX and SMOX, are important enzymes in the regulation of polyamine levels and can be involved in both normal and pathological conditions. As such, the ability to measure their activity both in vitro and in vivo is critical to understand their full influence on biological systems. Unfortunately, most assays currently used to measure both PAOX and SMOX activity use a coupled reaction with horse radish peroxidase (HRP) for either colorimetric or luminescent determinations. The coupled reactions have two significant shortcomings in certain instances. When examining compounds for their ability to selectively or specifically inhibit one of the polyamine oxidases, if the compound affects HRP activity, the resulting data do not reflect the compounds effects on the oxidases. A second issue arises when measuring oxidase activity in tissues. Since most polyamine oxidase activity assays indirectly measure the H_2O_2 produced, when attempting to measure their activity in tissues, the presence of iron or other divalent metal ions degrade the H_2O_2 produced by the polyamine oxidases by Fenton catalysis before it can react with the HRP, thus invalidating the assay (Ha et al., 1998; Kremer, 1962; Pedreño, López-Contreras, Cremades, & Peñafiel, 2005). In the HPLC method presented here, however, the spermidine product of the oxidases is measured and is not dependent on HRP activity. Consequently, this method is useful for most applications where the HRP-coupled reactions cannot be used.

3. Tissue sample preparation
3.1 Equipment
- Tissue homogenizer (Dremel, Mount Prospect, IL, USA)
- 2 mL microcentrifuge tubes (Eppendorf, Hamburg, Germany)
- 1.5 mL microcentrifuge tubes (Eppendorf, Hamburg, Germany)

3.2 Reagents
- Bradford protein assay reagent (Bio-Rad, Hercules, CA, USA)
- Glycine buffer (J.T. Baker, Philipsburg, NJ, USA). 0.5 M stock solution adjusted to pH 8.0 with NaOH should be made fresh. See Table 2.
- Aminoguanidine (Sigma-Aldrich, St. Louis, MO, USA). 100 mM stock solution (in HPLC-grade water) can be stored at −20°C. See Table 2.
- Pargyline (Sigma-Aldrich, St. Louis, MO, USA). 2.3 mg/mL stock solution (in HPLC-grade water) can be store at −20°C. See Table 2.
- Ice
- Optional: Dry ice-ethanol bath

Table 2 Lysis buffer ingredients, dilutions, and directions.

Reagent	Stock concentration	Stock storage conditions	Final concentration	volume per 480 μL reaction
Glycine Buffer	0.5 M, Adjusted to pH 8.0 with NaOH	None, make fresh	0.083 M	80 μL
Aminoguanidine	100 mM	−20 °C	1 mM	4.8 μL
Pargyline	2.3 mg/mL	−20 °C	0.019 mg/mL	4 μL
HPLC-grade water		RT		391.2 μL

3.3 Procedure

1. Harvest tissues of interest into 2 mL microcentrifuge tubes. Tissue can either be flash frozen in a dry ice-ethanol bath and stored at −80 °C or placed on ice for immediate use.
2. For immediate use, resuspend tissues in 480 µL ice-cold 0.083 M glycine buffer pH 8.0, 1 mM aminoguanidine, and 0.019 mg/mL pargyline (See Notes 1 and 2).
3. For frozen tissues, thaw on ice, and resuspend in enough volume of reagents in step 2 to cover the tissue (See Note 3).
4. Homogenize tissue quickly on ice (See Note 3).
5. Move ~10 µL lysate from each sample into a separate 1.5 mL microcentrifuge tube and determine total protein concentration in lysate using a Bradford assay or other preferred method.

3.4 Notes

1. For larger tissues, either section the tissue into a piece small enough to submerge or add enough buffer to cover the tissue.
2. Pargyline and aminoguanidine are mono and diamine oxidase inhibitors, respectively, and must be included when working with tissue or cellular samples that will be exposed to exogenous polyamines.
3. Enzymatic activity will be lost if samples are not kept on ice.

4. PAOX and SMOX reactions

4.1 Equipment

- Digital Block Heater (VWR, Radnor, PA, USA)
- Vortex Genie 2 sample mixer (Thermo Fisher Scientific, Waltham, MA, USA)
- 1.5 mL microcentrifuge tubes (Eppendorf, Hamburg, Germany)

4.2 Reagents

- 0.083 M Glycine buffer (1 part 0.5 M glycine buffer diluted in 5 parts HPLC grade water) (J.T. Baker, Phillipsburg, NJ, USA).
- Aminoguanidine (Sigma-Aldrich, St. Louis, MO, USA). 100 mM stock solution (in HPLC-grade water) can be stored at −20 °C. See Table 2.
- Pargyline (Sigma-Aldrich, St. Louis, MO, USA). 2.3 mg/mL stock solution (in HPLC-grade water) can be stored at −20 °C. See Table 2.
- Spermine tetrachloride (Sigma-Aldrich, St. Louis, MO, USA). 1.5 mM stock can be stored −20 °C.

- N^1-acetylspermine (Honeywell Fluka, Morris Plains, NJ, USA). 1.5 mM stock can be stored −20 °C.
- HPLC grade water (Thermo Fisher Scientific, Waltham, MA, USA)

4.3 Procedure

1. Move 50 μL of each tissue lysate from Section 3 into three new 1.5 mL microcentrifuge tubes on ice for each replicate (three recommended – see Note 1). Each tube will be incubated with a different substrate.
2. To each tube add 1.2 μL of 100 mM aminoguanidine and 1 μL of 2.3 mg/mL pargyline (see Table 3).
3. For each tissue: in tube one add 50 μL of 0.083 M glycine buffer (control). In the second tube add 50 μL of 1.5 mM spermine (as a substrate for SMOX) and add 50 μL of 1.5 mM N^1-acetylspermine (as a substrate for PAOX) into tube three (see Table 4).
4. Incubate all samples with gentle agitation in a 37 °C water bath for 2 h (see Note 2).

Table 3 Lysate tube components.

Reagent	Stock concentration	Final concentration	Volume per ~100 μL reaction
Aminoguanidine	100 mM	1 mM	1.2 μL
Pargyline	2.3 mg/mL	0.019 mg/mL	1 μL
Tissue Lysate			50 μL

Table 4 Enzyme activity workflow (one replicate).

Tube	Reaction	Component	Volume per 100 μL reaction
Tube 1	Control	Lysate tube from Table 3 0.083 M glycine	50 μL 50 μL
Tube 2	SMOX activity	Lysate tube from Table 3 1.5 mM spermine	50 μL 50 μL
Tube 3	PAOX activity	Lysate tube from Table 3 1.5 mM N^1-acetylspermine	50 μL 50 μL

4.4 Notes

1. For three replicates of each reaction, nine total tubes will need to receive 50 μL of tissue lysate (450 μL total).
2. Samples can then be snap frozen and stored at −80 °C until assayed for changes in spermidine product by HPLC.

5. HPLC analysis of polyamine oxidase activity

5.1 Equipment

1. PerkinElmer Series 200 HPLC (PerkinElmer, Waltham, MA, USA)
2. Vacuum elution manifold (Agilent, Santa Clara, CA, USA)
3. Refrigerated microcentrifuge (Eppendorf, Hamburg, Germany)
4. Digital Block Heater (VWR, Radnor, PA, USA)
5. Vortex Genie 2 sample mixer (Thermo Fisher Scientific, Waltham, MA, USA)
6. Sep-Pak™ C18 1-cc vacuum cartridges, 100 mg sorbent per cartridge, 55–105 μm (Waters, Milford, MA, USA)
7. Syncronis™ C18 reversed-phase HPLC column (5 μm, 150 × 4.6 mm) (Phenomenex, Torrance, CA, USA)
8. Flat-bottom HPLC vial inserts (Thermo Fisher Scientific, Waltham, MA, USA)
9. Screw cap HPLC vials, 2 mL (ThermoFisher, Waltham, MA, USA)
10. Caps for 2 mL HPLC vials (ThermoFisher, Waltham, MA, USA)
11. Septa inserts for HPLC vial caps (PTFE, 8 mm) (VWR, Radnor, PA, USA)
12. 1.5 mL microcentrifuge tubes (Eppendorf, Hamburg, Germany)

5.2 Reagents

- 0.083 M Glycine buffer (Sigma Aldrich, St. Louis, MO, USA)
- Perchloric acid (HClO$_4$) (J.T.Baker, Philipsburg, NJ, USA). 1:10 dilution of 12 N perchloric acid to yield a 1.2 N perchloric acid stock (See Note 1)
- Spermine tetrachloride (Sigma Aldrich, St. Louis, MO, USA)
- N^1-acetylspermine (Honeywell Fluka, Morris Plains, NJ, USA)
- HPLC-grade acetonitrile (ACN) (Thermo Fisher Scientific, Waltham, MA, USA)
- HPLC grade water (Thermo Fisher Scientific, Waltham, MA, USA)
- Dansyl chloride (MilliporeSigma, Burlington, MA, USA)
- Proline (MilliporeSigma, Burlington, MA, USA)
- Sodium carbonate (Na$_2$CO$_3$) (MilliporeSigma, Burlington, MA, USA)
- 1,7-Diaminoheptane (DAH) (MilliporeSigma, Burlington, MA, USA)

5.3 Procedure

Acid extraction and dansylation of polyamines.

1. Add 1 vol (100 µL) of 1.2 N perchloric acid to each 1.5 mL reaction tube from Section 4.
2. Vortex tubes briefly. Centrifuge for 10–15 min at 12,000 × g, 4 °C.
3. Move 50 µL of supernatant to a fresh 1.5 mL microcentrifuge tube.
4. To a new 1.5 mL tube, add 50 µL of 40 µM stock of endogenous polyamines (putrescine, spermidine, spermine, N1-acetylspermine) to serve as a standard
5. To each tube add 50 µL of 40 µM DAH to serve as an internal standard.
6. Add 200 µL saturated sodium carbonate (Na_2CO_3) to each tube.
7. Add 200 µL 10 mg/mL dansyl chloride (freshly diluted in acetone at time of labeling) to each reaction.
8. Vortex tubes 10–15 s
9. Incubate at 70 °C in heat block for 10 min.
10. Remove tubes from heat block and incubate at room temperature for 10 min.
11. Prepare proline solution (250 mg/mL) in HPLC water. Add 100 µL to each reaction tube.
12. Incubate tubes in the dark at room temperature for 10 min.
13. Set up elution vacuum apparatus. Wash each column with two column volumes of 100 % acetonitrile followed by two volumes of 35 % acetonitrile.
14. Load labeled samples & standard and draw through.
15. Wash columns with two volumes of 35 % acetonitrile.
16. Turn off vacuum and insert rack with 2 mL microcentrifuge tubes for collection.
17. Add 1 mL 100 % acetonitrile to columns and draw into collection tubes (See Note 2).

HPLC quantification of polyamines:

1. Load 100 µL of 100 % ACN to an HPLC vial for the blank.
2. Load 175 µL of standard and samples to individual HPLC vials.
3. Run HPLC with a constant flow of 2 mL/min using the following gradient of 100 % and 45 % ACN: 0–0.1 min at 45 % ACN; 0.1–13 min 100 % ACN; 13–18 min 100 % ACN, 18–23 min 45 %. Flush the system with 100 % ACN for 5 min between washes.
4. Record fluorescence at an excitation wavelength of 340 nm and an emission wavelength of 515 nm.

5. Quantify polyamine levels in samples relative to the prepared standard and normalize each to the area of the DAH internal standard peak.
   a. Set the area of each polyamine in the standard (STD) divided by the area of DAH in the standard equal to 0.1 nmol and calculate the conversion factor (X) for each polyamine normalized to DAH.
      i. (Area poly in STD/Area DAH in STD)*X = 0.1
   b. Calculate the raw amount of each polyamine in the samples based on the DAH internal standard conversion factor determined above.
      i. Raw amount (nmol) = (Area polyamine in sample/Area DAH in sample)*X
   c. Calculate the adjusted amount of each polyamine in the sample by dividing the raw amount (calculated in b above) by the mass (in mg) of protein (as determined by a Bradford Assay) and multiplying the result by 800.
      i. Adjusted amount = (raw amount/mg protein)* 800 = nmol polyamine/mg protein

SMOX and PAOX enzyme activity calculations:

1. SMOX calculation: Calculate the amount of spermidine produced by SMOX enzymatic activity during the reaction (Fig. 3A).
   a. The sample for this calculation is the spermine spiked sample (tube 2 from Table 4).
   b. Subtract the amount of spermidine (adjusted amount * mg protein) in tube 1 (see Note 3) from the amount of spermidine (adjusted amount * mg protein) in tube 2. This is the amount of spermidine produced by SMOX.
   c. Calculate SMOX enzymatic activity (in nmol/min) by dividing the amount of spermidine produced by the two-hour reaction time (See Note 4).
2. PAOX calculation: Calculate the amount of spermidine produced by PAOX enzymatic activity during the reaction (Fig. 3B-D).
   a. The sample for this calculation is the N^1-acetylspermine spiked sample (tube 3 from Table 4).
   b. Subtract the amount of spermidine (adjusted amount * mg protein) in tube 1 (see Note 3) from the amount of spermidine (adjusted amount * mg protein) in tube 3. This is the amount of spermidine produced by PAOX.
   c. Divide the amount of spermidine produced by PAOX by the two-hour reaction time to yield PAOX enzymatic activity (in nmol/min).

Quantification of spermine oxidase (SMOX) activity in tissues by HPLC   195

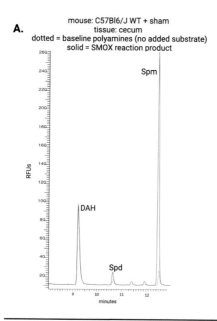

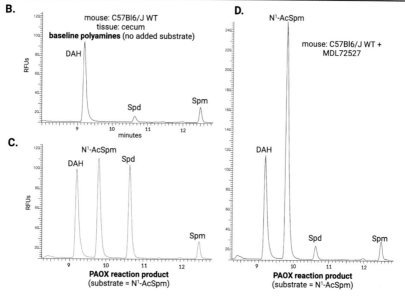

Fig. 3 Representative HPLC chromatograms of SMOX and PAOX activity in mouse cecum tissue. (A) Cecum tissue from an untreated mouse was evaluated with and without spermine addition to determine SMOX enzymatic activity. Peaks are labeled with their corresponding molecule. (B, C) Cecum tissue from the same untreated mouse was evaluated with and without N1-acetylspermine addition to determine PAOX enzymatic activity. (D) N1-acetylspermine was added to cecum tissue from a mouse treated with MDL72527 (a polyamine oxidase inhibitor). The chromatogram demonstrates an expected decreased spermidine peak following reduced PAOX activity due to enzymatic inhibition.

5.4 Notes

1. When diluting 12 N perchloric acid to 1.2 N, be sure to add the acid volume directly to the water volume and not the other way around (water to the acid).
2. Eluted samples can be stored at −20°C indefinitely.
3. This adjusts for baseline polyamine levels (no substrate was added to this tube).
4. SSAT is inactive in glycine buffer and adjustment for SSAT/PAOX activity is unnecessary in spermine-spiked samples unless there is direct evidence of N^1-acetylspermine present.

Acknowledgments

The authors would like to acknowledge Christina DeStafano Shields and Andrew Goodwin for their work on developing SMOX activity assays via HRP coupled reactions and HPLC. Figs. 2 and 3 were created using BioRender.

References

Babbar, N., & Casero, R. A. (2006). Tumor necrosis factor-alpha increases reactive oxygen species by inducing spermine oxidase in human lung epithelial cells: A potential mechanism for inflammation-induced carcinogenesis. *Cancer Research, 66*(23), 11125–11130. https://doi.org/10.1158/0008-5472.CAN-06-3174.

Babbar, N., Hacker, A., Huang, Y., & Casero, R. A. (2006). Tumor necrosis factor alpha induces spermidine/spermine N1-acetyltransferase through nuclear factor kappaB in non-small cell lung cancer cells. *The Journal of Biological Chemistry, 281*(34), 24182–24192. https://doi.org/10.1074/jbc.M601871200.

Babbar, N., Murray-Stewart, T., & Casero, R. A. (2007). Inflammation and polyamine catabolism: The good, the bad and the ugly. *Biochemical Society Transactions, 35*(Pt 2), 300–304. https://doi.org/10.1042/BST0350300.

Barone, S., Okaya, T., Rudich, S., Petrovic, S., Tenrani, K., Wang, Z., ... Soleimani, M. (2005). Distinct and sequential upregulation of genes regulating cell growth and cell cycle progression during hepatic ischemia-reperfusion injury. *American Journal of Physiology. Cell Physiology, 289*(4), C826–C835. https://doi.org/10.1152/ajpcell.00629.2004.

Casero, R. A., & Pegg, A. E. (1993). Spermidine/spermine N1-acetyltransferase–the turning point in polyamine metabolism. *The FASEB Journal, 7*(8), 653–661.

Cervelli, M., Averna, M., Vergani, L., Pedrazzi, M., Amato, S., Fiorucci, C., ... Marcoli, M. (2022). The involvement of polyamines catabolism in the crosstalk between neurons and astrocytes in neurodegeneration. *Biomedicines, 10*(7), https://doi.org/10.3390/biomedicines10071756.

Cervelli, M., Bellavia, G., D'Amelio, M., Cavallucci, V., Moreno, S., Berger, J., ... Mariottini, P. (2013). A new transgenic mouse model for studying the neurotoxicity of spermine oxidase dosage in the response to excitotoxic injury. *PLoS One, 8*(6), e64810. https://doi.org/10.1371/journal.pone.0064810.

Cervetto, C., Averna, M., Vergani, L., Pedrazzi, M., Amato, S., Pelassa, S., ... Cervelli, M. (2021). Reactive astrocytosis in a mouse model of chronic polyamine catabolism activation. *Biomolecules, 11*(9), https://doi.org/10.3390/biom11091274.

Chaturvedi, R., Asim, M., Romero-Gallo, J., Barry, D. P., Hoge, S., de Sablet, T., ... Wilson, K. T. (2011). Spermine oxidase mediates the gastric cancer risk associated with Helicobacter pylori CagA. *Gastroenterology, 141*(5), 1696–1708.e1691-1692. https://doi.org/10.1053/j.gastro.2011.07.045.

Chaturvedi, R., Cheng, Y., Asim, M., Bussière, F. I., Xu, H., Gobert, A. P., ... Wilson, K. T. (2004). Induction of polyamine oxidase 1 by Helicobacter pylori causes macrophage apoptosis by hydrogen peroxide release and mitochondrial membrane depolarization. *The Journal of Biological Chemistry, 279*(38), 40161–40173. https://doi.org/10.1074/jbc.M401370200.

Devereux, W., Wang, Y., Stewart, T. M., Hacker, A., Smith, R., Frydman, B., ... Casero, R. A. (2003). Induction of the PAOh1/SMO polyamine oxidase by polyamine analogues in human lung carcinoma cells. *Cancer Chemotherapy and Pharmacology, 52*(5), 383–390. https://doi.org/10.1007/s00280-003-0662-4.

Doğan, A., Rao, A. M., Baskaya, M. K., Hatcher, J., Temiz, C., Rao, V. L., & Dempsey, R. J. (1999). Contribution of polyamine oxidase to brain injury after trauma. *Journal of Neurosurgery, 90*(6), 1078–1082. https://doi.org/10.3171/jns.1999.90.6.1078.

Dumitriu, A., Latourelle, J. C., Hadzi, T. C., Pankratz, N., Garza, D., Miller, J. P., ... Myers, R. H. (2012). Gene expression profiles in Parkinson disease prefrontal cortex implicate FOXO1 and genes under its transcriptional regulation. *PLoS Genetics, 8*(6), e1002794. https://doi.org/10.1371/journal.pgen.1002794.

Fan, J., Chen, M., Wang, X., Tian, Z., Wang, J., Fan, D., ... Dai, X. (2019). Targeting smox is neuroprotective and ameliorates brain inflammation in cerebral ischemia/reperfusion rats. *Toxicological Sciences: an Official Journal of the Society of Toxicology, 168*(2), 381–393. https://doi.org/10.1093/toxsci/kfy300.

Gerner, E. W., & Meyskens, F. L. (2004). Polyamines and cancer: Old molecules, new understanding. *Nature Reviews. Cancer, 4*(10), 781–792. https://doi.org/10.1038/nrc1454.

Goodwin, A. C., Destefano Shields, C. E., Wu, S., Huso, D. L., Wu, X., Murray-Stewart, T. R., ... Casero, R. A. (2011). Polyamine catabolism contributes to enterotoxigenic Bacteroides fragilis-induced colon tumorigenesis. *Proceedings of the National Academy of Sciences of the United States of America, 108*(37), 15354–15359. https://doi.org/10.1073/pnas.1010203108.

Ha, H. C., Sirisoma, N. S., Kuppusamy, P., Zweier, J. L., Woster, P. M., & Casero, R. A. (1998). The natural polyamine spermine functions directly as a free radical scavenger. *Proceedings of the National Academy of Sciences of the United States of America, 95*(19), 11140–11145. https://doi.org/10.1073/pnas.95.19.11140.

Hacker, A., Marton, L. J., Sobolewski, M., & Casero, R. A. (2008). In vitro and in vivo effects of the conformationally restricted polyamine analogue CGC-11047 on small cell and non-small cell lung cancer cells. *Cancer Chemotherapy and Pharmacology, 63*(1), 45–53. https://doi.org/10.1007/s00280-008-0706-x.

Houen, G., Bock, K., & Jensen, A. L. (1994). HPLC and NMR investigation of the serum amine oxidase catalyzed oxidation of polyamines. *Acta Chemica Scandinavica, 48*(1), 52–60. https://doi.org/10.3891/acta.chem.scand.48-0052.

Hu, T., Sun, D., Zhang, J., Xue, R., Janssen, H. L. A., Tang, W., & Dong, L. (2018). Spermine oxidase is upregulated and promotes tumor growth in hepatocellular carcinoma. *Hepatology Research: The Official Journal of the Japan Society of Hepatology, 48*(12), 967–977. https://doi.org/10.1111/hepr.13206.

Hölttä, E. (1977). Oxidation of spermidine and spermine in rat liver: Purification and properties of polyamine oxidase. *Biochemistry, 16*(1), 91–100. https://doi.org/10.1021/bi00620a015.

Ivanova, S., Batliwalla, F., Mocco, J., Kiss, S., Huang, J., Mack, W., ... Tracey, K. J. (2002). Neuroprotection in cerebral ischemia by neutralization of 3-aminopropanal. *Proceedings of the National Academy of Sciences of the United States of America, 99*(8), 5579–5584. https://doi.org/10.1073/pnas.082609299.

Kremer, M. L. (1962). The promoting effect of cupric ions on the ferric ion catalyzed decomposition of hydrogen peroxide. *Journal of Catalysis, 1*, Elsevier, 351–355.

Murray Stewart, T., Dunston, T. T., Woster, P. M., & Casero, R. A. (2018). Polyamine catabolism and oxidative damage. *The Journal of Biological Chemistry, 293*(48), 18736–18745. https://doi.org/10.1074/jbc.TM118.003337.

Murray-Stewart, T., Wang, Y., Goodwin, A., Hacker, A., Meeker, A., & Casero, R. A. (2008). Nuclear localization of human spermine oxidase isoforms - possible implications in drug response and disease etiology. *The FEBS Journal, 275*(11), 2795–2806. https://doi.org/10.1111/j.1742-4658.2008.06419.x.

Pedreño, E., López-Contreras, A. J., Cremades, A., & Peñafiel, R. (2005). Protecting or promoting effects of spermine on DNA strand breakage induced by iron or copper ions as a function of metal concentration. *Journal of Inorganic Biochemistry, 99*(10), 2074–2080. https://doi.org/10.1016/j.jinorgbio.2005.07.005.

Shantz, L. M., & Levin, V. A. (2007). Regulation of ornithine decarboxylase during oncogenic transformation: Mechanisms and therapeutic potential. *Amino Acids, 33*(2), 213–223. https://doi.org/10.1007/s00726-007-0531-2.

Sharmin, S., Sakata, K., Kashiwagi, K., Ueda, S., Iwasaki, S., Shirahata, A., & Igarashi, K. (2001). Polyamine cytotoxicity in the presence of bovine serum amine oxidase. *Biochemical and Biophysical Research Communications, 282*(1), 228–235. https://doi.org/10.1006/bbrc.2001.4569.

Smirnova, O. A., Keinanen, T. A., Ivanova, O. N., Hyvonen, M. T., Khomutov, A. R., Kochetkov, S. N., ... Ivanov, A. V. (2017). Hepatitis C virus alters metabolism of biogenic polyamines by affecting expression of key enzymes of their metabolism. *Biochemical and Biophysical Research Communications, 483*(2), 904–909. https://doi.org/10.1016/j.bbrc.2017.01.032.

Snezhkina, A. V., Krasnov, G. S., Lipatova, A. V., Sadritdinova, A. F., Kardymon, O. L., Fedorova, M. S., ... Kudryavtseva, A. V. (2016). The dysregulation of polyamine metabolism in colorectal cancer is associated with overexpression of c-Myc and C/EBPβ rather than enterotoxigenic bacteroides fragilis infection. *Oxid Med Cell Longev, 2016*, 2353560. https://doi.org/10.1155/2016/2353560.

Tomitori, H., Usui, T., Saeki, N., Ueda, S., Kase, H., Nishimura, K., ... Igarashi, K. (2005). Polyamine oxidase and acrolein as novel biochemical markers for diagnosis of cerebral stroke. *Stroke; a Journal of Cerebral Circulation, 36*(12), 2609–2613. https://doi.org/10.1161/01.STR.0000190004.36793.2d.

Toprak, N. U., Yagci, A., Gulluoglu, B. M., Akin, M. L., Demirkalem, P., Celenk, T., & Soyletir, G. (2006). A possible role of Bacteroides fragilis enterotoxin in the aetiology of colorectal cancer. *Clinical Microbiology and Infection: the Official Publication of the European Society of Clinical Microbiology and Infectious Diseases, 12*(8), 782–786. https://doi.org/10.1111/j.1469-0691.2006.01494.x.

Vujcic, S., Liang, P., Diegelman, P., Kramer, D. L., & Porter, C. W. (2003). Genomic identification and biochemical characterization of the mammalian polyamine oxidase involved in polyamine back-conversion. *The Biochemical Journal, 370*(Pt 1), 19–28. https://doi.org/10.1042/BJ20021779.

Wang, Y., & Casero, R. A. (2006). Mammalian polyamine catabolism: A therapeutic target, a pathological problem, or both? *Journal of Biochemistry, 139*(1), 17–25. https://doi.org/10.1093/jb/mvj021.

Wang, Y., Devereux, W., Woster, P. M., Stewart, T. M., Hacker, A., & Casero, R. A. (2001). Cloning and characterization of a human polyamine oxidase that is inducible by polyamine analogue exposure. *Cancer Research, 61*(14), 5370–5373.

Wang, Y., Hacker, A., Murray-Stewart, T., Fleischer, J. G., Woster, P. M., & Casero, R. A. (2005). Induction of human spermine oxidase SMO(PAOh1) is regulated at the levels of new mRNA synthesis, mRNA stabilization and newly synthesized protein. *The Biochemical Journal, 386*(Pt 3), 543–547. https://doi.org/10.1042/BJ20041084.

Wu, T., Yankovskaya, V., & McIntire, W. S. (2003). Cloning, sequencing, and heterologous expression of the murine peroxisomal flavoprotein, N1-acetylated polyamine oxidase. *The Journal of Biological Chemistry, 278*(23), 20514–20525. https://doi.org/10.1074/jbc.M302149200.

Xu, H., Chaturvedi, R., Cheng, Y., Bussiere, F. I., Asim, M., Yao, M. D., ... Wilson, K. T. (2004). Spermine oxidation induced by Helicobacter pylori results in apoptosis and DNA damage: Implications for gastric carcinogenesis. *Cancer Research, 64*(23), 8521–8525. https://doi.org/10.1158/0008-5472.CAN-04-3511.

Zahedi, K., Barone, S., Destefano-Shields, C., Brooks, M., Murray-Stewart, T., Dunworth, M., ... Soleimani, M. (2017). Activation of endoplasmic reticulum stress response by enhanced polyamine catabolism is important in the mediation of cisplatin-induced acute kidney injury. *PLoS One, 12*(9), e0184570. https://doi.org/10.1371/journal.pone.0184570.

Zahedi, K., Barone, S., Kramer, D. L., Amlal, H., Alhonen, L., Jänne, J., ... Soleimani, M. (2010). The role of spermidine/spermine N1-acetyltransferase in endotoxin-induced acute kidney injury. *American Journal of, 299*(1), C164–C174. https://doi.org/10.1152/ajpcell.00512.2009.

Zahedi, K., Huttinger, F., Morrison, R., Murray-Stewart, T., Casero, R. A., & Strauss, K. I. (2010). Polyamine catabolism is enhanced after traumatic brain injury. *Journal of Neurotrauma, 27*(3), 515–525. https://doi.org/10.1089/neu.2009.1097.

Zahedi, K., Wang, Z., Barone, S., Prada, A. E., Kelly, C. N., Casero, R. A., ... Soleimani, M. (2003). Expression of SSAT, a novel biomarker of tubular cell damage, increases in kidney ischemia-reperfusion injury. *American Journal of, 284*(5), F1046–F1055. https://doi.org/10.1152/ajprenal.00318.2002.

CHAPTER TEN

Direct measurement of ATP13A2 polyamine-dependent ATPase activity following rapid purification of lysosomes

Christina Efthymiou, Sydney Drury, and Kenneth Lee*
Department of Cellular and Molecular Physiology, The Pennsylvania State University, College of Medicine, PA, United States
*Corresponding author. e-mail address: kenneth.lee@psu.edu

Contents

1. Introduction	202
2. LysoPure: rapid purification of lysosomes from mammalian cells	203
2.1 Equipment	203
2.2 Materials	203
2.3 Procedure and expected results	204
2.4 Considerations for the purification	205
3. Determination of lysosomal ATP13A2 spermine-dependent ATPase activity using a NADH-coupled ATPase assay	205
3.1 Equipment	205
3.2 Materials	205
3.3 Procedure and expected results	206
3.4 Considerations for the ATPase assay	208
Acknowledgments	208
References	208

Abstract

The P5B family P-type ATPase ATP13A2(PARK9) is a *bona fide* polyamine transporter resident in the endolysosomal compartment where it mediates the import of endocytosed polyamines from the lysosome lumen into the cytosol. Dysfunction of ATP13A2 can negatively impact cellular survival and genetic aberrations its coding gene are linked to a number of neurodegenerative disorders with devastating consequences. While there has been much progress in its structural characterization *in vitro*, our understanding of ATP13A2's mechanism of action and regulation in a native lysosomal setting remains incomplete. Here we describe our approach to measure the polyamine-dependent ATPase activity of lysosomal ATP13A2 following our newly developed method to rapidly capture and purify lysosomes from mammalian cells. This strategy enables the targeted functional interrogation of the lysosome-localized population of ATP13A2 specifically.

1. Introduction

To the best of our knowledge all lifeforms on this planet require polyamines to survive, proliferate and reproduce (Michael, 2016; Pegg & Casero, 2011). Pathogens, cancer cells and healthy cells depend crucially on the availability of polyamines (Huang, Zhang, Chen, & Zeng, 2020; Li, Meng, Wu, & Sun, 2020; Michael, 2016; Miller-Fleming, Olin-Sandoval, Campbell, & Ralser, 2015). Though targeted modulation of polyamine homeostasis appears to be a promising approach to improve human health, this strategy has yet to be widely translated into clinical practice (Huang et al., 2020; Madeo, Eisenberg, Pietrocola, & Kroemer, 2018; Minois, Carmona-Gutierrez, & Madeo, 2011; Seiler, 2003). Collectively, metabolism and transport processes maintain cellular polyamines in homeostatic balance (Abdulhussein & Wallace, 2014; Pegg & Casero, 2011). While polyamine metabolism is well understood, a large gap exists in our understanding of molecular mechanisms controlling polyamine transport.

Studies have shown that the P5B-family P-type ATP-powered ion pump, ATP13A2, mediates import of polyamines from the lysosome into the cytosol (van Veen et al., 2020; Vrijsen et al., 2020). Loss-of-function mutations in ATP13A2 (also known as PARK9) are associated with a spectrum of neurodegenerative diseases, most notably Parkinson's disease, which affects 10 million people worldwide (Bras, Verloes, Schneider, Mole, & Guerreiro, 2012; Di Fonzo et al., 2007; Estrada-Cuzcano et al., 2017; Klein & Westenberger, 2012; Lin et al., 2008; Lunati, Lesage, & Brice, 2018; Ramirez et al., 2006; Spataro et al., 2019; Usenovic & Krainc, 2012). There is a strong case supporting the development of novel ATP13A2 modulating compounds, but progress has been slow due to a lack of detailed information on the molecular basis of ATP13A2 function and regulation.

Substantive gains in our knowledge of human ATP13A2 mechanism were made possible by recent cryo-EM studies from our group and others (Chen et al., 2021; Li, Wang, Salustros, Gronberg, & Gourdon, 2021; Mu et al., 2023; Sim, von Bulow, Hummer, & Park, 2021; Tillinghast, Drury, Bowser, Benn, & Lee, 2021; Tomita et al., 2021). In addition to this body of work, much of what we currently know about the functional properties of ATP13A2 were derived from protein specimens purified with the use of detergents (van Veen et al., 2020; Van Veen, Martin, Schuermans, & Vangheluwe, 2021). While detergents remain essential tools for nascent structural investigations of membrane proteins, we must be cognizant of the potential that detergent use in purification procedures may strip important cellular factors from the target. Such factors, including and not limited to proteins and lipids, could potentially impinge on functional and regulatory properties that may be otherwise overlooked. In addition, it remains unclear whether findings from structural and functional studies of performed using ATP13A2 purified in detergent truly reflect its behavior in a lipid bilayer setting.

As a first step to address these issues, we introduce in this chapter a new method called LysoPure to directly purify and functionally characterize recombinant ATP13A2 embedded in its natural lysosomal membrane environment. Classical lysosome purification approaches require lengthy procedures that may result in loss of peripherally associated cellular factors. While the LysoIP protocol allows fast capture of lysosomes for multi-omics analysis it prohibits the recovery of intact lysosomes for downstream functional investigation (Abu-Remaileh et al., 2017). Our new LysoPure approach extends the LysoIP immune-isolation approach to allow rapid purification of lysosomes for *in vitro* biochemical functional studies. We will also describe a method to assay the polyamine-dependent ATPase activity of recombinant ATP13A2 (Tillinghast et al., 2021), following the LysoPure procedure.

2. LysoPure: rapid purification of lysosomes from mammalian cells

2.1 Equipment

- Multi-six Microcentrifuge Tube Magnetic Separator (RayBiotech, cat. # 801-205)
- 15/50 mL Tube Magnetic Separator (RayBiotech, cat. # 801-204)
- 2 mL dounce homogenizer (VWR, cat. # 89026-398 and 89026-386)
- AccuSpin Micro 17R centrifuge (Fisher Scientific, cat. # 13-100-675)
- Allegra X-I4R Centrifuge (Beckman Coulter)
- Fisherbrand™ Variable Speed Nutator (Fisher Scientific, cat. # S06622)
- Humidified CO_2 incubator (VWR, cat. # 10810-944)
- Cell scrapers (VWR, cat. # 10062-910)

2.2 Materials

- Dulbecco's Modification of Eagle's Medium (DMEM) medium (VWR, cat. # 45000-304)

- HyClone fetal bovine serum (Cytiva, cat. # SH30071.03)
- Pierce™ Anti-HA Magnetic Beads (ThermoFisher Scientific, cat # 88837)
- HEK293 cells (ATCC, cat. # CRL-1573)
- HBK buffer (20 mM Hepes-KOH, pH 7.5; 136 mM KCl)

2.3 Procedure and expected results

We generated a mammalian expression construct, TMEM192$_{SUMO-3xHA}$, encoding the established lysosomal marker TMEM192 fused to a C-terminal Small Ubiquitin-like Modifier (SUMO)-tag followed by haemagglutinin-tag (3xHA). This construct is designed to allow fast immune-capture of lysosomes using anti-HA antibody-coupled resin followed by proteolytic release of purified lysosomes under non-denaturing conditions. Additionally, we generated another mammalian expression plasmid encoding C-terminal GFP-tagged human ATP13A2 (hATP13A2$_{GFP}$) to enable tracking of the lysosome-resident population of hATP13A2$_{GFP}$ by fluorescence during the purification procedure.

Expression of TMEM192$_{SUMO-3xHA}$ and hATP13A2$_{GFP}$ was accomplished by transient transfection of HEK293 cells cultured in DMEM supplemented with 10 % FBS maintained at 37 °C in 5 % CO$_2$. Three 10 cm tissue culture dishes of HEK293 cells at ~85 % confluence were transfected with plasmids encoding TMEM192$_{SUMO-3xHA}$ and hATP13A2$_{GFP}$, respectively, using Lipofectamine 2000 according to the manufacturer's recommendations. As a control, three additional 10 cm dishes of HEK293 cells were transfected with only the hATP13A2$_{GFP}$ encoding plasmid.

At 48 h post-transfection, each dish of HEK293 cells was carefully washed once with 1.7 mL cold HBK. The washed cells were overlaid once more with 500uL mL cold HBK and dislodged from the tissue culture dish using a cell scraper. Cells transfected with the same plasmids were pooled and collected by centrifugation at 1000 × g for 2 min at 4 °C in microfuge tubes. Following removal of the supernatant by aspiration, the cell pellets were resuspended in 950uL cold HBK and homogenized with 25 strokes in a pre-chilled 2 mL Dounce homogenizer. The homogenates were spun at 1000 × g for 2 min 4 °C in microfuge tubes to remove large debris. The supernatants, containing lysosomes, were transferred to clean chilled microfuge tubes containing 3 mg of anti-HA magnetic beads, pre-washed with HBK. Following a 15 min mixing period by rocking 4 °C, the anti-HA beads were magnetically separated and washed three times with 3 mL ice-cold HBK in 15 mL conical tubes. The washed anti-HA beads were resuspended in 100uL ice-cold HBK and supplemented with 250 units of SUMO protease to elute bound lysosomes. After mixing at 4 °C for 1 h, the anti-HA beads were immobilized with a magnetic rack and the eluate, E1, was transferred to a clean chilled microfuge tube. 100uL of fresh ice-cold HBK was again mixed with the anti-HA beads by gentle pipetting and this second eluate, E2, was pooled with E1. Typically, ~200–300 μg of purified lysosomes (according to A_{280} measurements made using a Nanodrop UV–VIS spectrophotometer) can be obtained using the LysoPure procedure described above.

2.4 Considerations for the purification

a) Anti-HA magnetic beads should not be centrifuged. Always use a magnetic rack to collect the anti-HA beads during the LysoPure procedure.

b) in addition to HEK293 cells other cell lines can also be used for transient transfection (e.g. HeLa, CHO, COS-7). In such cases, it may be important to optimize transfection conditions to ensure the target proteins are expressed in sufficient amounts for downstream studies

c) The amount of hATP13A2$_{GFP}$ protein in the purified lysosomes can be estimated, if desired, using GFP fluorescence.

d) The combined E1 + E2 eluates can be snap frozen in LN_2 and stored at −80°C or immediately used for downstream analysis (for example, by SDS-PAGE or total internal reflection fluorescence (TIRF) microscopy, see Fig. 1).

3. Determination of lysosomal ATP13A2 spermine-dependent ATPase activity using a NADH-coupled ATPase assay

3.1 Equipment

- FlexStation 3 Multi-Mode Microplate Reader (Molecular Devices)
- 384 well UV transparent plates (Greiner Bio-One, Cat. # 781801)
- 20–200 uL 8-channel pipette (VWR, cat. # 89079-948)
- 1-10uL 8-channel pipette (VWR, cat. # 89079-944)

3.2 Materials

- Spermine tetrahydrochloride (Fisher Scientific, cat. # AAJ6306014)
- DTT (GoldBio, cat. # DTT500).

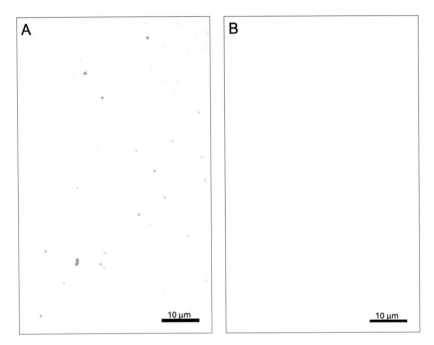

Fig. 1

- pyruvate kinase from rabbit muscle (Sigma Aldrich, cat. # 10128155001)
- β-nicotinamide adenine dinucleotide (NADH) (Sigma Aldrich, cat. # N8129)
- phosphoenolpyruvate (Sigma Aldrich, cat. # P7127)
- Adenosine 5′-triphosphate disodium salt hydrate (ATP) (Sigma Aldrich, cat. # A7699)

3.3 Procedure and expected results

We used a fluorimetric NADH-coupled assay to determine the polyamine-dependent ATPase activity of ATP13A2 associated with purified lysosomes produced using the LysoPure method (Tillinghast et al., 2021). Typically, purified lysosomes in HBK were diluted to a final concentration of 0.1 to 0.4 mg/mL into a reaction mix containing 50 mM Hepes-NaOH pH 7.5, 300 mM NaCl, 30 mM MgCl2, 10 mM DTT, 0.53 mM DDM:CHS (10:2, w/w), 0.041 mM spermine and 1 mM ATP. 0.015 mg/mL pyruvate kinase, 0.032 mg/mL lactate dehydrogenase, 0.33 mM NADH, and 4 mM phosphoenolpyruvate. Spermine is the preferred transport substrate of ATP13A2 and is needed to stimulate ATP13A2 ATPase activity. The detergent DDM:CHS is included to grant spermine access to the polyamine binding site located on side of ATP13A2 facing the lysosome lumen.

The coupled enzymatic reactions in this ATPase assay are setup such that ATP concentration is clamped at the starting concentration at the expense of NADH consumption. Under such conditions, the rate of ATP hydrolysis is constant until NADH is near depletion. The expenditure of NADH was monitored kinetically by tracking its fluorescence at $\lambda_{ex} = 340$ nm and $\lambda_{em} = 445$ nm using a FlexStation 3 Multi-Mode Microplate Reader (Molecular Devices). Rates of ATP hydrolysis were calculated by converting fluorescence loss to nmol NADH per minute using known standards of NADH.

Since hATP13A2$_{GFP}$ is co-purified with other lysosomal components in the LysoPure procedure, one should expect additional endogenous ATPase activities to be present (e.g. lysosomal V-type ATPase), as is observed to be the case (Fig. 2). However, under conditions of hATP13A2$_{GFP}$ overexpression, an ATPase activity significantly above the basal level can readily detected when spermine is present (Fig. 2). The amount of spermine-dependent hATP13A2$_{GFP}$ activity present can, therefore, be determined by subtracting the basal (no added spermine) activity component from the total ATPase activity obtained with added spermine. Importantly, the amount of spermine-dependent ATPase activity present, even when hATP13A2$_{GFP}$ is overexpressed, is below the detection limit when TMEM192$_{SUMO-3xHA}$ is omitted. This observation provides validation that enrichment of hATP13A2$_{GFP}$ activity due to non-specific interactions in the LysoPure method is minimal.

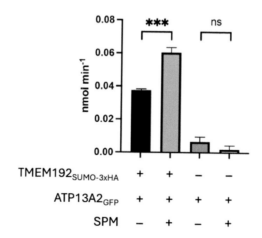

Fig. 2

3.4 Considerations for the ATPase assay

a) The reaction can be modified to setup titration studies, for example, by varying the initial concentration of ATP or spermine. Spermine may be replaced with other polyamines to assess their ability to stimulate ATP13A2 ATPase activity. Additional components including drugs and lipids may also be included in the reaction to assess their impact on ATP13A2 function.

b) The kinetic data can be fit by nonlinear regression to the Michaelis-Menten equation to calculate K_M and V_{max} values using GraphPad Prism or other commonly available software packages.

c) All reagents except for pyruvate kinase and lactate dehydrogenase are stored in aliquots at −20°C. After being thawed, they are stored on ice and used as soon as possible. Avoid re-freezing reagents.

d) We typically setup 25uL NADH-coupled ATPase assays in triplicate, which are read in 384-well UV transparent plates.

e) A common practice is to add ATP or SPM last during the reaction setup. The 384-well plates are immediately placed in the plate reader as soon as the last component is added and mixed using a multichannel pipette.

Acknowledgments

This work was supported by NIGMS grant R01GM145623.

References

Abdulhussein, A. A., & Wallace, H. M. (2014). Polyamines and membrane transporters. *Amino Acids, 46*(3), 655–660. https://doi.org/10.1007/s00726-013-1553-6.

Abu-Remaileh, M., Wyant, G. A., Kim, C., Laqtom, N. N., Abbasi, M., Chan, S. H., & Sabatini, D. M. (2017). Lysosomal metabolomics reveals V-ATPase- and mTOR-dependent regulation of amino acid efflux from lysosomes. *Science, 358*(6364), 807–813. https://doi.org/10.1126/science.aan6298.

Bras, J., Verloes, A., Schneider, S. A., Mole, S. E., & Guerreiro, R. J. (2012). Mutation of the parkinsonism gene ATP13A2 causes neuronal ceroid-lipofuscinosis. *Human Molecular Genetics, 21*(12), 2646–2650. https://doi.org/10.1093/hmg/dds089.

Chen, X., Zhou, M., Zhang, S., Yin, J., Zhang, P., Xuan, X., & Yang, M. (2021). Cryo-EM structures and transport mechanism of human P5B type ATPase ATP13A2. *Cell Discovery, 7*(1), 106. https://doi.org/10.1038/s41421-021-00334-6.

Di Fonzo, A., Chien, H. F., Socal, M., Giraudo, S., Tassorelli, C., Iliceto, G., & Bonifati, V. (2007). ATP13A2 missense mutations in juvenile parkinsonism and young onset Parkinson disease. *Neurology, 68*(19), 1557–1562. https://doi.org/10.1212/01.wnl.0000260963.08711.08.

Estrada-Cuzcano, A., Martin, S., Chamova, T., Synofzik, M., Timmann, D., Holemans, T., & Schule, R. (2017). Loss-of-function mutations in the ATP13A2/PARK9 gene cause complicated hereditary spastic paraplegia (SPG78). *Brain, 140*(2), 287–305. https://doi.org/10.1093/brain/aww307.

Huang, M., Zhang, W., Chen, H., & Zeng, J. (2020). Targeting polyamine metabolism for control of human viral diseases. *Infection and Drug Resistance, 13*, 4335–4346. https://doi.org/10.2147/IDR.S262024.

Klein, C., & Westenberger, A. (2012). Genetics of Parkinson's disease. *Cold Spring Harbor Perspectives in Medicine, 2*(1), a008888. https://doi.org/10.1101/cshperspect.a008888.

Li, J., Meng, Y., Wu, X., & Sun, Y. (2020). Polyamines and related signaling pathways in cancer. *Cancer Cell International, 20*(1), 539. https://doi.org/10.1186/s12935-020-01545-9.

Li, P., Wang, K., Salustros, N., Gronberg, C., & Gourdon, P. (2021). Structure and transport mechanism of P5B-ATPases. *Nature Communications, 12*(1), 3973. https://doi.org/10.1038/s41467-021-24148-y.

Lin, C. H., Tan, E. K., Chen, M. L., Tan, L. C., Lim, H. Q., Chen, G. S., & Wu, R. M. (2008). Novel ATP13A2 variant associated with Parkinson disease in Taiwan and Singapore. *Neurology, 71*(21), 1727–1732. https://doi.org/10.1212/01.wnl.0000335167.72412.68.

Lunati, A., Lesage, S., & Brice, A. (2018). The genetic landscape of Parkinson's disease. *Revue Neurologique (Paris), 174*(9), 628–643. https://doi.org/10.1016/j.neurol.2018.08.004.

Madeo, F., Eisenberg, T., Pietrocola, F., & Kroemer, G. (2018). Spermidine in health and disease. *Science, 359*(6374), https://doi.org/10.1126/science.aan2788.

Michael, A. J. (2016). Polyamines in eukaryotes, bacteria, and archaea. *Journal of Biological Chemistry, 291*(29), 14896–14903. https://doi.org/10.1074/jbc.R116.734780.

Miller-Fleming, L., Olin-Sandoval, V., Campbell, K., & Ralser, M. (2015). Remaining mysteries of molecular biology: The role of polyamines in the cell. *Journal of Molecular Biology, 427*(21), 3389–3406. https://doi.org/10.1016/j.jmb.2015.06.020.

Minois, N., Carmona-Gutierrez, D., & Madeo, F. (2011). Polyamines in aging and disease. *Aging (Albany NY), 3*(8), 716–732. https://doi.org/10.18632/aging.100361.

Mu, J., Xue, C., Fu, L., Yu, Z., Nie, M., Wu, M., & Liu, Z. (2023). Conformational cycle of human polyamine transporter ATP13A2. *Nature Communications, 14*(1), 1978. https://doi.org/10.1038/s41467-023-37741-0.

Pegg, A. E., & Casero, R. A., Jr. (2011). Current status of the polyamine research field. *Methods in Molecular Biology, 720*, 3–35. https://doi.org/10.1007/978-1-61779-034-8_1.

Ramirez, A., Heimbach, A., Grundemann, J., Stiller, B., Hampshire, D., Cid, L. P., & Kubisch, C. (2006). Hereditary parkinsonism with dementia is caused by mutations in ATP13A2, encoding a lysosomal type 5 P-type ATPase. *Nature Genetics, 38*(10), 1184–1191. https://doi.org/10.1038/ng1884.

Seiler, N. (2003). Thirty years of polyamine-related approaches to cancer therapy. Retrospect and prospect. Part 2. Structural analogues and derivatives. *Current Drug Targets, 4*(7), 565–585. https://doi.org/10.2174/1389450033490876.

Sim, S. I., von Bulow, S., Hummer, G., & Park, E. (2021). Structural basis of polyamine transport by human ATP13A2 (PARK9). *Molecular Cell, 81*(22), 4635–4649.e4638. https://doi.org/10.1016/j.molcel.2021.08.017.

Spataro, R., Kousi, M., Farhan, S. M. K., Willer, J. R., Ross, J. P., Dion, P. A., & Katsanis, N. (2019). Mutations in ATP13A2 (PARK9) are associated with an amyotrophic lateral sclerosis-like phenotype, implicating this locus in further phenotypic expansion. *Human Genomics, 13*(1), 19. https://doi.org/10.1186/s40246-019-0203-9.

Tillinghast, J., Drury, S., Bowser, D., Benn, A., & Lee, K. P. K. (2021). Structural mechanisms for gating and ion selectivity of the human polyamine transporter ATP13A2. *Molecular Cell, 81*(22), 4650–4662.e4654. https://doi.org/10.1016/j.molcel.2021.10.002.

Tomita, A., Daiho, T., Kusakizako, T., Yamashita, K., Ogasawara, S., Murata, T., & Nureki, O. (2021). Cryo-EM reveals mechanistic insights into lipid-facilitated polyamine export by human ATP13A2. *Molecular Cell, 81*(23), 4799–4809.e4795. https://doi.org/10.1016/j.molcel.2021.11.001.

Usenovic, M., & Krainc, D. (2012). Lysosomal dysfunction in neurodegeneration: The role of ATP13A2/PARK9. *Autophagy, 8*(6), 987–988. https://doi.org/10.4161/auto.20256.

Van Veen, S., Martin, S., Schuermans, M., & Vangheluwe, P. (2021). Polyamine transport assay using reconstituted yeast membranes. *Bio-Protocol, 11*(2), e3888. https://doi.org/10.21769/BioProtoc.3888.

van Veen, S., Martin, S., Van den Haute, C., Benoy, V., Lyons, J., Vanhoutte, R., & Vangheluwe, P. (2020). ATP13A2 deficiency disrupts lysosomal polyamine export. *Nature, 578*(7795), 419–424. https://doi.org/10.1038/s41586-020-1968-7.

Vrijsen, S., Besora-Casals, L., van Veen, S., Zielich, J., Van den Haute, C., Hamouda, N. N., & Vangheluwe, P. (2020). ATP13A2-mediated endo-lysosomal polyamine export counters mitochondrial oxidative stress. *Proceedings of the National Academy of Sciences of the United States of America, 117*(49), 31198–31207. https://doi.org/10.1073/pnas.1922342117.

CHAPTER ELEVEN

Analysis of translational regulation using polysome profiling and puromycin incorporation

Quinton W. Wood[a] and Teresa L. Mastracci[a,b,c,]*

[a]Department of Biology, Indiana University Indianapolis, Indianapolis, IN, United States
[b]Department of Biochemistry and Molecular Biology, Indiana University School of Medicine, Indianapolis, IN, United States
[c]Center for Diabetes and Metabolic Diseases, Indiana University School of Medicine, Indianapolis, IN, United States
*Corresponding author. e-mail address: tmastrac@iu.edu

Contents

1. Introduction	212
2. Before you begin	215
3. Key resources table	216
4. Materials and equipment	217
5. Step-by-step method details	217
5.1 Preparing the sucrose gradient	217
5.2 Preparing cell and tissue samples (lysis)	218
5.3 Ultracentrifugation of samples through sucrose gradient	219
5.4 Collection of polysome profiles	219
6. Expected outcomes	221
7. Quantification and statistical analysis	222
8. Advantages	222
9. Limitations	222
10. Optimization and troubleshooting	223
10.1 Improperly layered sucrose gradient	223
10.2 Potential solution for an improperly layered sucrose gradient	223
10.3 Insufficient cycloheximide treatment of cells and tissue	223
10.4 Potential solution for an insufficient cycloheximide treatment of cells and tissue	223
11. Safety considerations and standards	224
12. Before you begin	224
13. Key resources table	224
14. Materials and equipment	225
15. Step-by-step method details	225
15.1 Dissecting sample of interest	225
15.2 Puromycin treatment	226

Methods in Enzymology, Volume 715
ISSN 0076-6879, https://doi.org/10.1016/bs.mie.2025.01.035
Copyright © 2025 Elsevier Inc. All rights are reserved, including those for text and data mining, AI training, and similar technologies.

15.3 Preparation of lysate for western	226
15.4 Protein quantification and western blot	226
16. Expected outcomes	226
17. Quantification and statistical analysis	227
18. Advantages	227
19. Limitations	228
20. Optimization and troubleshooting	228
20.1 Insufficient puromycin treatment of samples	228
20.2 Potential solution to insufficient puromycin treatment of samples	228
21. Safety considerations and standards	228
22. Conclusion	228
Acknowledgments	229
Competing interests	229
References	229

Abstract

Translation is the process of decoding an mRNA transcript to permit the synthesis of a protein. This process occurs in three steps: initiation, elongation, and termination. Each step of translation is regulated by translation factors. By regulating translation, the quantity and quality of proteins can be controlled. When translation becomes dysfunctional, disease can ensue, making translational regulation an important avenue of research. Polysome profiling and puromycin incorporation are experimental techniques used in concert to analyze the translational state of cells or tissues. Polysome profiling evaluates the state of translation by quantifying mRNAs based on the abundance of associated ribosomes. Puromycin incorporation measures the amount of newly synthesized protein. Together these methodologies can decipher stark and subtle changes in the rate and efficiency of translation, and provide the opportunity to dissect alterations to the translation of specific transcripts.

1. Introduction

For a protein to be made, the cell must initiate mRNA translation. Translation is the process of decoding mRNAs to synthesize proteins. This process is cyclic and includes four steps: initiation, elongation, termination, and ribosome recycling (Fig. 1). Like all essential processes in a cell, mRNA translation is regulated, which permits the cell to control the rate and efficiency of protein synthesis (Hershey, Sonenberg, & Mathews, 2012). Therefore, protein synthesis is regulated at all stages of mRNA translation, which occurs as a result of specialized factors binding to sequences in the transcript to enhance or inhibit polypeptide formation (Reynolds, 2002; Sonenberg & Hinnebusch, 2009). Translation can also be regulated globally

Analysis of translational regulation using polysome profiling and puromycin incorporation   213

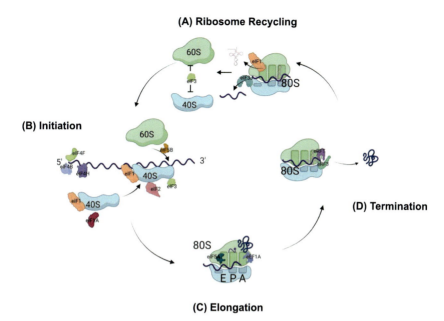

Fig. 1 An overview of mRNA translation. (A) In ribosomal recycling, euakryotic initiation factors (eIFs; eIF1 and eIF3j) dissociate the ribosomes resulting in the release of the 60S ribosomes, 40S ribosomes, tRNA, and mRNA. Reassosciation is prevented by eIF3. (B) In initiation, eIFs (eIF1, eIF1A, eIF4F, eIF4B, eIF4H, eIF2, eIF3, and eIF5B) aid in unraveling the 5′ end of mRNA, binding of the 40S subunit, scanning for a start codon, and binding of the 60S subunit. (C) In elongation, eukaryotic elongation factors (eEFs; eEF1A) assisted in certain circumstances by eIF5A, aid in binding of peptides and positioning of tRNAs and amino acids. (D) In termination, eukaryotic release factors (eRF1 and eRF3) facilitate the release of the polypeptide.

through activation or inhibition of the translation machinery (Gebauer, Preiss, & Hentze, 2012). The translation machinery consists of ribosomes, tRNAs, and translation factors. Translation factors, including the eukaryotic initiation factors (eIFs), the eukaryotic elongation factors (eEFs), and the eukaryotic release factors (eRFs), are central to translation control as these factors can alter the quantitative and qualitative yield of proteins (Yuan, Zhou, & Xu, 2024). Translation can be impaired in certain settings including disease. Therefore, techniques such as polysome profiling and puromycin incorporation can be used to analyze the rate and efficiency of mRNA translation. The polysome profiling technique quantifies mRNAs based on the abundance of associated ribosomes. Additionally, puromycin incorporation measures the amount of newly synthesized protein.

The process of mRNA translation begins with initiation. Initiation involves the proper assembly of ribosomes on an mRNA thus permitting the formation of the first peptide bond (Marintchev & Wagner, 2004). This process begins when post-termination ribosomal complexes are dissociated from 80S ribosomes to 60S subunits and 40S subunits. This dissociation is tightly regulated by initiation factors, eukaryotic initiation factor 1 (eIF1) and eukaryotic initiation factor 3 subunit J (eIF3j), which promote the dissociation of tRNA and mRNA from the 40S subunit, respectively. Eukaryotic initiation factor 3 (eIF3) prevents the reassociation of the free 40S subunit with free 60S subunit (Jackson, Hellen, & Pestova, 2010) (Fig. 1A). The binding of eIF1 and eukaryotic initiation factor 1 A (eIF1A) to the 40S subunit opens the mRNA entry channel, which allows the subunit to attach to mRNA. However, this attachment cannot occur until eukaryotic initiation factor 4F (eIF4F) and eukaryotic initiation factor 4B (eIF4B) or eukaryotic initiation factor 4 H (eIF4H) unwind the 5' cap of mRNA to prepare it for ribosomal attachment (Jackson et al., 2010) (Fig. 1B). The 40S subunit then scans the mRNA, along with many initiation factors including eIF1, eukaryotic initiation factor 2 (eIF2), and eukaryotic initiation factor 3 (eIF3), until an initiation codon is encountered (Jackson et al., 2010). Once encountered, a 60S subunit joins the 40S subunit, mediated by eukaryotic initiation factor 5B (eIF5B), creating an 80S ribosomal complex that can begin forming peptide bonds between amino acids (Fig. 1B). In general, translation initiation requires many "initiation factors" and the availability and/or efficiency of these factors can be controlled by stimuli such as hormones or growth factors (Roux & Topisirovic, 2018).

After completion of translation initiation, the stage of translation elongation will decode the mRNA transcript and add further amino acids to the polypeptide chain growing at the P-site (Xie et al., 2019). The elongating aminoacyl-tRNA is delivered to the A-site and eukaryotic elongation factor 1 A (eEF1A) assists in its binding (Dever, Dinman, & Green, 2018) (Fig. 1C). Peptide bond formation then occurs between an amino acid in the aminoacyl-tRNA and an amino acid in the peptidyl-tRNA in the P-site. In specialized circumstances, the translation factor eukaryotic initiation factor 5 A (eIF5A), which is bound in the E-site, can assist with peptide bond formation when challenging stretches of sequence are being decoded (Dever et al., 2018). Following peptide bond formation, the bound amino acid in the P-site is transferred to the A-site to form an extended peptidyl-tRNA. This process continues, adding amino acids to the peptide chain until a stop codon is encountered. Elongation factors such as eEF1A and eukaryotic elongation factor 2 (eEF2) ensure proper binding and positioning of different elements of translation elongation.

When a stop codon is encountered, the final stage of termination begins. The stop codon is decoded at the A-site which subsequently induces a conformational change in the A-site allowing for release factors to bind (Nakamura, Ito, & Isaksson, 1996). Once the release factor eukaryotic release factor 1 (eRF1) is bound to the stop codon, it is able to promote the binding of GTP to eukaryotic release factor 3 (eRF3) (Hellen, 2018). As a result of the binding of eRF3 to GTP, eRF1 adopts a conformation that activates peptide release, while eRF3 stops codon reading (Hellen, 2018) (Fig. 1D).

Many human diseases have been linked to impaired mRNA translation (Lu, 2022; Tahmasebi, Khoutorsky, Mathews, & Sonenberg, 2018). For example, mutations identified in eEF2 result in spinocerebellar ataxia, eukaryotic elongation factor 1A2 (eEF1A2) in epileptic encephalopathy, eukaryotic elongation factor 1B2 (eEF1B2) in non-syndromic intellectual disability, eukaryotic initiation factor 2B (eIF2B) in leukoencephalopathy with vanishing white matter, and eIF5A in craniofacial-neurodevelopmental disorder (Faundes et al., 2021; Tahmasebi et al., 2018). Dysfunction of translational regulation is also a common mechanism in tumor development (Ruggero, 2013). Given the link with human disease, the study of translation has become critically important for therapeutic development. As such, the creation of experimental techniques that can analyze the state of translation in cells or tissues is necessary to advance this research. Described herein are two methodologies: polysome profiling and puromycin incorporation. These techniques can identify impairments in translation as well as changes to overall protein synthesis.

METHOD: Polysome Profiling.

2. Before you begin

1. 10% Sucrose Solution
    a) The night before polysome profiling, in a 50 mL Falcon tube add 5 g of sucrose (Fisher, BP2201).
    b) Add RNase free water (Fisher, 10-977-015) up to the 40 mL marker.
    c) Add 1 mL 1 M Tris-HCl pH 7.5 (Fisher, BP152-1; Fisher, SA49), 1 mL 5 M NaCl (ThermoFisher, 327300010), 250 μL 1 M MgCl₂ (Fisher, BP214-500), and 7.5 mL RNase free water (Fisher, 10-977-015).
    d) Gently mix the solution at 4 °C on a rocking/nutating shaker overnight.

2. 50% Sucrose Solution
   a) The night before polysome profiling, in a 50 mL Falcon tube add 25 g of sucrose (Fisher, BP2201).
   b) Add RNase free water (Fisher, 10-977-015) up to the 40 mL marker.
   c) Add 1 mL 1 M Tris-HCl pH 7.5 (Fisher, BP152-1; Fisher, SA49), 1 mL 5 M NaCl (ThermoFisher, 327300010), 250 µL 1 M MgCl$_2$ (Fisher, BP214-500), and 7.5 mL RNase free water (Fisher, 10-977-015).
   d) Gently mix the solution at 4 °C on a rocking/nutating shaker overnight.
3. Lysis Buffer
   a) The morning of polysome profiling in a 50 mL Falcon tube combine 46.5 mL RNase free water (Fisher, 10-977-015), 500 µL Triton X-100 (Fisher, BP151-500), 500 µL 10% deoxycholate (Fisher, BP349-100), 1 mL 1 M Tris-HCl pH 7.5 (Fisher, BP152-1; Fisher, SA49), 1 mL 5 M NaCl (ThermoFisher, 327300010), 500 µL 1 M MgCl$_2$ (Fisher, BP214-500), and a Protease Inhibitor Cocktail Tablet (Roche, 1183617001).
4. PBS with cycloheximide wash solution
   a) The morning of polysome profiling, in a 50 mL Falcon tube, combine 25 µL of cycloheximide (Sigma, C4859-mL) and 50 mL of PBS.
   Note: Cycloheximide is an inhibitor of eukaryotic translation elongation, used to freeze ribosomes on the RNA in their *in vivo* positional distribution.

3. Key resources table

REAGENT or RESOURCE	SOURCE	IDENTIFIER
Chemicals, Peptides, and Recombinant Proteins		
Cycloheximide	Sigma	C4859-mL
HCl	Fisher	SA49
Deoxycholic acid	Fisher	BP349-100
MgCl$_2$	Fisher	BP214-500
NaCl	Fisher	327300010
Protease Inhibitor Cocktail	Roche	1183617001
RNase free water	Fisher	10-977-015

RNasin	Fisher	PR-N2515
Sucrose	Fisher	BP2201
Tris	Fisher	BP152-1
Triton X-100	Fisher	BP151-500

4. Materials and equipment

- Gradient Station Base Unit (Biocomp Instruments, part #153)
- Gradient Forming Attachments (Biocomp Instruments, part #105-914A-IR)
- Fractionating Tube Mount (Biocomp Instruments, part #151-114 A)
- Gilson Fraction FC-203B Collector #171011 G (Biocomp Instruments, part #153-203)
- Gilson Fraction FC-203B Rack (Biocomp Instruments, part #151-203-1.5E)
- Triax flow cell for dual UV (260 nm) and fluorescence scans (FC Software included) (Biocomp Instruments, part FC-2-UV/VIS-260)
- Set of LED, Ex filter, 2 Em filters mounted on Hamamatsu PDs for mCherry detection (Biocomp Instruments, part #FC-2-FK-mCherry)
- Polyclear™ Open Top UltracentrifugeTubes for SW41 centrifuge rotor (Biocomp Instruments, part#151-514 A) or 14 x 89mm Centrifuge Tubes (Beckman Coulter, #331372)
- SW41 Ti Swinging-Bucket Ultracentrifuge Rotor (Beckman Coulter, part #331362)
- Eppendorf Centrifuge (Eppendorf, 5420000245)
- Rubber stoppers (MilliporeSigma, Z164437)
- Blunt ended 14 G needle, 20 mL syringe (MilliporeSigma, CAD7941)
- Bubble level (ThermoFisher, BBL-1)
- 25 G needles (Fisher, 14-826-88)

5. Step-by-step method details
5.1 Preparing the sucrose gradient

1. To the 10% and 50% sucrose solutions made the night before, as described in 'before you begin', add 50 µL of 50 mg/mL Cycloheximide (Sigma, C4859-mL) to each.

2. Prepare one centrifuge tube for each sample to be analyzed. Place each tube in a rack and mark the midline of the centrifuge tube with a permanent marker.
3. Using a serological pipet, add 10% sucrose solution to the marked midline of each centrifuge tube; add this sucrose slowly down the side of the tube.
4. Using a blunt ended 14 G needle attached to a 20 mL syringe, slowly pull 50% sucrose solution into the syringe without introducing bubbles. Push a small droplet of 50% sucrose solution out of the needle to ensure there is no air in the needle. Touch the end of the needle to a paper towel to remove excess solution.
5. Add the 50% sucrose solution to centrifuge tubes already filled to the midline with 10% sucrose solution. Do this by quickly pushing the needle to the bottom of the centrifuge tube while keeping the needle along the wall of the tube. Slowly add the 50% sucrose solution to the bottom of the tube until the tube begins to overflow.

   Note: There should be no visible air bubbles in the sucrose-filled centrifuge tube.
6. Cap each sucrose-filled centrifuge tube with a rubber stopper.
7. Ensure the platform on the Gradient Master is perfectly level using a bubble level.
8. Place the tubes in the tube holder on the Gradient Master and initiate the machine to create the gradient.

5.2 Preparing cell and tissue samples (lysis)

1. Precool Eppendorf centrifuge to 4 °C for spin at step 9.
2. Harvest cells or tissue of interest.

   Note: Both cells and tissue can be used for this method; however, the protocol must be optimized for that specific cell/tissue type as abundance and cellular content vary. Embryonic day 14.5 mouse pancreas tissue, 3-week-old mouse brain tissue and human-derived lymphoblast cell lines are examples of samples that have been analyzed with this method (Padgett et al., 2021; Padgett et al., 2023).
3. For cells, add cycloheximide to the media in the culture dish for a final concentration of 50 µg/mL, then incubate for 10 min at 37 °C. For tissue, the amount of cycloheximide should be adjusted based on the size of the tissue and amount of culture media within which it will be incubated.
4. For cells and tissue, aspirate media from dishes and wash the cells/tissue twice with PBS with cycloheximide wash solution (50 µg/mL), as described in 'before you begin'.

5. For tissues, following treatment and washes, mince the tissue and transfer to a 10 cm plate.
6. To prepare the final lysis buffer, add 50 μL of 50 mg/mL of cycloheximide and 62.5 μL of RNasin (Fisher, PR-N2515) to the Lysis Buffer, previously described in 'before you begin'.
7. Add 500 μL of lysis buffer to each 10 cm plate, scrape cells/tissue to one side of the plate and add the lysed cell or tissue solution to a pre-chilled 1.5 mL Eppendorf tube.
8. Shear each lysed sample 7 times through a 25 G needle.
9. Vortex each tube and then incubate on ice for 10 min.
10. Centrifuge each sample for 10 min at 10,000 rpm at 4 °C.
11. Pour off supernatant into a new pre-chilled 1.5 mL Eppendorf tube.

5.3 Ultracentrifugation of samples through sucrose gradient

1. Remove 300 μL of sucrose solution from the top of each sucrose gradient centrifuge tube.
2. Add 400 μL of purified cell/tissue lysate to the top of the sucrose gradient.
3. To extract RNA from the input, add RNA purification reagents to the remaining supernatant. This will be the polysome profiling 'input'.
   Note: An RNA purification kit can be used for this step.
4. Place each centrifuge tube loaded with a cell/tissue lysate into a swinging bucket in the SW41 ultracentrifuge rotor. Ensure the rotor is balanced and that tubes are securely hung.
5. Centrifuge the tubes at 40,000 rpm for 2 h at 4 °C in the ultracentrifuge.
6. At the end of the spin, remove each swinging bucket from the rotor and place it gently on a rotor rack. The tubes must be handled very gently hereon as the sucrose gradient can be disturbed if the tubes are tipped or moved in a rigorous way.
7. Store the centrifuged sample sucrose gradient tubes at 4 °C; remove one at a time for analysis.

5.4 Collection of polysome profiles

1. Turn on the computer, UV monitor, Gradient Station and Profiler.
2. Perform a test run of the fraction collector.
   a. First wash the fraction collection tubing by filling a test centrifuge tube with milliQ water, mounting the tube on the Gradient Station and bringing the piston down to the bottom of the tube to allow it to flush the milliQ water through the line.

b. Refill the centrifuge tube with milliQ water and repeat with 3 more washes.
   c. Use the Gradient Station air feature to clear the line of water.
   d. Fill a centrifuge tube with milliQ water to the same volume as the sucrose gradients and position the piston at the top of the water level and begin to flow the water into the line to prime it.
   e. Use the Gradient Station RINSE function and then auto zero the UV monitor to set up the machine to collect.
   f. Put 1.5 mL Eppendorf tubes into the fraction collector to collect milliQ water "fractions" as a simulation of fraction collection.
   g. Initiate the profile and fraction collection engaging all components of the Gradient Station and fraction collector.
   h. Ensure that fractions are properly collected into the Eppendorf tubes and that the test profile can be saved on the computer.
3. If the test run performs correctly, begin sample collection.
4. To begin sample collection, remove one of the sample centrifuged sucrose gradient tubes from the 4°C, mount it on the Gradient Station and lock the cap in place.
5. Bring the piston down to the top of the sucrose gradient similarly to the test collection so that the piston begins to flow sample through the line to prime it.
6. Load 1.5 mL Eppendorf tubes on to the fraction collector, and make sure the rotation wheel is in place.
7. Double check that the computer and the UV monitor are set to the same wavelength.
8. Initiate the profile and fraction collection engaging all components of the Gradient Station and fraction collector.
9. Once the last fraction is collected, the profile will be complete. Save this to the computer.
10. To save each fraction for RNA extraction later, combine the fraction with RNA preservation reagents and store samples at −20 °C.
    Note: An RNA purification kit can be used for RNA preservation.
11. The Gradient Station can now be set up for analysis of the next sample. Be sure to wash the tubing thoroughly with milliQ water.
12. Once the equipment is washed, the next sample can be analyzed. Return to Step 4 above. Repeat until all samples are analyzed.

6. Expected outcomes

Centrifugation of a lysate through the sucrose gradient stratifies mRNA transcripts based on the density of the associated ribosomes, with monoribosome-containing mRNAs at the top followed by polyribosome-containing transcripts, which sediment lower due to increasing ribosome density (Fig. 2A). Measuring absorbance (at 254 nm) from the top to the bottom of the sucrose gradient then generates a polysome profile graph, which reveals distinct peaks that represent the abundance of monoribosome- and polyribosome-containing transcripts (Fig. 2B).

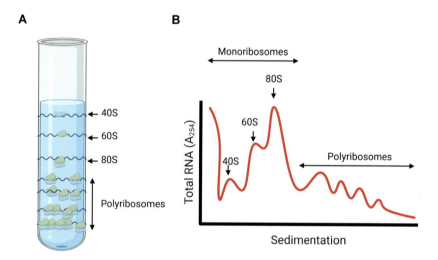

Fig. 2 Polysome profiling. (A) A sucrose gradient is created by mixing 10 % and 50 % sucrose solutions. Centrifugation of a lysate through the sucrose gradient then stratifies the mRNAs from the sample based on the density of ribosomes on each transcript. Monoribosome-containing transcripts are lighter and at the top. Transcripts with 2, 3, 4, etc ribosomes (polyribosome-containing transcripts) will sediment in subsequently lower fractions. (B) A polysome profile is generated as the outcome of the experiment. The y-axis of this graph displays the quantity of total RNA (as measured by absorbance; A_{254}) and the x-axis shows sedimentation stratified by the fractions collected. Distinct peaks represent the abundance of mRNAs that are associated with monoribosome and increasing numbers of polyribosomes.

7. Quantification and statistical analysis

Once the data is obtained from polysome profiling, it should be quantified by calculating the area under the curve of the monoribosome peaks and the area under the curve of the polyribosomes peaks, which will then permit the calculation of a polysome:monosome ratio. If there are two groups, then a Student's *t*-test can be used for statistical analysis to determine significance. The polysome:monosome ratio will identify if some part of the process of translation is impaired (Padgett et al., 2021). Additionally, if a baseline profile is set with a control sample, the distinct peaks formed by the monoribosomes can be analyzed to infer efficiency of translation initiation (Saini, Eyler, Green, & Dever, 2009). To determine changes to specific transcripts, RNA can be extracted from each individual fraction and analyzed by RT-PCR. Quantifying the relative abundance of a specific mRNAs in each fraction can identify if that transcript is being actively translated or if its translation was interrupted (Chassé, Boulben, Costache, Cormier, & Morales, 2017).

8. Advantages

Polysome profiling is a great tool to analyze global translation efficiency using the ratio of polysomes:monosomes. Efficiency of specific steps of translation such as initiation, elongation, and termination can be inferred by analysis of distinct monosome and polysome peaks in the profile. Translation of specific mRNAs can also be determined by evaluation of the fractions collected in this protocol. Additionally, the Gradient Master used in this method forms the sucrose gradient, fractionates, and serially measures absorbance, to streamline execution of the protocol.

9. Limitations

The greatest limitation is cell and tissue quantity; more cells or tissue will produce more accurate data (Liang et al., 2018). This technique is also laborious and time consuming, making it an unrealistic method for large studies (Liang et al., 2018). The availability of the specialized equipment can be cost prohibitive and therefore challenging for laboratories with limited resources (Sobhany & Stanley, 2021).

10. Optimization and troubleshooting

10.1 Improperly layered sucrose gradient

If a sucrose gradient is improperly layered, the RNA will not sediment correctly. This can occur when adding the 50% sucrose solution to the 10% sucrose solution or when adding the supernatant to the sucrose gradient. If either of these are performed too quickly, or inaccurately, the gradient may not be properly layered. No air bubbles should be seen when adding the 50% sucrose solution to the 10% sucrose solution or when adding the supernatant to the sucrose gradient.

10.2 Potential solution for an improperly layered sucrose gradient

An improperly layered sucrose gradient will not be noticed until a profile is made. At this point, the sucrose gradient will have to be remade, and special attention should be taken when combining the 10% and 50% sucrose solutions, and when adding the supernatant.

10.3 Insufficient cycloheximide treatment of cells and tissue

The optimal dose and incubation time of cycloheximide can vary from cell type to cell type and from tissue to tissue. Cycloheximide freezes the ribosomes on the RNA in their *in vivo* positional distribution, and if this does not occur the data obtained from polysome profiling may not reflect the true state of translation within your sample. Additionally, too high a concentration of cycloheximide can result in cell death.

10.4 Potential solution for an insufficient cycloheximide treatment of cells and tissue

Step 3 of 'Preparing Cells and Tissue Samples' should be optimized for cell confluency, concentration of cycloheximide, and incubation time with cycloheximide. To optimize the treatment of cells or tissues with cycloheximide, treat your sample with different concentrations of cycloheximide for different lengths of time to identify the dose that can be used without degrading your sample.

11. Safety considerations and standards

- Cycloheximide, Hydrochloric acid, and Deoxycholic acid are hazardous and should be used in a chemical fume hood with eye protection and gloves. Avoid contact with skin or eyes and inhalation.
  METHOD: Puromycin Incorporation.

12. Before you begin

1. Lysis Buffer
   a. The morning of the puromycin incorporation experiment, in a 50 mL falcon tube combine 2.5 mL 1 M Tris pH 8 (Fisher, BP152-1), 1.5 mL 5 M NaCl (Fisher, 327300010), 250 µL 10% deoxycholate (Fisher, BP349-100), 50 µL IGEPAL (NP-40) (Fisher, AAJ61055AP), 500 µL 10% SDS (ThermoFisher, 15553027), 1 mL 10% sarcosyl (MilliporeSigma, L7414-mL), 5 mL glycerol (Fisher, G33-500), 50 µL DTT (Fisher, BP172-5), 100 µL 0.5 M EDTA pH 8 (Fisher, PRV4233), 21 mg NaF (Fisher, S299-500), and 39.05 mL H$_2$O.
2. Cycloheximide wash solution
   a. The morning of the puromycin incorporation experiment, in a 50 mL Falcon tube combine 50 µL of cycloheximide (Sigma, C4859-mL) and 50 mL of PBS.

13. Key resources table

REAGENT or RESOURCE	SOURCE	IDENTIFIER
Antibodies		
Rat Anti-Puromycin	MilliporeSigma	MABE341
Chemicals, Peptides, and Recombinant Proteins		
Benzonase	MilliporeSigma	70746-3
Cycloheximide	Sigma	C4859-mL
Deoxycholic acid	Fisher	BP349-100
DTT	Fisher	BP172-5

EDTA	Fisher	PRV4233
FBS	HyClone	SH3091003
Glycerol	Fisher	G33-500
IGEPAL (NP-40)	Fisher	AAJ61055AP
MgCl$_2$	Fisher	BP214-500
NaCl	Fisher	327300010
NaF	Fisher	S299-500
Phosphatase Inhibitor Cocktail	Roche	4906845001
Protease Inhibitor Cocktail	Roche	1183617001
Puromycin	MilliporeSigma	P8833
RPMI 1640 media	Fisher	11-875-119
Sarcosyl	MilliporeSigma	L7414-mL
SDS	ThermoFisher	15553027
Tris	Fisher	BP152-1
Critical Commercial Assays		
DC Protein Assay Kit II	BioRad	5000112
Software and Algorithms		
Image Studio Software	LI-COR Biosciences	

14. Materials and equipment

- Handheld Homogenizer (Grainger, 39J349)
- Insulin syringe (Fisher, 14-826-79)
- Eppendorf Centrifuge (Eppendorf, 5420000245)
- DC Protein Assay Kit II (Bio-Rad, 5000112)

15. Step-by-step method details
15.1 Dissecting sample of interest

1. Harvest tissue of interest.

    Note: This protocol can also be performed using cell lines or primary cells grown in culture.

2. Place tissue into a 12-well plate containing cold RPMI 1640 media (Fisher, 11-875-119) supplemented with 10% FBS (HyClone, SH3091003).
3. Incubate at 37 °C for 30 min. Incubation time should be optimized for type and quantity of tissue.

15.2 Puromycin treatment
1. Remove plate from the incubator and add Puromycin (MilliporeSigma, P8833) at a final concentration of 20 μg/mL.
2. Incubate for 20 min at 37 °C.
3. Remove plate from incubator and carefully remove media using a pipet.
4. Wash each tissue sample twice with 1 mL of ice-cold PBS containing 100 μg/mL cycloheximide, as described in 'before you begin'.
5. Use clean forceps to move the explant to a 1.5 mL Eppendorf tube.
6. Tissue can now be snap frozen or immediately prepared for a western blot.

15.3 Preparation of lysate for western
1. To the lysis buffer previously made, as described in 'before you begin', add Protease inhibitors (Roche, 11836170001), Phosphatase inhibitors (Roche, 4906845001), 100 μL $MgCl_2$ (Fisher, BP214-500), and 5 μL Benzonase (MilliporeSigma, 70746-3).
2. Homogenize tissue in lysis buffer using a handheld homogenizer followed by shearing with an insulin syringe (10 times). The volume of lysis buffer should be scaled for the amount of tissue.
3. Vortex the tube containing the sample and let it sit on ice for 10 min.
4. Spin in an Eppendorf Centrifuge (Eppendorf, 5420000245) at 4 °C for 10 min at 13,000 rpm.
5. If there is a pellet – transfer supernatant to a new tube and freeze at −80 °C or immediately proceed to protein quantification and western blot.

15.4 Protein quantification and western blot
1. Quantify the amount of protein in the sample.
   Note: The DC Protein Assay Kit II (Bio-Rad, #5000112) can be used for this step.
2. Perform a western blot using Rat anti-puromycin primary antibody (MilliporeSigma, MABE341).

16. Expected outcomes
Following puromycin treatment, the tissue sample is homogenized and a western blot is performed (Fig. 3A–C). As depicted in (Padgett et al., 2021) the resulting western blot using an anti-puromycin primary antibody

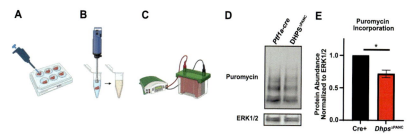

Fig. 3 Puromycin incorporation. (A) The tissue of interest is treated with puromycin. (B) Following treatment, the tissue is homogenized. (C) A western blot is performed to measure the abundance of newly synthesized proteins. (D) Adapted from Padgett et al. 2021: an example of a western blot probed for puromycin and a control protein (ERK1/2), which facilitates normalization of the blot. Densitometry is then used to quantify protein abundance. (E) The resulting quantification graph identifies a reduction in the amount of newly synthesized protein in the experimental sample compared with the control sample.

will show the expression of puromycin and therefore the abundance of newly synthesized proteins. Immunoblot for a control protein is also necessary to facilitate normalization (Fig. 3D). Densitometric quantification of puromycin and the control then permits the evaluation of relative abundance of protein to determine if an experimental sample is altered compared with a control (Fig. 3E).

17. Quantification and statistical analysis

Fluorescently labeled secondary antibodies can be quantified by densitometry using Image Studio Software (LI-COR Biosciences); however, densitometric analysis can also be performed with colorimetric readouts. Quantified protein abundance should be graphed and statistically analyzed using software. Identifying the abundance of newly synthesized protein between a control and experimental tissue can clarify if, for example, an increase in the polyribosomes represents an increase in protein synthesis or a block in translation leading to a buildup of mRNA with many stalled ribosomes.

18. Advantages

Other techniques to study protein synthesis involve radiolabeling amino acids and measuring protein synthesis by quantifying radioactivity (Marciano, Leprivier, & Rotblat, 2018). An advantage to the puromycin incorporation technique described herein is that protein synthesis can be measured by densitometry of an immunoblot without the use of radiolabels, making this technique much safer.

19. Limitations

Puromycin Incorporation gives a broad overview of protein synthesis for an entire sample. However, if the abundance of the sample is limiting it is possible a western blot cannot be accurately performed.

20. Optimization and troubleshooting

20.1 Insufficient puromycin treatment of samples

The optimal dose and incubation time of puromycin can vary from sample to sample. Puromycin inhibits protein synthesis by incorporating into elongating amino acid chains resulting in premature translation termination, and if puromycin is not fully incorporated into your sample, the data obtained may not reflect the true state of protein synthesis within your sample. Additionally, puromycin can result in cell death if used at too high a concentration.

20.2 Potential solution to insufficient puromycin treatment of samples

Step 1 and 2 of 'Puromycin Treatment' should be optimized for concentration of puromycin, and incubation time with puromycin. Test the incubation of tissue/cells with different concentrations of puromycin for varying lengths of time to identify the optimal dose.

21. Safety considerations and standards

- Cycloheximide, Deoxycholic acid, and EDTA are hazardous and should be worked with in a chemical fume hood with eye protection and gloves. Avoid contact to skin or eyes and inhalation.

22. Conclusion

Translation can be analyzed using the techniques polysome profiling and puromycin incorporation. Polysome Profiling can be used to gather data on translation. This technique sediments mRNAs, from lysed cells/tissues, in a sucrose gradient based on the abundance of bound ribosomes. The subsequent analysis by absorbance quantifies the abundance of mRNAs in each fraction. The resultant "polysome profile" can be analyzed for a polysome:monosome ratio, which can identify changes in translation. Puromycin Incorporation can be used to quantify newly synthesized proteins. Puromycin acts as a molecular mimic of an aminoacyl-tRNA, allowing it to be incorporated into growing polypeptide chains during translation. This will effectively label nascent proteins, which can then be detected using western blot and antibodies specific to puromycin. This technique provides a measure of the rate of protein synthesis within a cell population or living tissue sample. Coupling polysome profiling with a puromycin incorporation assay provides a means to evaluate changes in mRNA translation.

Acknowledgments
We would like to thank Mollie Shinkle, Catharina Villaca and Danielle Overton for helpful discussions. This work was supported by funding to TLM from the National Institutes of Health (NIH) (R01DK121987). All figures were created in BioRender: Mastracci, T. (2024) https://BioRender.com/y92s151; Mastracci, T. (2024) https://BioRender.com/f09j505; Mastracci, T. (2024) https://BioRender.com/c61o867.

Competing interests
The authors declare no competing interests.

References
Chassé, H., Boulben, S., Costache, V., Cormier, P., & Morales, J. (2017). Analysis of translation using polysome profiling. *Nucleic Acids Research, 45*(3), e15. https://doi.org/10.1093/nar/gkw907.
Dever, T. E., Dinman, J. D., & Green, R. (2018). Translation elongation and recoding in eukaryotes. *Cold Spring Harbor Perspectives in Biology, 10*(8), a032649. https://doi.org/10.1101/cshperspect.a032649.
Faundes, V., Jennings, M. D., Crilly, S., Legraie, S., Withers, S. E., Cuvertino, S., ... Banka, S. (2021). Impaired eIF5A function causes a Mendelian disorder that is partially rescued in model systems by spermidine. *Nature Communications, 12*, 833. https://doi.org/10.1038/s41467-021-21053-2.
Gebauer, F., Preiss, T., & Hentze, M. W. (2012). From cis-regulatory elements to complex RNPs and back. *Cold Spring Harbor Perspectives in Biology, 4*(7), a012245. https://doi.org/10.1101/cshperspect.a012245.
Hellen, C. U. T. (2018). Translation termination and ribosome recycling in eukaryotes. *Cold Spring Harbor Perspectives in Biology, 10*(10), a032656. https://doi.org/10.1101/cshperspect.a032656.
Hershey, J. W. B., Sonenberg, N., & Mathews, M. B. (2012). Principles of translational control: An Overview. *Cold Spring Harbor Perspectives in Biology, 4*(12), a011528. https://doi.org/10.1101/cshperspect.a011528.

Jackson, R. J., Hellen, C. U. T., & Pestova, T. V. (2010). The mechanism of eukaryotic translation initiation and principles of its regulation. *Nature Reviews. Molecular Cell Biology, 11*(2), 113–127. https://doi.org/10.1038/nrm2838.

Liang, S., Bellato, H. M., Lorent, J., Lupinacci, F. C. S., Oertlin, C., van Hoef, V., ... Larsson, O. (2018). Polysome-profiling in small tissue samples. *Nucleic Acids Research, 46*(1), e3. https://doi.org/10.1093/nar/gkx940.

Lu, B. (2022). Translational regulation by ribosome-associated quality control in neurodegenerative disease, cancer, and viral infection. *Frontiers in Cell and Developmental Biology, 10*, 970654. https://doi.org/10.3389/fcell.2022.970654.

Marciano, R., Leprivier, G., & Rotblat, B. (2018). Puromycin labeling does not allow protein synthesis to be measured in energy-starved cells. *Cell Death & Disease, 9*(2), 1–3. https://doi.org/10.1038/s41419-017-0056-x.

Marintchev, A., & Wagner, G. (2004). Translation initiation: Structures, mechanisms and evolution. *Quarterly Reviews of Biophysics, 37*(3–4), 197–284. https://doi.org/10.1017/S0033583505004026.

Nakamura, Y., Ito, K., & Isaksson, L. A. (1996). Emerging understanding of translation termination. *Cell, 87*(2), 147–150. https://doi.org/10.1016/S0092-8674(00)81331-8.

Padgett, L. R., Robertson, M. A., Anderson-Baucum, E. K., Connors, C. T., Wu, W., Mirmira, R. G., & Mastracci, T. L. (2021). Deoxyhypusine synthase, an essential enzyme for hypusine biosynthesis, is required for proper exocrine pancreas development. *The FASEB Journal, 35*(5), e21473. https://doi.org/10.1096/fj.201903177R.

Padgett, L. R., Shinkle, M. R., Rosario, S., Stewart, T. M., Foley, J. R., Casero, R. A., ... Mastracci, T. L. (2023). Deoxyhypusine synthase mutations alter the post-translational modification of eukaryotic initiation factor 5A resulting in impaired human and mouse neural homeostasis. *Human Genetics and Genomics Advances, 4*(3), 100206. https://doi.org/10.1016/j.xhgg.2023.100206.

Reynolds, P. (2002). In sickness and in health: The importance of translational regulation. *Archives of Disease in Childhood, 86*(5), 322–324. https://doi.org/10.1136/adc.86.5.322.

Roux, P. P., & Topisirovic, I. (2018). Signaling pathways involved in the regulation of mRNA translation. *Molecular and Cellular Biology, 38*(12), e00070. https://doi.org/10.1128/MCB.00070-18.

Ruggero, D. (2013). Translational control in cancer etiology. *Cold Spring Harbor Perspectives in Biology, 5*(2), a012336. https://doi.org/10.1101/cshperspect.a012336.

Saini, P., Eyler, D. E., Green, R., & Dever, T. E. (2009). Hypusine-containing protein eIF5A promotes translation elongation. *Nature, 459*(7243), 118–121. https://doi.org/10.1038/nature08034.

Sobhany, M., & Stanley, R. E. (2021). Polysome profiling without gradient makers or fractionation systems. *Journal of Visualized Experiments (JoVE), 172*, e62680. https://doi.org/10.3791/62680.

Sonenberg, N., & Hinnebusch, A. G. (2009). Regulation of translation initiation in eukaryotes: mechanisms and biological targets. *Cell, 136*(4), 731–745. https://doi.org/10.1016/j.cell.2009.01.042.

Tahmasebi, S., Khoutorsky, A., Mathews, M. B., & Sonenberg, N. (2018). Translation deregulation in human disease. *Nature Reviews. Molecular Cell Biology, 19*(12), 791–807. https://doi.org/10.1038/s41580-018-0034-x.

Xie, J., de Souza Alves, V., von der Haar, T., O'Keefe, L., Lenchine, R. V., Jensen, K. B., ... Proud, C. G. (2019). Regulation of the elongation phase of protein synthesis enhances translation accuracy and modulates lifespan. *Current Biology, 29*(5), 737–749.e5. https://doi.org/10.1016/j.cub.2019.01.029.

Yuan, S., Zhou, G., & Xu, G. (2024). Translation machinery: The basis of translational control. *Journal of Genetics and Genomics, 51*(4), 367–378. https://doi.org/10.1016/j.jgg.2023.07.009.

CHAPTER TWELVE

Targeting polyamine metabolism in an ex vivo prostatectomy model

Hayley C. Affronti[a], Aryn M. Rowsam[b], Spencer R. Rosario[c], and Dominic J. Smiraglia[d],*

[a]Department of Biology, Vassar College, Poughkeepsie, NY, United States
[b]Department of Biology, Hartwick College, Oneonta, NY, United States
[c]Department of Biostatistics and Bioinformatics, Roswell Park Comprehensive Cancer Center, Buffalo, NY, United States
[d]Department of Cell Stress Biology, Roswell Park Comprehensive Cancer Center, Buffalo, NY, United States
*Corresponding author. e-mail address: dominic.smiraglia@roswellpark.org

Contents

1. Introduction — 232
2. Materials and reagents — 234
   - 2.1 Key materials (see Table 1) — 234
   - 2.2 Graphical overview (see Fig. 2) — 235
3. Protocols — 236
   - 3.1 Tissue explant culture — 236
   - 3.2 Preparation for metabolic analysis — 236
   - 3.3 Ex Vivo sample metabolomics analysis — 237
   - 3.4 IHC analysis — 238
References — 238

Abstract

Ex vivo models allow for testing drug efficacy and patient response, yet it remains a challenge to develop representative 3D cultures for prostate cancer. Tissue explant models offer a more clinically relevant alternative to organoids due to their ability to provide adequate tissue quantities, maintain tumor-stromal interactions and metabolic activity, and their relatively inexpensive culturing conditions. In this chapter we outline a protocol for culturing patient prostatectomy tumors for up to 7 days on dental sponges soaked in either control or drug containing media for evaluating drug efficacy. Further, we describe the preparation of tissue samples for downstream immunohistochemistry and metabolic analysis. We have tested the efficacy of a combination therapy targeting polyamine metabolism, which is dysregulated in prostate cancer, using this patient tumor explant model. We found that activating polyamine catabolism in combination with inhibition of methionine salvage was effective at inducing target protein expression, reducing intratumoral polyamines, and inducing apoptosis in a majority of the patient samples tested. Additionally, we were able to confirm drug induced effects were specific to the malignant prostate epithelial cells. This *ex vivo* prostatectomy model lends itself to both targeted metabolite analyses as well as more comprehensive metabolomic analyses. This method can be applied to strategies aiming to target metabolic pathways in solid tumor diseases.

1. Introduction

Ex vivo model systems are powerful tools for predicting patient response and evaluating the efficacy of cancer therapies in patient derived samples. However, there are few models that exist, some of which can be very costly and technically challenging to develop and maintain. The most common *ex vivo* models derived from patient samples are cell lines, 3 dimensional (3D) organoids and tissue explants.

Organoids can be derived from embryonic, normal, and tumor tissue. They can be developed in a number of ways including but not limited to matrix based methods, organ-on-a-chip and 3D bioprinting (Lv et al., 2024; Yang et al., 2023). For some tissue types, organoids can maintain a similar histological and molecular phenotype as that of the original tumor (Gao et al., 2014). Additionally, they are excellent for evaluating drug response for a range of doses in hopes of predicting patient response to treatments. Organoids can also be genetically modified to evaluate drug targets and tumor progression. Although very expensive to maintain and culture, these provide a clinically relevant model for studying biology, tumorigenesis and drug treatments. However, for some tissue types such as prostate cancer (CaP), organoids can be difficult to develop that accurately represent the original tumor tissue (Servant et al., 2021).

In contrast to 3D organoids, tissue explant models utilize intact pieces of tissue, maintaining the original architecture of the tumor. With this system, a piece of tissue from a solid tumor is cut up and placed on dental sponges in tissue culture media containing various vehicles and drugs (Centenera, Raj, Knudsen, Tilley, & Butler, 2013). To account for effects of the culture conditions themselves, one piece of tissues is saved immediately (frozen and/or FFPE) with no culturing to compare against the vehicle control culturing. These tissues can be maintained in drug and media for seven days, and immunohistochemistry and biochemical analysis can be performed on the tissue. This represents one advantage over organoid culture because CaP organoids grow very slowly and therefore it is difficult to develop large enough quantities to perform biochemical analysis or measure metabolites. Additionally, culture conditions, including the dental sponges, are relatively inexpensive compared to organoid culturing. Nevertheless, both organoid culture and tissue explant approaches provide additional, and more clinically relevant models for studying CaP and predicting patient response to therapies.

In our work, we utilized tissue explants as a model for assessing the efficacy in clinical samples of a combination therapy that targets polyamine metabolism (Affronti et al., 2020). Polyamines are small alkylamines that due to their cationic nature, interact with negatively charged macromolecules in various cellular processes. Putrescine, spermidine, and spermine are the natural polyamines in mammals which are at higher concentrations in the prostate than any other tissue. Due to their essential role in reproduction (Pegg, 2014, 2016), the prostate synthesizes high levels of polyamines to be exported from the cell into the prostatic fluid. The catabolic enzyme, spermidine/spermine N1-acetyltransferase (SSAT), acetylates the polyamines which promotes their export and prevents accumulation of intracellular polyamines (Casero & Pegg, 1993). This heightened polyamine secretion adds strain on prostate cells to replace intracellular polyamine levels through the polyamine biosynthetic pathway. S-adenosylmethionine (SAM) is consumed from the methionine cycle, first to decarboxylated SAM, which then directly contributes to polyamine synthesis by donating its aminopropyl group. The one-carbon unit is lost in the form of 5-methylthioadenosine (MTA). The methionine salvage pathway serves as a means to recycle MTA in order to salvage the one-carbon unit and regenerate methionine and ultimately maintain homeostatic regulation of SAM pools (Bistulfi et al., 2016).

This pathway can be leveraged by targeting the main enzymes involved in polyamine homeostasis (Fig. 1). The combined effect of activated SSAT via the polyamine analog, N^1,N^{11}-bisethylnorspermine (BENSpm), with the MTAP inhibitor, MT-DADMe-ImmA (MTDIA), synergistically inhibits prostate cancer cell line growth, and additively blocked androgen independent xenograft growth. Applying this drug combination in the ex vivo prostatectomy setting allowed us to ask whether the strategy (1) is capable of inducing apoptosis in human prostate cancer, (2) if the effect is limited to the epithelial compartment, or broadly affecting the entire tissue including stroma, (3) if the target enzyme, SSAT, is induced in human tissue as predicted, and (4) if predicted metabolic effects are observed.

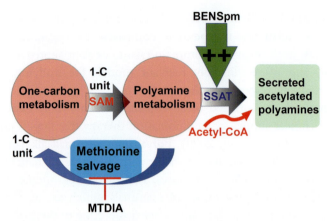

Fig. 1 Schematic overview of BENSpm and MTDIA therapeutic approach. The one carbon unit containing metabolite, S-adenosylmethionine (SAM), synthesized during one-carbon metabolism is utilized in polyamine biosynthesis but can be recycled back via the methionine salvage pathway. Methylthio-DADMe-Immucillin-A (MTDIA) is used to inhibit the rate limiting enzyme in the methionine salvage pathway thereby blocking the cells' ability to recycle the one-carbon unit lost to polyamine biosynthesis. This strain can be enhanced by N^1,N^{11}-bisethylnorspermine (BENSpm) which increases the activity of spermidine/spermine N^1-acetyltransferase (SSAT) which utilizes acetyl-CoA and enhances acetylated polyamine secretion, thereby driving increased polyamine biosynthesis to replenish intracellular polyamine pools.

2. Materials and reagents

2.1 Key materials (see Table 1)

Table 1 Key materials used for assessing efficacy of combination drug treatments in an *ex vivo* prostatectomy model.

Name	Source	Identifier
6-Well plates	VWR	10062-892
RPMI	ThermoFisher Scientific	11875093
FBS	ThermoFisher Scientific	16000044
Pen Strep	ThermoFisher Scientific	15140122
Gelatin Surgifoam Sponges	VWR	10611585

Formalin	Sigma	HT501128
BENSpm	Synthesis Med Chem	N/A
MTA	Sigma	D5011
MTDIA	Synthesized by Dr. Jim Philips at Cleveland Clinic Taussig Cancer Institute	N/A
Fresh prostatectomy specimen	RPCCC Pathology Core	N/A

2.2 Graphical overview (see Fig. 2)

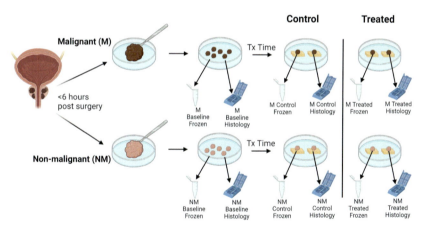

Fig. 2 Graphical depiction of the *ex vivo* model from radical prostatectomy to sample collection. Tissue samples collected during radical prostatectomy are analyzed by a pathologist. Pieces of non-malignant (NM) and malignant (M; > 40 % neoplastic involvement) tissues are separated and preserved in complete tissue culture media at room temperature. Within 6 h of surgery, samples are received and brought to the lab for *ex vivo* analysis. The tissue is cut up in a tissue culture dish containing media and baseline samples are snap frozen or collected in cassettes for histology. The remaining tissue samples are placed on sponges in media containing vehicle (control) or treatments. After the treatment (Tx) is complete samples are again frozen or collected in cassettes for histology. *Created in BioRender. Affronti, H. (2024) BioRender.com/y96w392.*

3. Protocols
3.1 Tissue explant culture

Note: The method below must be performed in a sterile tissue culture hood.
1. Obtain fresh tissue following surgery to resect solid tumor masses that are confirmed by a pathologist as having a high percent of neoplastic involvement.
   a. For radical prostatectomy samples, tumors should have greater than 40% neoplastic involvement.
   b. It is also ideal to obtain matched, adjacent non-tumor tissue.
   c. Tumor tissue was collected and used within 6 h of prostatectomy
2. Place tissue in a tissue culture dish containing RPMI 1640 10% FBS and 1% penicillin streptomycin and cut into approximately 10 mg pieces.
3. Snap freeze a small piece of tissue for baseline measurements.
4. Place 1 cm^3 Gelatin Surgifoam Sponges in sterile tubes containing RPMI 1640 10% FBS and 1% penicillin with drug or vehicle of interest.
5. Incubate sponges in media at 37 °C for 1 h, with gentle mixing every 10 min to ensure sponges are thoroughly soaked.
6. Remove dental sponges from tubes and place in a 6-well tissue culture dish containing media with vehicle or drug, two sponges per well; one for histology and one for biochemical assays.
7. Place one 10 mg piece of tissue on top of each dental sponge.
8. Cover with lid and place in an incubator at 37 °C and 5% CO_2.
9. Refresh media with drugs or vehicle every 48 h.
10. Snap freeze tissue for biochemical analysis or place in 10% formalin for IHC.

3.2 Preparation for metabolic analysis
1. Homogenize tissue using mortar and pestle on dry ice.
2. Aliquot 10 mg of homogenized tissue for UPLC analysis.
3. Sonicate sample in 100 μL of 0.6 M PCA for 30 s (15 s on, 15 s off, 15 s on).
4. Centrifuge sample at 4 °C for 15 min at 10,000 × g.
5. Ultra-Performance Liquid Chromatography experiments were performed on samples as previously described (Affronti et al., 2017, 2020; Bistulfi et al., 2009; Rosario et al., 2023).

3.3 Ex Vivo sample metabolomics analysis

For the analysis of small tissues, we implemented use of the Biocrates MxP® Quant 500 assay. It allowed for assessment of up to 630 metabolites from 26 biochemical classes. The assay offers reliable quantification of a broad range of metabolites, from a very small tissue quantity (>10 mg, in this case). Table 2 indicates the number of detectable metabolites we have observed from various tissue sources. Samples were prepared and handled in accordance with the Biocrates optimized methodologies. Briefly:

1. Snap freeze tissue samples in liquid nitrogen, and store in −80°C until ready for processing.
2. Homogenize tissue samples in a ratio of 1 mg of tissue to 3 µL of solvent (25 % ethanol and 75 % 0.01 M phosphate buffer) using optimized settings on the Omni-Bead Ruptor 24 (Omni International, Kennesaw, GA).
3. Centrifuge the homogenate to obtain supernatant.
4. Add 10 µL of each sample supernatant, quality control (QC) samples, blank, zero sample, or calibration standard on the filterspot (already containing internal standard) in the appropriate wells of a 96-well plate.
5. Dry the plate under a gentle stream of nitrogen.
6. Derivatize the samples with phenyl isothiocyanate (PITC) for the amino acids and biogenic amines, and then dry under a gentle stream of nitrogen again.
7. Elute the sample extract using 5 mM ammonium acetate in methanol.
8. Dilute the sample extracts with either water for the HPLC-MS/MS analysis (1:1) or kit running solvent (Biocrates Life Sciences AG) for flow injection analysis (FIA)-MS/MS (50:1), using a Shimadzu HPLC system interfaced with a Sciex 5500 mass spectrometer.
9. Process data using the MetIDQ software (Biocrates Life Sciences AG).

Table 2 Detectable metabolites from various tissue sources using the Biocrates MxP Quant 500 assay.

Sample type	Number of metabolites detected
Cell Media	527
Cell Pellets	323
Ex Vivo Samples (<1 mg)	368
Mouse Tumors (>5 mg)	540
Mouse Tissues (>5 mg)	549

3.4 IHC analysis

1. Using high precision forceps, gently remove tissue and place in a tissue cassette with 10 % formalin for at least 24 h.
2. Preserve samples by paraffin-embedding.
3. Prepare slides and incubate with primary and secondary antibodies of choice.
4. Analyze the entire stained slide by scoring (1–5) the percent positively stained cells and (0–4) the intensity of the stained cells.
   a. Determine the percent positively stained cells where a 1 indicates < 5 %, 2 is between 5–25 %, 3 is between 26–50 %, 4 is between 51–75 % and 5 is > 75 % positively stained cells.
   b. Identify the intensity of the staining for the positively stained cells where 0 indicates an absence of staining, 1 is weak intensity, 2 is moderate intensity, 3 is strong and 4 is very strong intensity.
   c. For each slide, calculate the index score by multiplying the percent positively stained cells by intensity.

References

Affronti, H. C., Long, M. D., Rosario, S. R., Gillard, B. M., Karasik, E., Boerlin, C. S., & Smiraglia, D. J. (2017). Dietary folate levels alter the kinetics and molecular mechanism of prostate cancer recurrence in the CWR22 model. *Oncotarget, 8*(61), 103758–103774. https://doi.org/10.18632/oncotarget.21911.

Affronti, H. C., Rowsam, A. M., Pellerite, A. J., Rosario, S. R., Long, M. D., Jacobi, J. J., & Smiraglia, D. J. (2020). Pharmacological polyamine catabolism upregulation with methionine salvage pathway inhibition as an effective prostate cancer therapy. *Nature Communications, 11*(1), 52. https://doi.org/10.1038/s41467-019-13950-4.

Affronti, H. (2024) BioRender.com/y96w392.

Bistulfi, G., Affronti, H. C., Foster, B. A., Karasik, E., Gillard, B., Morrison, C., & Smiraglia, D. J. (2016). The essential role of methylthioadenosine phosphorylase in prostate cancer. *Oncotarget, 7*(12), 14380–14393. https://doi.org/10.18632/oncotarget.7486.

Bistulfi, G., Diegelman, P., Foster, B. A., Kramer, D. L., Porter, C. W., & Smiraglia, D. J. (2009). Polyamine biosynthesis impacts cellular folate requirements necessary to maintain S-adenosylmethionine and nucleotide pools. *The FASEB Journal, 23*(9), 2888–2897. https://doi.org/10.1096/fj.09-130708.

Casero, R. A., Jr., & Pegg, A. E. (1993). Spermidine/spermine N1-acetyltransferase–the turning point in polyamine metabolism. *The FASEB Journal, 7*(8), 653–661.

Centenera, M. M., Raj, G. V., Knudsen, K. E., Tilley, W. D., & Butler, L. M. (2013). Ex vivo culture of human prostate tissue and drug development. *Nature Reviews. Urology, 10*(8), 483–487. https://doi.org/10.1038/nrurol.2013.126.

Gao, P., Postiglione, M. P., Krieger, T. G., Hernandez, L., Wang, C., Han, Z., & Shi, S. H. (2014). Deterministic progenitor behavior and unitary production of neurons in the neocortex. *Cell, 159*(4), 775–788. https://doi.org/10.1016/j.cell.2014.10.027.

Lv, J., Du, X., Wang, M., Su, J., Wei, Y., & Xu, C. (2024). Construction of tumor organoids and their application to cancer research and therapy. *Theranostics, 14*(3), 1101–1125. https://doi.org/10.7150/thno.91362.

Pegg, A. E. (2014). The function of spermine. *IUBMB Life, 66*(1), 8–18. https://doi.org/10.1002/iub.1237.

Pegg, A. E. (2016). Functions of polyamines in mammals. *The Journal of Biological Chemistry, 291*(29), 14904–14912. https://doi.org/10.1074/jbc.R116.731661.

Rosario, S. R., Jacobi, J. J., Long, M. D., Affronti, H. C., Rowsam, A. M., & Smiraglia, D. J. (2023). JAZF1: A metabolic regulator of sensitivity to a polyamine-targeted therapy. *Molecular Cancer Research: MCR, 21*(1), 24–35. https://doi.org/10.1158/1541-7786.MCR-22-0316.

Servant, R., Garioni, M., Vlajnic, T., Blind, M., Pueschel, H., Muller, D. C., & Le Magnen, C. (2021). Prostate cancer patient-derived organoids: detailed outcome from a prospective cohort of 81 clinical specimens. *The Journal of Pathology, 254*(5), 543–555. https://doi.org/10.1002/path.5698.

Yang, S., Hu, H., Kung, H., Zou, R., Dai, Y., Hu, Y., & Li, F. (2023). Organoids: The current status and biomedical applications. *Medicine Communications (2020), 4*(3), e274. https://doi.org/10.1002/mco2.274.

CHAPTER THIRTEEN

Development and characterization of a *Drosophila* model of Snyder-Robinson syndrome

Xianzun Tao and R. Grace Zhai*

Department of Neurology, University of Chicago, Chicago, IL, United States
*Corresponding author. e-mail address: rgzhai@uchicago.edu

Contents

1. Introduction	242
2. Rational	244
3. Generation of *dSms*-deficient flies	245
3.1 Equipment	245
3.2 Reagents	246
3.3 Fly lines	246
3.4 Procedure	246
3.5 Notes	246
4. *dSms* mRNA level measurement	248
4.1 Equipment	248
4.2 Reagents	248
4.3 Procedure	248
5. Polyamine levels measurement	248
5.1 Equipment	248
5.2 Reagents	248
5.3 Procedure	249
6. SMS re-expression	249
6.1 Fly lines	249
6.2 Procedure	249
6.3 Notes	249
7. Embryo viability measurement	250
7.1 Equipment	250
7.2 Procedure	250
8. Adult lifespan measurement	250
8.1 Equipment	250
8.2 Procedure	250
9. Climbing performance analysis	251
9.1 Equipment	251
9.2 Procedure	251

10. Eye morphology analysis	251
10.1 Equipment	252
10.2 Procedure	252
11. Eye activity analysis - electroretinography	253
11.1 Equipment	253
11.2 Procedure	253
12. Discussion	253
Acknowledgments	254
References	254

Abstract

Snyder-Robinson Syndrome (SRS) is an X-linked disorder characterized by intellectual disability, skeletal abnormalities, and immune system dysfunction. SRS is caused by mutations in the spermine synthase (SMS) gene, leading to polyamine dysregulation, and a myriad of cellular dysfunctions. This chapter presents a methodology for developing and characterizing a *Drosophila* model of SRS that recapitulates phenotypes of SMS deficiency. The protocol covers the generation of the *Drosophila* model, phenotypic characterization, data analysis methods, and troubleshooting to enhance reproducibility and rigor.

1. Introduction

Snyder-Robinson Syndrome (SRS) (Snyder & Robinson, 1969) is a rare X-linked recessive disorder caused by mutations in the *spermine synthase* (SMS) gene (Cason et al., 2003), a key enzyme in the polyamine biosynthetic pathway. Polyamines, including spermidine and spermine, are crucial for various cellular functions, such as DNA stabilization, protein synthesis and homeostasis, and ion channel regulation (Pegg, 2016). SRS manifests with symptoms including intellectual disability, skeletal abnormalities, hypotonia, and seizures, indicating the systemic effects of SMS mutation (Schwartz, Peron, & Kutler, 1993).

Three types of SMS mutations have been observed in SRS patients: (1) point mutations in splice sites leading to alternative splicing (Cason et al., 2003), (2) nucleotide deletions resulting in translational frameshifts that result in either protein truncation or nonsense-mediated mRNA decay (Larcher et al., 2020), and (3) missense mutations causing enzyme activity impairment or protein instability (Abela et al., 2016; Albert et al., 2015; Becerra-Solano et al., 2009; de Alencastro et al., 2008; Dontaine et al., 2021; Lemke et al., 2012; Li et al., 2017; Mouskou et al., 2021; Peng, Norris, Schwartz, & Alexov, 2016; Peron et al., 2013; Qazi et al., 2020; Zhang, Teng, Wang, Schwartz, & Alexov, 2010; Zhang et al., 2013). Most of the tested mutations result in a significant reduction in protein levels or enzymatic activity. In silico modeling and in vitro studies suggest these mutations alter SMS's structural integrity, leading to misfolding and degradation (Akinyele et al., 2024; Peng et al., 2016; Zhang et al., 2010). Together, these findings highlight loss of SMS function as the key mechanism in SRS pathogenesis.

The fruit fly *Drosophila melanogaster* is a powerful model organism for studying human diseases due to several advantages: (1) its genetic similarity to humans, with approximately 75 % of human disease-related genes having homologs in *Drosophila* (Reiter, Potocki, Chien, Gribskov, & Bier, 2001); (2) the conservation of fundamental biological pathways involved in development, signaling, and metabolism; (3) its well-characterized nervous system, which, despite being relatively simple, provides an effective platform for studying neurological diseases and neurodegenerative disorders (Bellen, Tong, & Tsuda, 2010); (4) the availability of versatile genetic tools, including transgene expression, gene knockouts, RNA interference, and CRISPR/Cas9, which enable precise genetic manipulation (Housden, Lin, & Perrimon, 2014; St Johnston, 2002); (5) the ease of phenotypic analysis, allowing for high-throughput screening of developmental, behavioral, and physiological traits (Venken, Simpson, & Bellen, 2011); (6) a short life cycle and rapid reproduction, facilitating the observation of genetic effects and disease progression within a short timeframe (Jennings, 2011); and (7) low maintenance costs, allowing for large-scale genetic screens and drug testing (Pandey & Nichols, 2011).

In recent years, substantial progress has been made in modeling and studying rare diseases using *Drosophila*. These *Drosophila* models have contributed significantly to our understanding of the genetic and molecular mechanisms underlying various rare diseases. Moreover, they provide a valuable platform for in vivo drug screening and therapeutic development, accelerating the discovery of new treatments for conditions that currently have limited options.

In this chapter, we present a *Drosophila* model for SRS. *Drosophila* has a single predicted SMS gene (CG4300, referred to as *dSms*), which produces two annotated transcripts, dSms-RA and dSms-RB, corresponding to protein isoforms PA and PB, respectively. These protein isoforms share identical catalytic and binding residues, essential for their enzymatic function. The protein sequence of *Drosophila* SMS-PA is highly conserved across species, showing significant homology with both mouse and human SMS. Notably, the missense mutations identified in SRS patients are enriched in identical amino acids conserved across *Drosophila*, mouse, and human SMS (Fig. 1) (Li et al., 2017), highlighting the relevance of *dSms* deficiency in modeling human SRS.

Fig. 1 Alignment of SMS protein sequences in human, mouse and *Drosophila*. The identity and similarity of the full protein sequences among these three species are shown as previously described (Li et al., 2017). The numbers on the right indicate amino acid positions in the human SMS sequence. Red stars mark amino acids that are mutated in SRS patients and are identical across human, mouse, and *Drosophila* (n = 11). Blue stars indicate amino acids that are mutated in SRS patients but differ between human and *Drosophila* (n = 8).

2. Rational

Unlike human or mouse SMS gene, *Drosophila* SMS gene is located on an autosome. Flies with a single copy of the *dSms* genes mutated ($dSms^{+/-}$, heterozygous) appear normal, partially mimicking the condition of mothers or sisters of SRS patients who carry one mutant SMS allele on the X chromosome but are unaffected. In contrast, flies with both copies of the *dSms* genes mutated ($dSms^{-/-}$, homozygous) exhibit reduced survival, poor health, and infertility, closely resembling the severe SMS deficiency observed in SRS patients (Li et al., 2017; Tao et al., 2022). To create a Drosophila model of SRS, *dSms* heterozygous flies are crossed, and homozygous progenies are collected for analysis.

Two key aspects must be clarified to confirm that the mutant flies exhibit true *dSms* deficiency without interference from other genetic defects: (1) the suppression of *dSms* expression and the consequent loss of enzyme activity, and (2) the rescue effect of re-expressing SMS on the major phenotypes. To address these, the first step is measuring the *dSms* mRNA levels and polyamine concentrations in homozygous flies, comparing them with those of control and heterozygous flies. Second, re-express either dSms or human SMS in the homozygous flies and assesse the effects on embryonic viability, adult lifespan, climbing performance, eye morphology and activity.

To re-express SMS, the UAS-Gal4 system, which is a powerful genetic tool used in Drosophila melanogaster to control the spatial and temporal expression of genes, could be used. It involves two key components: (1) Gal4 driver line: Gal4 is a transcriptional activator derived from yeast. In a Drosophila Gal4 line, Gal4 is expressed under the control of a tissue-specific or inducible promoter. Different driver lines express Gal4 in specific tissues, cell types, or at certain developmental stages. (2) UAS responder line: The UAS (Upstream Activating Sequence) is a specific DNA sequence recognized by Gal4. A gene of interest is placed downstream of the UAS, and it remains silent until Gal4 binds to the UAS, activating its transcription. In this system, when a Drosophila carrying a Gal4 driver is crossed with a fly containing a UAS responder, the progeny fly carrying both Gal4 and UAS will express Gal4 in a tissue specific manner and drive the expression of the gene of interest linked to UAS in the tissues or cells where Gal4 is expressed. This allows precise control of gene expression, making it a valuable tool for studying gene function, cell behavior, and disease models in Drosophila.

3. Generation of *dSms*-deficient flies

3.1 Equipment

1. Fly incubator
2. Microscope

3.2 Reagents
1. Cornmeal-molasses-yeast medium (Genesee Scientific #62–106)
2. Yeast powder

3.3 Fly lines
Various techniques have been employed to mutate *dSms* gene, including transposable element insertion and CRISPR-based mutagenesis. These insertions in the *dSms* locus can disrupt normal *dSms* transcription (Bellen et al., 2004; Hayashi et al., 2002; Kanca et al., 2019; Molnar, Lopez-Varea, Hernandez, & de Celis, 2006; Ryder et al., 2004; Thibault et al., 2004). Here, we provide a list of available fly strains in which the *dSms* locus has been modified by either a transposable element or a CRISPR cassette (Table 1).

3.4 Procedure
1. The *dSms* mutant lines are backcrossed to the control strain to minimize genetic background difference. For example, in our study, the BL86001 line was backcrossed to the *yw* flies.
2. Flies are maintained on a cornmeal-molasses-yeast medium (15 g/L) at 25 °C, 65 % humidity, 12 h light/12 h dark cycle.
3. To set up the cross, 5-6 female dSms+/− flies are paired with 4-5 male dSms +/− flies in a vial containing a small amount of yeast powder.
4. After flies reach adulthood, sort the adult flies by their phenotypic markers to distinguish between homozygous mutants (dSms−/−) and heterozygous controls (dSms+/−).

3.5 Notes
1. The survival rate of homozygous fly ($dSms^{-/-}$) embryos is approximately 10 % (Li et al., 2017), consistent with observations in a mouse model (Akinyele et al., 2024) and the ultra-rare occurrence of the disease in humans. Therefore, a cross scale approximately 10 times larger than usual is necessary to obtain a sufficient number of homozygous progenies for experiments.
2. The strains listed are derived from large-scale *Drosophila* gene mutation projects. However, the *dSms* expression levels and phenotypes of these strains have not been systematically characterized. In our experiments, we utilized and validated the BL86001 line.

Table 1 Available fly strains with dSms deletion.

Stock #	Genotype	Insertion[c]	References
K126079[a]	w[1118];P{w[+mW.Scer\FRT.hs]=RS5}cCG4300[5-SZ-3469]	TE[c]	Ryder et al. (2004)
K105274	y[*] w[*];P{w[+mW.hs]=GawB}CG4300[NP6588]/TM6, P{w[-]=UAS-lacZ.UW23-1}UW23-1	TE	Hayashi et al. (2002)
BL15526[b]	y[1] w[67c23];P{y[+mDint2] w[+mC]=EPgy2}Sms[EY01645]/TM3, Sb[1]	TE	Bellen et al. (2004)
BL18410	w[1118];PBac{w[+mC]=WH}Sms[f01110]/TM6B, Tb[1]/TM2	TE	Thibault et al. (2004)
BL43396	w[*];P{w[+mGS]=GSV1}Sms[C909]/TM2	TE	Molnar et al. (2006)
BL86001	w[1118];PBac{w[+mC]=RB}Sms[e00382]/TM6B, Tb[1]	TE	Thibault et al. (2004)
BL97639	y[1] w[*];TI{GFP[3xP3.cLa]=CRIMIC.TG4.0} Sms[CR02839-TG4.0]/TM3, Sb[1] Ser[1]	CRISPR	Kanca et al. (2019)

[a]K, Kyoto stock center.
[b]BL, Bloomington stock center.
[c]TE, transposable element.

4. *dSms* mRNA level measurement

4.1 Equipment
1. Homogenizer (Motorized pestle or Bullet blender)
2. Mini vortex mixer
3. Microcentrifuge
4. Quantitative PCR (qPCR) thermocycler

4.2 Reagents
1. TRIzol
2. Reverse transcription (RT) kit
3. qPCR kit
4. *dSms* primers

4.3 Procedure
1. Ten fly heads are homogenized in 100 uL of TRIzol using a motorized pestle or bullet blender to ensure thorough tissue disruption for optimal RNA yield.
2. Add 900 uL of TRIzol to above sample to make a 1 mL in total volume and vortex briefly.
3. Total RNA is extracted from the lysate, following the manufacturer's protocol.
4. Reverse transcribes the RNA to cDNA using RT kit, following the manufacturer's protocol.
5. Measure cDNA level using the qPCR kit and qPCR thermocycler, following the manufacturer's protocol.

5. Polyamine levels measurement

5.1 Equipment
1. Homogenizer (Motorized pestle or Bullet blender)
2. Mini vortex mixer
3. Microcentrifuge

5.2 Reagents
1. 1xPBS
2. Bradford protein assay (Bio-Rad # 5000201)
3. 1.2 N perchloric acid

5.3 Procedure
1. Ten flies are homogenized in 200 uL of 1xPBS using a motorized pestle or bullet blender to ensure thorough tissue disruption.
2. Centrifuge samples for 10 min at 12,000 g, 4 °C.
3. Measure protein levels in lysate supernatant with Bradford protein assay, following the manufacturer's protocol.
4. Add 50 μL of lysate supernatant to 50 μL 1.2 N perchloric acid, vertex briefly, and store the mixture at −80 °C.
5. Polyamine content is determined using the pre-column dansylation high-performance liquid chromatography method, with 1,7 diaminoheptane as the internal standard. The details refer to Ashley Nwafor et al. in this issue.

6. SMS re-expression
6.1 Fly lines
1. SMS transgene lines: $UAS\text{-}dSms^{RA}$, $UAS\text{-}hSms^{wt}$ and $UAS\text{-}hSms^{443}$.
2. *Actin-Gal4* driver line could be used for ubiquitous expression in the whole fly.

6.2 Procedure
1. Cross *UAS-SMS* transgene flies with $dSms^-$/*balancer* flies to recombine the *UAS-SMS* transgene chromosome and the $dSms^-$ chromosome and crossing *Gal4* driver flies with $dSms^-$/*balancer* flies to recombine the *Gal4* driver chromosome and the $dSms^-$ chromosome.
2. Cross the recombined *Gal4 driver;$dSms^-$/balancer* flies with *UAS-SMS;$dSms^-$/balancer* flies to get $dSms^{-/-}$ flies with SMS re-expression.
3. After flies reach adulthood, sort the adult flies by their phenotypic markers to pick $dSms^{-/-}$ flies expressing the transgene.

6.3 Notes
1. The mutation c.443 A > G (hSMS443, p.Gln148Arg), which affects the highly conserved 5′-MTA-binding site, was found in two brothers with severe and expanded SRS phenotypes (Albert et al., 2015).
2. Different transgene flies should be crossed to $dSms^-$/*balancer* flies in parallel to reduce the genetic background difference.

7. Embryo viability measurement

7.1 Equipment
1. Apple juice plates
2. Paintbrushes for embryo collection
3. CO_2 tank
4. CO_2 fly pad

7.2 Procedure
1. Place a group of 10-15 well-fed females and 5-7 males in a laying chamber (e.g., a bottle or cage with a mesh-covered opening).
2. Add a fresh apple juice plate smeared with yeast paste at the bottom of the chamber to encourage egg-laying.
3. Allow flies to lay eggs for 2-4 h to obtain a synchronized batch of embryos.
4. Using a moistened paintbrush, gently transfer 80-100 embryos from the plate to each vial containing fresh food.
5. After flies reach adulthood, sort the adult flies by their phenotypic markers to distinguish the target genotype from other genotypes.
6. Count the number of target genotype flies and total flies in each vial.
7. Calculate the embryo viability using the following formula: compare the percentage of target genotype flies to the theoretical Mendelian ratio expected for your cross.

8. Adult lifespan measurement

8.1 Equipment
1. Humidified chamber or incubator set at a consistent temperature (e.g., 25 °C)
2. Microscope
3. CO_2 tank
4. CO_2 fly pad

8.2 Procedure
1. Newly enclosed flies are collected and 20 flies of the same sex from each group are kept in fresh vials.
2. Place the vials in a 25 °C incubator or at room temperature, depending on your experimental setup.
3. Flies are transferred to new vials every 3 days and the number of live flies is counted every other day.

4. Calculate the median lifespan (the age at which 50% of the flies have died) and maximum lifespan for each group.
5. Use Kaplan-Meier survival analysis and statistical tests (e.g., log-rank test) to compare survival curves between groups.

9. Climbing performance analysis
9.1 Equipment
1. Plastic vials longer than 12 cm
2. CO_2 tank
3. CO_2 fly pad

9.2 Procedure
1. Ten age-matched male or female flies from each genotype are anesthetized and placed in a vial marked with a line 8 cm from the bottom surface.
2. Let the flies recover from the anesthesia for about 30 min.
3. The flies are gently tapped onto the bottom and given 10 s to climb.
4. After 10 s, the number of flies that successfully climbed above the 8 cm mark is recorded and divided by the total number of flies.
5. The assay is repeated 10 times, and 10 independent groups from each genotype are tested.

10. Eye morphology analysis

Due to its highly organized structure and easily observable phenotypes, *Drosophila* eye provides a powerful, accessible, and versatile system for studying the genetic and molecular basis of human diseases. The *Drosophila* compound eye is composed of about 800 repeating units called ommatidia, each containing photoreceptor cells organized in a precise pattern. This highly ordered structure and stereotyped architecture makes it easy to identify morphological abnormalities or disruptions caused by genetic mutations.

The *Drosophila* eye contains pigment granules that give it a characteristic color. Loss of pigmentation can result from genetic mutations or disruptions in metabolic pathways, often linked to defects in melanin synthesis, vesicle trafficking, and cellular stress. Pigmentation loss can be used as a visible marker for cellular dysfunctions, such as lysosomal storage diseases, neurodegenerative conditions, or metabolic disorders. Pigmentation changes are easy to observe, making it a rapid and cost-effective screening tool for identifying genetic mutations or drug effects that may contribute to disease pathogenesis (Fig. 2).

10.1 Equipment
4. Microscope
5. CO_2 tank
6. CO_2 fly pad
7. Digital or microscope camera
8. Diffused lighting board

10.2 Procedure
1. Flies are anesthetized and glued to a glass slide under a microscope, with the thorax, legs, and wings immobilized.
2. Place flies on a diffused lighting board.
3. Use a digital camera attached to a microscope or a microscope camera to take images of fly eyes from different angles.

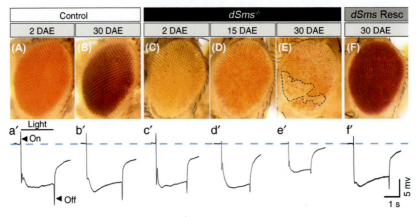

Fig. 2 Retinal degeneration in $dSms^{-/-}$ flies. Eye exterior morphology (A–F) of flies at indicated ages. The black dashed outline in panel (E) highlights the area with pigment loss. Panels a'–f' show traces from ERG recordings, depicting light-induced depolarization and on/off responses (example indicated by black arrowheads in panel a'). *Adapted from Li et al. (2017).*

11. Eye activity analysis - electroretinography

Electroretinography (ERG) is used to measure the electrical activity of photoreceptor cells in response to light stimuli. In *Drosophila*, researchers can record ERG signals by placing electrodes on the eyes of immobilized flies and stimulating them with light pulses. ERG helps assess retinal function and synaptic activity between photoreceptors and other neurons. It is particularly useful for studying neurodegenerative diseases with visual impairment. ERG provides a quantitative assessment of visual function, allowing researchers to evaluate the impact of genetic changes, disease progression, or therapeutic interventions on neuronal activity in the eye.

11.1 Equipment
1. Microscope
2. CO_2 tank
3. CO_2 fly pad
4. ERG recording system

11.2 Procedure
1. Flies are anesthetized and glued to a glass slide under a microscope, with the thorax, legs, and wings immobilized.
2. A recording electrode filled with 3 M KCl solution is placed on the compound eye, while a reference electrode is inserted into the thorax.
3. After 1-2 min of dark adaptation, the flies are exposed to a white light stimulus, and their responses are recorded and analyzed using AxoScope 10.5 software (Fig. 2).

12. Discussion

In this chapter, we introduced methods for establishing and characterizing a *Drosophila* model of SRS. *Drosophila* serves as a powerful model due to its genetic tractability, short life cycle, and suitability for drug screening and testing. It is worth noting that the physiological and metabolic differences between flies and humans necessitate cautious interpretation of the results. To fully understand the disorder and assess the translational potential of therapeutic interventions, findings in flies often need to be complemented with studies in more complex models, such as mammalian systems.

An important consideration when working with *Drosophila* models is the backcrossing of mutant or transgenic flies with control flies to standardize the genetic background. The genetic background can significantly influence phenotypic outcomes, especially in the study of complex traits like metabolic and neurodegenerative phenotypes. By backcrossing the transgenic *dSms*-deficient flies with control flies for several generations, researchers can minimize genetic variability unrelated to the *dSms* mutation. This approach ensures that observed phenotypes are primarily due to the specific genetic modifications rather than background genetic differences, thereby enhancing the validity of the findings.

Acknowledgments

We would like to thank Chong Li and Yi Zhu (University of Miami) for key method development, and Jackson R. Foley, Tracy Murray Stewart, and Robert A. Casero, Jr. (Johns Hopkins School of Medicine) for fly extract polyamine measurement. The study is supported by National Institutes of Health grant R01NS109640 and RF1NS109640 to RGZ.

References

Abela, L., Simmons, L., Steindl, K., Schmitt, B., Mastrangelo, M., Joset, P., ... Plecko, B. (2016). N(8)-acetylspermidine as a potential plasma biomarker for Snyder-Robinson syndrome identified by clinical metabolomics. *Journal of Inherited Metabolic Disease, 39*(1), 131–137. https://doi.org/10.1007/s10545-015-9876-y.

Akinyele, O., Munir, A., Johnson, M. A., Perez, M. S., Gao, Y., Foley, J. R., ... Kemaladewi, D. U. (2024). Impaired polyamine metabolism causes behavioral and neuroanatomical defects in a mouse model of Snyder-Robinson syndrome. *Disease Models & Mechanisms, 17*(6), dmm050639. https://doi.org/10.1242/dmm.050639.

Albert, J. S., Bhattacharyya, N., Wolfe, L. A., Bone, W. P., Maduro, V., Accardi, J., ... Boerkoel, C. F. (2015). Impaired osteoblast and osteoclast function characterize the osteoporosis of Snyder - Robinson syndrome. *Orphanet Journal of Rare Diseases, 10*, 27. https://doi.org/10.1186/s13023-015-0235-8.

Becerra-Solano, L. E., Butler, J., Castaneda-Cisneros, G., McCloskey, D. E., Wang, X., Pegg, A. E., ... Garcia-Ortiz, J. E. (2009). A missense mutation, p.V132G, in the X-linked spermine synthase gene (SMS) causes Snyder-Robinson syndrome. *American Journal of Medical Genetics. Part A, 149A*(3), 328–335. https://doi.org/10.1002/ajmg.a.32641.

Bellen, H. J., Levis, R. W., Liao, G. C., He, Y. C., Carlson, J. W., Tsang, G., ... Spradling, A. C. (2004). The BDGP gene disruption project: Single transposon insertions associated with 40% of Drosophila genes. *Genetics, 167*(2), 761–781. https://doi.org/10.1534/genetics.104.026427.

Bellen, H. J., Tong, C., & Tsuda, H. (2010). 100 years of Drosophila research and its impact on vertebrate neuroscience: A history lesson for the future. *Nature Reviews. Neuroscience, 11*(7), 514–522. https://doi.org/10.1038/nrn2839.

Cason, A. L., Ikeguchi, Y., Skinner, C., Wood, T. C., Holden, K. R., Lubs, H. A., ... Schwartz, C. E. (2003). X-linked spermine synthase gene (SMS) defect: The first polyamine deficiency syndrome. *European Journal of Human Genetics: EJHG, 11*(12), 937–944. https://doi.org/10.1038/sj.ejhg.5201072.

de Alencastro, G., McCloskey, D. E., Kliemann, S. E., Maranduba, C. M., Pegg, A. E., Wang, X., ... Sertie, A. L. (2008). New SMS mutation leads to a striking reduction in spermine synthase protein function and a severe form of Snyder-Robinson X-linked recessive mental retardation syndrome. *Journal of Medical Genetics, 45*(8), 539–543. https://doi.org/10.1136/jmg.2007.056713.

Dontaine, P., Kottos, E., Dassonville, M., Balasel, O., Catros, V., Soblet, J., ... Vilain, C. (2021). Digestive involvement in a severe form of Snyder-Robinson syndrome: Possible expansion of the phenotype. *European Journal of Medical Genetics, 64*(1), 104097. https://doi.org/10.1016/j.ejmg.2020.104097.

Hayashi, S., Ito, K., Sado, Y., Taniguchi, M., Akimoto, A., Takeuchi, H., ... Goto, S. (2002). GETDB, a database compiling expression patterns and molecular locations of a collection of Gal4 enhancer traps. *Genesis : the Journal of Genetics and Development, 34*(1-2), 58–61. https://doi.org/10.1002/gene.10137.

Housden, B. E., Lin, S., & Perrimon, N. (2014). Cas9-based genome editing in Drosophila. *Methods in Enzymology, 546,* 415–439. https://doi.org/10.1016/B978-0-12-801185-0.00019-2.

Jennings, B. H. (2011). A versatile model in biology & medicine. *Materials Today, 14*(5), 190–195. https://doi.org/10.1016/S1369-7021(11)70113-4.

Kanca, O., Zirin, J., Garcia-Marques, J., Knight, S. M., Yang-Zhou, D., Amador, G., ... Bellen, H. J. (2019). An efficient CRISPR-based strategy to insert small and large fragments of DNA using short homology arms. *Elife, 8.* https://doi.org/10.7554/eLife.51539.

Larcher, L., Norris, J. W., Lejeune, E., Buratti, J., Mignot, C., Garel, C., ... Whalen, S. (2020). The complete loss of function of the SMS gene results in a severe form of Snyder-Robinson syndrome. *European Journal of Medical Genetics, 63*(4), 103777. https://doi.org/10.1016/j.ejmg.2019.103777.

Lemke, J. R., Riesch, E., Scheurenbrand, T., Schubach, M., Wilhelm, C., Steiner, I., ... Biskup, S. (2012). Targeted next generation sequencing as a diagnostic tool in epileptic disorders. *Epilepsia, 53*(8), 1387–1398. https://doi.org/10.1111/j.1528-1167.2012.03516.x.

Li, C., Brazill, J. M., Liu, S., Bello, C., Zhu, Y., Morimoto, M., ... Zhai, R. G. (2017). Spermine synthase deficiency causes lysosomal dysfunction and oxidative stress in models of Snyder-Robinson syndrome. *Nature Communications, 8*(1), 1257. https://doi.org/10.1038/s41467-017-01289-7 (pii).

Molnar, C., Lopez-Varea, A., Hernandez, R., & de Celis, J. F. (2006). A gain-of-function screen identifying genes required for vein formation in the Drosophila melanogaster wing. *Genetics, 174*(3), 1635–1659. https://doi.org/10.1534/genetics.106.061283.

Mouskou, S., Katerelos, A., Doulgeraki, A., Leka-Emiri, S., Manolakos, E., Papoulidis, I., ... Voudris, K. (2021). Novel hemizygous missense variant of spermine synthase (SMS) gene causes Snyder-Robinson syndrome in a four-year-old boy. *Molecular Syndromology, 12*(3), 194–199. https://doi.org/10.1159/000514122.

Pandey, U. B., & Nichols, C. D. (2011). Human disease models in and the role of the fly in therapeutic drug discovery. *Pharmacological Reviews, 63*(2), 411–436. https://doi.org/10.1124/pr.110.003293.

Pegg, A. E. (2016). Functions of polyamines in mammals (pii) *The Journal of Biological Chemistry, 291*(29), 14904–14912. https://doi.org/10.1074/jbc.R116.731661 (pii).

Peng, Y., Norris, J., Schwartz, C., & Alexov, E. (2016). Revealing the effects of missense mutations causing Snyder-Robinson syndrome on the stability and dimerization of spermine synthase. *International Journal of Molecular Sciences, 17*(1), https://doi.org/10.3390/ijms17010077.

Peron, A., Spaccini, L., Norris, J., Bova, S. M., Selicorni, A., Weber, G., ... Mastrangelo, M. (2013). Snyder-Robinson syndrome: A novel nonsense mutation in spermine synthase and expansion of the phenotype. *American Journal of Medical Genetics. Part A, 161A*(9), 2316–2320. https://doi.org/10.1002/ajmg.a.36116.

Qazi, T. J., Wu, Q., Aierken, A., Lu, D., Bukhari, I., Hussain, H. M. J., ... Qing, H. (2020). Whole-exome sequencing identifies a novel mutation in spermine synthase gene (SMS) associated with Snyder-Robinson syndrome. *BMC Medical Genetics, 21*(1), 168. https://doi.org/10.1186/s12881-020-01095-x.

Reiter, L. T., Potocki, L., Chien, S., Gribskov, M., & Bier, E. (2001). A systematic analysis of human disease-associated gene sequences in Drosophila melanogaster. *Genome Research, 11*(6), 1114–1125. https://doi.org/10.1101/gr.169101.

Ryder, E., Blows, F., Ashburner, M., Bautista-Llacer, R., Coulson, D., Drummond, J., ... Russell, S. (2004). The DrosDel collection: A set of P-element insertions for generating custom chromosomal aberrations in Drosophila melanogaster. *Genetics, 167*(2), 797–813. https://doi.org/10.1534/genetics.104.026658.

Schwartz, C. E., Peron, A., & Kutler, M. J. (1993). Snyder-Robinson syndrome. In M. P. Adam, H. H. Ardinger, R. A. Pagon, S. E. Wallace, L. J. H. Bean, K. W. Gripp, G. M. Mirzaa, & A. Amemiya (Eds.). *GeneReviews (R)*Seattle (WA): University of Washington. https://www.ncbi.nlm.nih.gov/pubmed/23805436.

Snyder, R. D., & Robinson, A. (1969). Recessive sex-linked mental retardation in the absence of other recognizable abnormalities. Report of a family. *Clinical Pediatrics, 8*(11), 669–674. https://doi.org/10.1177/000992286900801114.

St Johnston, D. (2002). The art and design of genetic screens. *Nature Reviews. Genetics, 3*(3), 176–188. https://doi.org/10.1038/nrg751.

Tao, X., Zhu, Y., Diaz-Perez, Z., Yu, S. H., Foley, J. R., Stewart, T. M., ... Zhai, R. G. (2022). Phenylbutyrate modulates polyamine acetylase and ameliorates Snyder-Robinson syndrome in a Drosophila model and patient cells. *JCI Insight, 7*(13), https://doi.org/10.1172/jci.insight.158457.

Thibault, S. T., Singer, M. A., Miyazaki, W. Y., Milash, B., Dompe, N. A., Singh, C. M., ... Margolis, J. (2004). A complementary transposon tool kit for Drosophila melanogaster using P and piggyBac. *Nature Genetics, 36*(3), 283–287. https://doi.org/10.1038/ng1314.

Venken, K. J., Simpson, J. H., & Bellen, H. J. (2011). Genetic manipulation of genes and cells in the nervous system of the fruit fly. *Neuron, 72*(2), 202–230. https://doi.org/10.1016/j.neuron.2011.09.021.

Zhang, Z., Norris, J., Kalscheuer, V., Wood, T., Wang, L., Schwartz, C., ... Van Esch, H. (2013). A Y328C missense mutation in spermine synthase causes a mild form of Snyder-Robinson syndrome. *Human Molecular Genetics, 22*(18), 3789–3797. https://doi.org/10.1093/hmg/ddt229.

Zhang, Z., Teng, S., Wang, L., Schwartz, C. E., & Alexov, E. (2010). Computational analysis of missense mutations causing Snyder-Robinson syndrome. *Human Mutation, 31*(9), 1043–1049. https://doi.org/10.1002/humu.21310.

… CHAPTER FOURTEEN

Monitoring ODC activity and polyamines in Bachmann-Bupp syndrome patient biological samples

Chad R. Schultz[a,b], Elizabeth A. VanSickle[b,c], Caleb P. Bupp[a,b,c], and André S. Bachmann[a,b],*

[a]Department of Pediatrics and Human Development, College of Human Medicine, Michigan State University, Grand Rapids, MI, United States
[b]International Center for Polyamine Disorders, Grand Rapids, MI, United States
[c]Division of Medical Genetics, Corewell Health/Helen DeVos Children's Hospital, Grand Rapids, MI, United States
*Corresponding author. e-mail address: bachma26@msu.edu

Contents

1. Introduction — 258
2. Rationale — 259
3. Experimental section — 260
   3.1 RBC preparation — 260
   3.2 Protein quantification assay — 260
   3.3 ODC activity assay — 262
   3.4 Dansylating polyamines in the RBC preparations — 263
   3.5 Running samples on the HPLC — 266
Acknowledgments — 267
Competing interests — 268
References — 268

Abstract

Polyamines are aliphatic molecules that include putrescine, spermidine, and spermine. Polyamines are present in most living organisms including humans. These positively charged molecules play important roles in cell physiology and pathology by contributing to embryonic cell development, regulation of cell division and, if overproduced, the stimulation of cancer cell proliferation and tumorigenesis. We recently discovered Bachmann-Bupp Syndrome (BABS); a rare neurodevelopmental disorder linked to *de novo* mutations in the ornithine decarboxylase 1 (*ODC1*) gene. *ODC1* gene mutations that are linked to BABS always produce C-terminally truncated versions of the enzyme ornithine decarboxylase (ODC). These shortened ODC proteins remain enzymatically active and are not cleared by the proteasome, therefore leading to ODC protein accumulation in cells. ODC is a key enzyme of

Methods in Enzymology, Volume 715
ISSN 0076-6879, https://doi.org/10.1016/bs.mie.2025.01.071
Copyright © 2025 Elsevier Inc. All rights reserved, including those for text and data mining, AI training, and similar technologies.

polyamine biosynthesis by converting ornithine to putrescine, and if accumulated, can lead to high putrescine levels in human cells including red blood cells (RBCs) and primary dermal fibroblasts.

Here we describe how to quantitatively measure ODC enzymatic activity and the polyamines by a radiolabeled ^{14}C-ornithine assay and by reverse phase (RP)-HPLC, respectively. While these methods have been developed decades ago, many publications provide incomplete protocols with omission of experimental details, which inadvertently can lead to mistakes, inconclusive results, and failed experiments. There is a growing number of laboratories that have become interested in exploring polyamines (in part due to metabolomics analyses in human health-related studies). The detailed protocols of this chapter provide step-by-step guidance detailing how to measure ODC activity and polyamines in human RBCs.

1. Introduction

Polyamines were discovered several centuries ago, in the year 1677 by Antonie van Leeuwenhoek (Lewenhoeck, 1677). They are polycationic molecules that interact with negatively charged molecules including DNA, RNA, proteins, and lipids. Polyamines are ubiquitous and found in almost all organisms in both prokaryotes and eukaryotes including bacteria, parasites, plants, and mammals. In humans, polyamines have been primarily studied in the context of cancer and infectious disease (Bachmann & Geerts, 2018; Casero & Marton, 2007; Casero, Murray Stewart, & Pegg, 2018; Gerner & Meyskens, 2004; Holbert, Cullen, Casero, & Stewart, 2022). Most prominently, polyamines and treatments were explored for West African sleeping sickness (trypanosomiasis) (Alirol et al., 2013; Alirol et al., 2012; Bacchi et al., 1983; Priotto et al., 2009) and several types of cancer including colorectal and prostate cancer (Gerner & Meyskens, 2004; Meyskens et al., 2008; Raj et al., 2013; Simoneau, Gerner, Phung, McLaren, & Meyskens, 2001) as well as the pediatric cancer neuroblastoma (Bachmann, Geerts, & Sholler, 2012; Geerts et al., 2010; Hogarty et al., 2024; Oesterheld et al., 2023; Saulnier Sholler et al., 2015; Sholler et al., 2018; Wallick et al., 2005). The irreversible ODC inhibitor alpha-difluoromethylornithine (DFMO, also known as Eflornithine) received FDA approval in 1990 (intravenous formulation) to treat trypanosomiasis (in combination with nifurtimox). In 2000, the FDA approved a topical formulation that includes 13.9 % DFMO to treat hirsutism (Wolf et al., 2007). In 2023, the FDA approved DFMO (oral formulation, available as tablets) now known as Iwilfin for the treatment and relapse prevention of adult and pediatric patients with high-risk neuroblastoma. In addition, many other diseases are currently being studied to understand the importance of polyamines in pathophysiology and to explore opportunities to treat such diseases with specific polyamine pathway inhibitors (Bachmann & Levin, 2012).

In 2018, we discovered a neurodevelopmental disorder referred to as Bachmann-Bupp Syndrome (BABS) that is linked to C-terminally truncated forms of ODC (Bupp, Schultz, Uhl, Rajasekaran, & Bachmann, 2018). BABS was the second polyamine disorder, or polyaminopathy, to be identified in the polyamine pathway, after Snyder-Robinson Syndrome (SRS) (Arena et al., 1996; Cason et al., 2003; Schwartz, Peron, & Kutler, 2020; Snyder & Robinson, 1969). We repurposed DFMO to treat BABS patients (Bachmann, VanSickle, Michael, Vipond, & Bupp, 2024). We showed that untreated BABS patients have excessive baseline amounts of accumulated ODC due to the lack of proper clearance via the proteasome, leading to higher ODC enzymatic activity and high levels of putrescine in human red blood cells (RBCs) and primary dermal fibroblasts (Schultz, Bupp, Rajasekaran, & Bachmann, 2019). Treatment of the first BABS patients with DFMO rapidly reduced N^1-acetylputrescine levels in patient plasma compared to levels before treatment and led to remarkable overall patient phenotype improvements, including hair regrowth and developmental progress (Rajasekaran et al., 2021). Currently, a total of 17 BABS patients have been identified worldwide and six BABS patients are being treated with DFMO, leading to similar patient responses (unpublished data).

2. Rationale

Although ODC activity assays and polyamine analyses are described in many prior publications, these protocols often omit important experimental details which inadvertently can lead to mistakes, inconclusive results, and failed experiments. Moreover, most ODC and polyamine determinations are made using cell lines and not human biological samples. Given the importance of polyamines and their recent association with a plethora of new biological processes, human diseases, genetic disorders, and various medical conditions (sometimes identified "by accident" via broad spectrum metabolomics analyses), we are publishing our detailed protocols to measure ODC enzymatic activity and polyamines in human RBCs. The blood was collected from patients with BABS or healthy controls and RBCs prepared as described in a separate chapter of this book (see VanSickle et al.).

3. Experimental section

3.1 RBC preparation

3.1.1 Equipment
- Handheld homogenizer (VWR, Radnor PA, USA)
- Refrigerated Microcentrifuge (Eppendorf, Hamburg, Germany)

3.1.2 Reagents
- Liquid nitrogen
- 1.8 ml microcentrifuge tubes (Thermo Fisher Scientific, Waltham, MA, USA)
- ODC Breaking Buffer – 25 mM Tris HCL, 0.1 mM EDTA, 2.5 mM DTT

3.1.3 Procedure
1. Pipet 50 µl of each RBC sample into a microcentrifuge tube.
2. Add 200 µl of ODC Breaking Buffer to the RBCs.
3. Snap freeze the mixture with liquid nitrogen and allow the sample to thaw on ice.
4. Homogenize the mixture for 10–15 s
5. Centrifuge the samples for 10 min at 12,000 rpm.
6. Pipet the supernatant into a fresh microcentrifuge tube to use for polyamine preparations and ODC activity assay.

3.1.4 Notes
1. DTT should be added fresh (day of) to the ODC breaking buffer.
2. All of the steps of the procedure should be done on ice.
3. The RBC samples are a thick slurry, so it is helpful to cut the end of the pipet tip in order to draw up the 50 µl of sample.

3.2 Protein quantification assay

3.2.1 Equipment
- CLARIOstar Plate Reader (BMG Labtech, Ortenberg, Germany)

3.2.2 Reagents
- Costar Assay Plate, 96-well (#9017 Corning, Corning, NY)
- Albumin Standard, 2 mg/ml (#23209 Thermo Fisher Scientific, Waltham, MA)
- Protein Assay Dye Concentrate (#5000006, Bio-Rad Laboratories, Hercules, CA)

3.2.3 Procedure
1. Dilute the albumin standard from 2 mg/ml to 100 µg/ml (1:20) using deionized water.

2. Setup a standard curve in the top row of a 96-well plate using the diluted albumin standard as follows (Table 1):
3. For each RBC preparation sample pipet into a 96-well plate well 48 μl of water plus 2 μl of sample.
4. Add 200 μl of 1:3 diluted Protein Assay Dye Concentrate to all the wells containing the albumin standard or RBC preparation and incubate at room temperature for 10 min.
5. Measure the absorbance value at OD595 on the CLARIOstar plate reader.
6. Using Microscoft Excel, plot the standard curve with the concentration of albumin on the Y-axis and the OD595 readings on the X-axis.
7. Use the formula for the standard curve line to determine the unknown concentrations of the RBC preparations.
8. Based on the determined concentration calculate the mg of protein in 50 μl of each RBC preparation.

3.2.4 Notes

1. Each concentration of the standard curve as well as each RBC preparation sample should be done at least in duplicate.
2. The RBC preparations are typically very high in protein concentration, so you may have to dilute the preparations 1:5 or 1:10 prior to using in the protein quantification assay so that your OD595 reading will be within the value range of the standard curve.
3. The amount of protein (mg) in 50 μl of sample will be used to normalize the ODC activity values and polyamine amounts across samples.

Table 1 Diluted albumin standards for protein quantification assay.

100 μg/ml Albumin (μl)	Water (μl)	[Albumin] μg/ml
0	0	0
3.13	46.87	6.25
6.25	43.75	12.5
12.5	37.5	25
25	25	50
50	0	100

3.3 ODC activity assay

3.3.1 Equipment
- Isotemp 37 °C shaking waterbath (Fisher Scientific, Waltham, MA)
- TriCarb 4910 TR Liquid Scintillation Counter (Perkin Elmer, Waltham, MA)

3.3.2 Reagents
- 0.5 M Tris – HCl pH 7.5
- 2 mM pyridoxal – 5 – phosphate
- 25 mM DTT
- 20 mM L-ornithine
- 55 mCi/mmol L- [1–14C] ornithine (American Radiolabeled Chemicals, St. Louis, MO)
- 0.1 N sodium hydroxide
- 5 M sulfuric acid
- 1.8 ml microcentrifuge tubes (Thermo Fisher Scientific, Waltham, MA, USA)
- Beckman PolyQ scintillation vials (#566350, Fisher Scientific, Waltham MA)
- Circular Filter Paper, 20 mm diameter (#1001-020, Cytiva Life Sciences, Marlborough, MA)
- Research Products International Bio-Safe II complete counting cocktail (#111195, Fisher Scientific, Waltham, MA)

3.3.3 Procedure
1. Prepare the ODC activity assay reaction buffer mix as follows:
   a. Per reaction:
      i. 2.5 µl of 0.5 M Tris-HCl pH 7.5
      ii. 1 µl 20 mM L-ornithine
      iii. 5 µl of 2 mM pyridoxal – 5- phosphate
      iv. 12.5 µl of 25 mM DTT
      v. 1 µl L-[1-^{14}C] ornithine (55 mCi/mmol)
      vi. 178 µl deionized water
   b. Allow the mix to sit for 30 min at room temperature.
2. Pipet 50 µl of RBC preparation samples from section one into a 1.8 ml microcentrifuge tube.
3. For each sample prepare a Beckman PolyQ scintillation vial with a 20 mm circular piece of filter paper soaked with 200 µl of 0.1 N sodium hydroxide to capture the ^{14}C labeled carbon dioxide produced in the reaction.
4. Add 200 µl of ODC activity assay reaction buffer to the 50 µl of RBC preparation sample and place the 1.8 ml microcentrifuge tube with the cap open into the prepared scintillation vial.

5. Place the cap on the vial and incubate the samples in the 37 °C water bath while shaking at 30 rpm for 2 h.
6. Remove the samples from the water bath and add 250 μl of 5 M sulfuric acid to the reaction mix to end the reaction.
7. Incubate the samples for an additional 30 min in the 37 °C water bath while shaking at 30 rpm.
8. Remove the reaction mix from the microcentrifuge tube with a 1 ml pipet and put it into a liquid radiation waste container.
9. Using forceps, grab the opened cap of the microcentrifuge tube and remove it from the scintillation vial and place it in a solid radiation waste container.
10. Add 5 ml of Research Products International Bio-Safe II complete counting cocktail to each vial.
11. Cap the vial and centrifuge briefly (2–3 s) and allow the vials to sit overnight.
12. Run the vials in the TriCarb 4910 TR Liquid Scintillation Counter, capturing the disintegrations per minute (DPM) and microCurie (μCi) for 1 min per sample.
13. Use the original activity of the L-[1-^{14}C] ornithine of 55 mCi/mmol to convert the μCi values to molar values of CO_2 liberated in the reaction.
14. ODC activity values are reported as pmol CO_2/120 min/mg protein (Fig. 1A).

3.3.4 Notes
1. Samples are run in duplicate in the ODC activity assay.
2. Aliquots of 2 mM pyridoxal-5-phosphate, 25 mM DTT, and 20 mM L-ornithine are made in advance and kept frozen at −20 °C. Aliquots are used only once.

3.4 Dansylating polyamines in the RBC preparations
3.4.1 Equipment
- Refrigerated Microcentrifuge (Eppendorf, Hamburg, Germany)
- Microcentrifuge Tube Thermomixer R (Eppendorf, Hamburg, Germany)

3.4.2 Reagents
- 1.8 ml microcentrifuge tubes (Thermo Fisher Scientific, Waltham, MA, USA)
- Perchloric acid buffer (0.2 M perchloric acid/1 M sodium chloride)
- 0.15 mM 1,7 diaminoheptane
- 1 M sodium carbonate
- 5 mg/ml dansyl choride dissolved in acetone
- 1 M proline

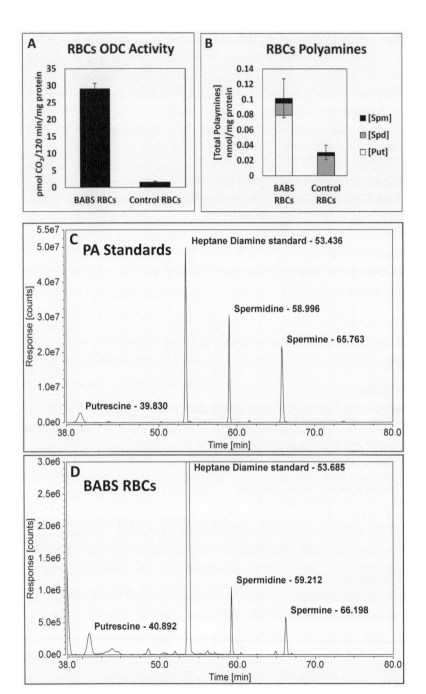

Fig. 1 (A) Ornithine decarboxylase (ODC) enzyme activity measurement in red blood cells (RBCs) using the ^{14}C-ornithine radiolabeled assay. ODC activity in Bachmann-

- 4 ml amber glass vials with cap (C4015-17AW, Thermo Fisher Scientific, Waltham, MA)
- Methylene chloride
- Methanol
- Hypersep C18, 25 mg cartridges (#60108-376, Thermo Fisher Scientific, Waltham, MA)
- 2 ml glass vials w/septa for autosampler (#C5000-180, Thermo Fisher Scientific, Waltham, MA)

3.4.3 Procedure

1. Pipet 50 μl of each RBC preparation sample into a 1.8 ml microcentrifuge tube.
2. Add 100 μl of perchloric acid buffer to each sample and vortex briefly. This will protonate the polyamines and precipitate out the protein in the sample.
3. Pellet the precipitated protein by spinning the samples at 12,000 rpm for 10 min at room temperature using the microcentrifuge.
4. Transfer the supernatant to a clean 1.8 ml microcentrifuge tube.
5. Add 30 μl of 0.15 mM 1,7 diaminoheptane internal standard to each sample.
6. Add 200 μl of 1 M sodium carbonate to each sample.
7. Add 400 μl of 5 mg/ml dansyl chloride in acetone to each sample and vortex for a few seconds.
8. Incubate the samples at 37 °C while shaking at 300 rpm in the thermomixer R for 1 h.
9. Add 100 μl of 1 M proline to each sample.
10. Incubate the samples at 37 °C while shaking at 300 rpm in the thermomixer R for 20 min.
11. Transfer the samples to the 4 ml amber glass vials with caps.
12. In a fume hood add 1 ml of methylene chloride to each sample and mix by inverting ten times. This will extract out the dansylated polyamines into the clear bottom layer.

Bupp Syndrome (BABS) RBCs was significantly higher compared to control RBCs. (B) Polyamine (PA) concentration of putrescine (Put), spermidine (Spd), and spermine (Spm) measured by reverse phase (RP)-HPLC indicating a significant increase in putrescine levels in BABS RBCs compared to control RBCs. (C) Chromatogram showing the peaks of internal PA standards that are used on every RP-HPLC run. (D) Chromatogram showing PA peaks in a BABS RBC sample. Numbers represent the retention time of each metabolite. To enhance the visibility of the PA peaks, the scale was reduced, which caused the standard peak to be cut off.

13. Using a 1 ml pipet tip, go to the bottom of the vial and extract as much of the clear bottom layer without drawing up the upper milky white layer and transfer it to a clean 4 ml amber vial.
14. Leave the cap off of the vial with the extracted polyamines and allow the methylene chloride to evaporate overnight in the fume hood.
15. Add 1 ml of methanol to each sample and vortex to reconstitute the polyamines.
16. To preclear the samples so that they do not cause a clog in the HPLC system, the samples are passed through Hypersep C18 25 mg cartridges. Pipet the 1 ml of sample onto the cartridge. Using a hose attached to a lab air line, use the air to push the sample through the cartridge into a clean uncapped 2 ml glass vial that will be used on the HPLC autosampler.
17. Screw the cap with septa onto the 2 ml glass vials and load onto the autosampler trays to run on the HPLC.

3.4.4 Notes
1. Each RBC preparation sample should be processed in duplicate.
2. Accurate pipetting of the 1,7 diaminoheptane standard is crucial as the relative molar response of the polyamine peaks to the 1,7 diaminoheptane standard peak will be used to quantify the amount of polyamines in a sample.
3. Step 15 above can be shortened if there is access to a vacufuge to evaporate the methylene chloride.

3.5 Running samples on the HPLC
3.5.1 Equipment
1. Dionex Ultimate HPLC System with an autosampler and fluorescence detector (Thermo Fisher Scientific, Waltham, MA)
2. Syncronis C18 (250 × 4.6 mm, 5 μm particle size) reverse phase (RP) column (Thermo Fisher Scientific, Waltham, MA)

3.5.2 Reagents
1. Buffer A: 10 mM Sodium −1- heptane sulfonate in 10% acetonitrile, pH 3.5
2. Buffer B: 100% Acetonitrile

3.5.3 HPLC running conditions
1. Set the flow rate to 0.7 ml per minute.
2. Equilibrate the column with 55% buffer B/45% buffer A for 7 min prior to injecting the first sample.
3. Inject 20 μl of dansylated polyamine sample into the HPLC system.

4. Program the buffer gradient to elute the polyamines as follows:
   a. Isocratic solution of 55% buffer B/45% buffer A for 28 min.
   b. Slowly ramp to 95% buffer B/5% buffer A for 20 min. Hold this gradient for 20 min to elute the spermine peak.
   c. Drop back to 55% buffer B/45% buffer A for 7 min.
   d. Re-equilibrate the column in 55% buffer B/45% buffer A for 7 min prior to the next sample being injected.
5. The approximate elution times using the above gradients are as follows (Fig. 1B–D):
   a. Putrescine: 40 min
   b. 1,7 diaminoheptane internal standard: 54 min
   c. Spermidine: 59 min
   d. Spermine: 66 min

3.5.4 HPLC analysis

1. For each HPLC system, a relative molar response between each of the polyamines and the 1,7 diaminoheptane internal standard must be determined. This is done by dansylating equal amounts of each individual polyamine with an equal amount of 1,7 diaminoheptane and determining the peak area ratio for each polyamine in comparison to the internal standard.
2. Determine the polyamine concentration in a sample by measuring the peak areas for each polyamine and 1,7 diaminoheptane standard. Then use the relative molar response determined above to determine the number of moles of each polyamine and normalize to the amount of protein in each sample. Polyamine (putrescine, spermidine, spermine) concentration values are reported as nmol PA/mg protein (Fig. 1B).

3.5.5 Notes

1. The run time for each sample is over 1 h, so the use of an autosampler is highly recommended to allow for samples to be run when personnel is not in the lab.
2. Variations in the retention time of the polyamine peaks can occur if the acetonitrile percentage in Buffer A differs slightly between runs. It is important to include a polyamines standard sample to each run to ensure proper identification of each peak.

Acknowledgments

We acknowledge the National Institutes of Health grant award R01 HD110500 (A.S.B., C.P.B.) for the provided funds to study BABS and SRS and we thank the MSU-Spectrum Health Alliance Corporation (SH-MSU-ACF RG101298) for the provided support to generate the data shown in this book chapter. We also thank the late Dr. Patrick Woster (Medical University of South Carolina, Charleston, South Carolina) for his countless contributions to the polyamine field and for providing us and the polyamine community with DFMO over many years.

Competing interests

A.S.B. and C.P.B. are listed inventors of two U.S. patents (US 11,273,137 B2 and US 12,194,010 B2) issued on March 15, 2022 and January 14, 2025, respectively, entitled "Methods and compositions to prevent and treat disorders associated with mutations in the *ODC1* gene" and Michigan State University and Corewell Health have an exclusive licensing agreement with Orbus Therapeutics, Inc. C.P.B. provides consulting services for Orbus Therapeutics. A.S.B. is sole inventor of a U.S. patent (US 9,072,778) issued on July 7, 2015, entitled "Treatment regimen for N-Myc, C-Myc, and L-Myc amplified and overexpressed tumors". No potential conflicts of interest were disclosed by the other authors.

References

Alirol, E., Schrumpf, D., Amici Heradi, J., Riedel, A., de Patoul, C., Quere, M., & Chappuis, F. (2013). Nifurtimox-eflornithine combination therapy for second-stage gambiense human African trypanosomiasis: Medecins Sans Frontieres experience in the Democratic Republic of the Congo. *Clinical Infectious Diseases: an Official Publication of the Infectious Diseases Society of America, 56*(2), 195–203. https://doi.org/10.1093/cid/cis886.

Alirol, E., Shrumpf, D., Heradi, J. A., Riedel, A., de Patoul, C., Quere, M., & Chappuis, F. (2012). Nifurtimox-Eflornithine Combination Therapy (NECT) for second-stage gambiense human African trypanosomiasis: MSF experience in the Democratic Republic of the Congo. *Clinical Infectious Diseases: An Official Publication of the Infectious Diseases Society of America*. https://doi.org/10.1093/cid/cis886.

Arena, J. F., Schwartz, C., Ouzts, L., Stevenson, R., Miller, M., Garza, J., & Lubs, H. (1996). X-linked mental retardation with thin habitus, osteoporosis, and kyphoscoliosis: Linkage to Xp21.3-p22.12. *American Journal of Medical Genetics, 64*(1), 50–58. https://doi.org/10.1002/(SICI)1096-8628(19960712)64:1<50::AID-AJMG7>3.0.CO;2-V.

Bacchi, C. J., Garofalo, J., Mockenhaupt, D., McCann, P. P., Diekema, K. A., Pegg, A. E., & Hutner, S. H. (1983). In vivo effects of alpha-DL-difluoromethylornithine on the metabolism and morphology of Trypanosoma brucei brucei. *Molecular and Biochemical Parasitology, 7*(3), 209–225.

Bachmann, A. S., & Geerts, D. (2018). Polyamine synthesis as a target of MYC oncogenes. *The Journal of Biological Chemistry, 293*(48), 18757–18769. https://doi.org/10.1074/jbc.TM118.003336.

Bachmann, A. S., Geerts, D., & Sholler, G. (2012). Neuroblastoma: Ornithine decarboxylase and polyamines are novel targets for therapeutic intervention. In In. M. A. Hayat (Vol. Ed.), *Pediatric cancer, neuroblastoma: diagnosis, therapy, and prognosis: Vol. 1*, (pp. 91–103). Springer.

Bachmann, A. S., & Levin, V. A. (2012). Clinical applications of polyamine-based therapeutics. In P. M. Woster, & R. A. CaseroJr. (Eds.). *Polyamine drug discovery* (pp. 257–276) Royal Society of Chemistry.

Bachmann, A. S., VanSickle, E. A., Michael, J., Vipond, M., & Bupp, C. P. (2024). Bachmann-Bupp syndrome and treatment. *Developmental Medicine and Child Neurology, 66*(4), 445–455. https://doi.org/10.1111/dmcn.15687.

Bupp, C. P., Schultz, C. R., Uhl, K. L., Rajasekaran, S., & Bachmann, A. S. (2018). Novel de novo pathogenic variant in the ODC1 gene in a girl with developmental delay, alopecia, and dysmorphic features. *American Journal of Medical Genetics. Part A, 176*(12), 2548–2553. https://doi.org/10.1002/ajmg.a.40523.

Casero, R. A., Jr., & Marton, L. J. (2007). Targeting polyamine metabolism and function in cancer and other hyperproliferative diseases. *Nature Reviews. Drug Discovery, 6*(5), 373–390.

Casero, R. A., Jr., Murray Stewart, T., & Pegg, A. E. (2018). Polyamine metabolism and cancer: Treatments, challenges and opportunities. *Nature Reviews. Cancer, 18*, 681–695. https://doi.org/10.1038/s41568-018-0050-3.

Cason, A. L., Ikeguchi, Y., Skinner, C., Wood, T. C., Holden, K. R., Lubs, H. A., & Schwartz, C. E. (2003). X-linked spermine synthase gene (SMS) defect: The first polyamine deficiency syndrome. *European Journal of Human Genetics: EJHG, 11*(12), 937–944. https://doi.org/10.1038/sj.ejhg.5201072.

Geerts, D., Koster, J., Albert, D., Koomoa, D. L., Feith, D. J., Pegg, A. E., & Bachmann, A. S. (2010). The polyamine metabolism genes ornithine decarboxylase and antizyme 2 predict aggressive behavior in neuroblastomas with and without MYCN amplification. *International Journal of Cancer. Journal International du Cancer, 126*(9), 2012–2024. https://doi.org/10.1002/ijc.25074.

Gerner, E. W., & Meyskens, F. L., Jr. (2004). Polyamines and cancer: Old molecules, new understanding. *Nature Reviews. Cancer, 4*(10), 781–792.

Hogarty, M. D., Ziegler, D. S., Franson, A., Chi, Y. Y., Tsao-Wei, D., Liu, K., & Marachelian, A. (2024). Phase 1 study of high-dose DFMO, celecoxib, cyclophosphamide and topotecan for patients with relapsed neuroblastoma: A New Approaches to Neuroblastoma Therapy trial. *British Journal of Cancer.* https://doi.org/10.1038/s41416-023-02525-2.

Holbert, C. E., Cullen, M. T., Casero, R. A., Jr., & Stewart, T. M. (2022). Polyamines in cancer: Integrating organismal metabolism and antitumour immunity. *Nature Reviews. Cancer, 22*(8), 467–480. https://doi.org/10.1038/s41568-022-00473-2.

Lewenhoeck, D. A. (1677). Observationes D. Anthonii Lewenhoeck, de natis e semine genitali aminalculis. *Philosophical Transactions, 12*, 1040–1046.

Meyskens, F. L., Jr., McLaren, C. E., Pelot, D., Fujikawa-Brooks, S., Carpenter, P. M., Hawk, E., & Gerner, E. W. (2008). Difluoromethylornithine plus sulindac for the prevention of sporadic colorectal adenomas: A randomized placebo-controlled, double-blind trial. *Cancer Prevention Research (Phila), 1*(1), 32–38. https://doi.org/10.1158/1940-6207.CAPR-08-0042.

Oesterheld, J., Ferguson, W., Kraveka, J. M., Bergendahl, G., Clinch, T., Lorenzi, E., & Saulnier Sholler, G. L. (2023). Eflornithine as postimmunotherapy maintenance in high-risk neuroblastoma: Externally controlled, propensity score-matched survival outcome comparisons. *Journal of Clinical Oncology: Official Journal of the American Society of Clinical Oncology,* JCO2202875. https://doi.org/10.1200/JCO.22.02875.

Priotto, G., Kasparian, S., Mutombo, W., Ngouama, D., Ghorashian, S., Arnold, U., & Kande, V. (2009). Nifurtimox-eflornithine combination therapy for second-stage African Trypanosoma brucei gambiense trypanosomiasis: A multicentre, randomised, phase III, non-inferiority trial. *Lancet, 374*(9683), 56–64. https://doi.org/10.1016/S0140-6736(09)61117-X.

Raj, K. P., Zell, J. A., Rock, C. L., McLaren, C. E., Zoumas-Morse, C., Gerner, E. W., & Meyskens, F. L. (2013). Role of dietary polyamines in a phase III clinical trial of difluoromethylornithine (DFMO) and sulindac for prevention of sporadic colorectal adenomas. *British Journal of Cancer, 108*(3), 512–518. https://doi.org/10.1038/bjc.2013.15.

Rajasekaran, S., Bupp, C. P., Leimanis-Laurens, M., Shukla, A., Russell, C., Junewick, J., & Bachmann, A. S. (2021). Repurposing eflornithine to treat a patient with a rare ODC1 gain-of-function variant disease. *Elife, 10*. https://doi.org/10.7554/eLife.67097.

Saulnier Sholler, G. L., Gerner, E. W., Bergendahl, G., MacArthur, R. B., VanderWerff, A., Ashikaga, T., & Bachmann, A. S. (2015). A phase I trial of DFMO targeting polyamine addiction in patients with relapsed/refractory neuroblastoma. *PLoS One, 10*(5), e0127246. https://doi.org/10.1371/journal.pone.0127246.

Schultz, C. R., Bupp, C. P., Rajasekaran, S., & Bachmann, A. S. (2019). Biochemical features of primary cells from a pediatric patient with a gain-of-function ODC1 genetic mutation. *The Biochemical Journal, 476*(14), 2047–2057. https://doi.org/10.1042/BCJ20190294.

Schwartz, C. E., Peron, A., & Kutler, M. J. (2020). Snyder-robinson syndrome. In M. P. Adam, G. M. Mirzaa, R. A. Pagon, S. E. Wallace, L. J. H. Bean, K. W. Gripp, & A. Amemiya (Eds.), *GeneReviews* (R). Seattle, WA, 1993–2024.

Sholler, G. L. S., Ferguson, W., Bergendahl, G., Bond, J. P., Neville, K., Eslin, D., & Kraveka, J. M. (2018). Maintenance DFMO increases survival in high risk neuroblastoma. *Scientific Reports, 8*(1), 14445. https://doi.org/10.1038/s41598-018-32659-w.

Simoneau, A. R., Gerner, E. W., Phung, M., McLaren, C. E., & Meyskens, F. L., Jr. (2001). Alpha-difluoromethylornithine and polyamine levels in the human prostate: Results of a phase IIa trial. *Journal of the National Cancer Institute, 93*(1), 57–59.

Snyder, R. D., & Robinson, A. (1969). Recessive sex-linked mental retardation in the absence of other recognizable abnormalities. Report of a family. *Clinical Pediatrics (Phila), 8*(11), 669–674. https://doi.org/10.1177/000992286900801114.

Wallick, C. J., Gamper, I., Thorne, M., Feith, D. J., Takasaki, K. Y., Wilson, S. M., & Bachmann, A. S. (2005). Key role for p27Kip1, retinoblastoma protein Rb, and MYCN in polyamine inhibitor-induced G1 cell cycle arrest in MYCN-amplified human neuroblastoma cells. *Oncogene, 24*(36), 5606–5618.

Wolf, J. E., Jr., Shander, D., Huber, F., Jackson, J., Lin, C. S., Mathes, B. M., & Eflornithine, H. S. G. (2007). Randomized, double-blind clinical evaluation of the efficacy and safety of topical eflornithine HCl 13.9% cream in the treatment of women with facial hair. *International Journal of Dermatology, 46*(1), 94–98. https://doi.org/10.1111/j.1365-4632.2006.03079.x.

CHAPTER FIFTEEN

Gene replacement therapy to restore polyamine metabolism in a Snyder-Robinson syndrome mouse model

Oluwaseun Akinyele, Krystal B. Tran, Marie A. Johnson, and Dwi U. Kemaladewi[*,1]

Division of Genetic and Genomic Medicine, Department of Pediatrics, University of Pittsburgh School of Medicine, UPMC Children's Hospital of Pittsburgh, Pittsburgh, PA, United States
[*]Corresponding author. e-mail address: dwk24@pitt.edu

Contents

1. Introduction	272
2. Materials	275
2.1 Adeno-associated viral vector (AAV)	275
2.2 Animals and genotyping	275
2.3 Temporal vein injection in neonatal mice	276
2.4 Tail vein injection in juvenile mice	276
2.5 Tissue isolation and biodistribution evaluation	276
2.6 Expression analysis	276
3. Methods	277
3.1 AAV design	277
3.2 Preparation of AAV aliquots	279
3.3 Mouse genotyping	281
3.4 Temporal vein injection in neonatal mice	283
3.5 Tail vein injection in juvenile mice	285
3.6 Tissue isolation and biodistribution evaluation	286
3.7 Analysis of SMS protein expression using western blot	287
4. Conclusion	289
Acknowledgments	289
References	290

Abstract

Polyamines, including putrescine, spermidine, and spermine, are organic cations essential for cell growth, proliferation, and tissue regeneration. Their levels are tightly

---

[1] Present address: Gene Transfer and Immunogenicity Branch, Office of Therapeutic Products, Center for Biologics Evaluation & Research, U.S. Food and Drug Administration (FDA), 10903 New Hampshire Ave., Silver Spring, MD 20993

Methods in Enzymology, Volume 715
ISSN 0076-6879, https://doi.org/10.1016/bs.mie.2025.01.068
Copyright © 2025 Elsevier Inc. All rights reserved, including those for text and data mining, AI training, and similar technologies.

regulated by a set of enzymes controlling their biosynthesis, catabolism, and inter-conversion. Dysregulation of polyamine metabolism is associated with a group of rare genetic neurodevelopmental disorders collectively known as "polyaminopathies", including Snyder-Robinson Syndrome (SRS).

SRS is an X-linked recessive disorder caused by mutations in the *SMS* gene, which encodes the spermine synthase enzyme. The lack of spermine synthase leads to aberrant polyamine levels and neurological impairments, as observed in patients and animal models. Currently, there are no available treatment options for SRS.

Due to its monogenic nature, SRS is an excellent candidate for gene replacement therapy. The recent success of Zolgensma in treating children with Spinal Muscular Atrophy and the establishment of Platform Vector Gene Therapy (Pave-GT) initiative at the National Institute of Health (NIH) offer a framework to adapt-and-apply the same gene delivery system for multiple rare disease gene therapies.

This chapter outlines strategies for delivering a functional copy of the *SMS* gene using an adeno-associated viral (AAV) vector, as well as methods to evaluate the molecular efficacy of this approach in an SRS mouse model. Our ultimate goal is to establish a versatile platform for genetic interventions targeting SRS and other polyaminopathies.

1. Introduction

Polyamines, including putrescine, spermidine, and spermine, are vital for cellular processes such as chromatin stabilization, protein synthesis, and cell proliferation. Several studies have revealed critical role of polyamine metabolism in five distinct genetic disorders, collectively termed "polyaminopathies" (Fig. 1).

The first member of polyaminopathies identified is Snyder-Robinson Syndrome (SRS; MIM: 309583), which is caused by mutations in *SMS* gene encoding spermine synthase protein, leading to elevated spermidine/spermine ratio (Cason et al., 2003; Larcher et al., 2020; Leung et al., 2024; Marhabaie et al., 2021; Mouskou et al., 2021; Peron et al., 2013; Qazi et al., 2020; Zhang et al., 2013). Individuals with SRS are presented with intellectual disability, seizures, thin habitus with low muscle mass, osteoporosis, kyphoscoliosis, and facial dysmorphism.

Bachman-Bupp Syndrome (BABS; MIM: 619075) is caused by gain-of-function mutations in *ODC* encoding the ornithine decarboxylase-1 protein, which clinically manifest as intellectual disability, developmental delay, low muscle mass, and non-congenital alopecia (Bupp, Schultz, Uhl, Rajasekaran, & Bachmann, 2018; Rodan et al., 2018). Elevated ODC enzymatic activity results in increased putrescine level (Schultz, Bupp,

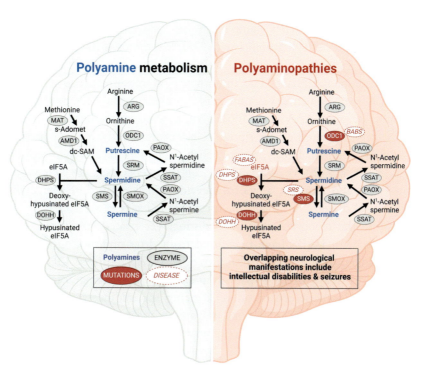

Fig. 1 **Polyamine metabolism and polyaminopathies.** The left panel illustrates the polyamine metabolism pathway, highlighting key metabolites (blue) such as putrescine, spermidine, and spermine, as well as enzymes (gray ovals) involved in their biosynthesis, interconversion, and catabolism. The right panel depicts polyaminopathies (white ovals), which arise from mutations in genes encoding enzymes critical for polyamine metabolism (red ovals). These mutations disrupt normal metabolic processes, leading to intellectual disabilities, seizures, and other neurological manifestations. Essential processes such as the formation of hypusinated eIF5A (eukaryotic translation initiation factor 5A) are shown. ARG, Arginase; ODC1, Ornithine decarboxylase 1; SRM, Spermidine synthase; SMS, Spermine synthase; SMOX, Spermine oxidase; PAOX, Polyamine oxidase; SSAT, Spermidine/Spermine N1-Acetyltransferase 1; MAT, Methionine adenosyltransferase; s-Adomet, S-Adenosylmethionine; dcSAM, Decarboxylated S-Adenosylmethionine; AMD1, S-Adenosylmethionine Decarboxylase; eIF5A, Eukaryotic Translation Initiation Factor 5A; DHPS, Deoxyhypusine Synthase; DOHH, Deoxyhypusine Hydroxylase; FABAS, Faundes–Banka Syndrome. *Figures are made using BioRender.*

Rajasekaran, & Bachmann, 2019), which can be restored to physiological baseline using eflornithine, an ODC inhibitor (Rajasekaran et al., 2021).

Subsequent research identified genetic mutations leading to perturbation in hypusination process of eIF5A, which is mediated by spermidine. Decreased enzymatic activity of deoxyhypusine synthase (DHPS) due to biallelic variants in *DHPS* gene results in reduced eIF5A hypusination (Ganapathi et al., 2019). Patients with DHPS disorder (MIM: 618480) are presented with neurodevelopmental delay, seizures, and in some cases, short stature, and microcephaly. The importance of spermine-mediated eIF5A hypusination in human development is also evidence in cases of Faundes-Banka Syndrome (FABAS, MIM: 619376), in which frameshift mutations in *EIF5A* cause impaired eIF5A-ribosome interactions and variable combinations of developmental delay, microcephaly, micrognathia and dysmorphism (Faundes et al., 2021). Variants in *DOHH* encoding deoxyhypusine hydroxylase also manifest as global developmental delay, intellectual disability, facial dysmorphism, and microcephaly (Ziegler et al., 2022).

Despite the recognition of these distinct yet related conditions, only individuals with BABS have benefitted from drug repurposing of eflornithine, an Food and Drug Administration (FDA)-approved ODC inhibitor used to treat African trypanosomiasis. For SRS, therapeutic approaches in development center on rebalancing spermidine/spermine levels using polyamine supplementation, polyamine analogs, and enzyme inhibition with varying outcomes in patient-derived and Drosophila models of SRS (Murray Stewart et al., 2020; Stewart et al., 2023; Tantak, Sekhar, Tao, Zhai, & Phanstiel, 2021; Tao et al., 2022). In addition to these efforts, we reason that re-establishing normal polyamine levels could be achieved by introducing a functional copy of the *SMS* gene.

Adeno-associated virus (AAV) vectors have become a key platform for delivering therapeutic genes to target cells in organs such as eye, muscle, liver, and brain, in mice, nonhuman primates, and humans (Wang, Tai, & Gao, 2019). As of January 2025, the U.S. FDA has approved AAV-based gene therapy products for treating Leber congenital amaurosis, Hemophilia A, Hemophilia B, Duchenne muscular dystrophy, and spinal muscular atrophy (SMA).

AAV vectors are inherently disease-agnostic, meaning their suitability for a specific disease is determined by factors like their biodistribution, capsid serotype, route of administration, dose, the genetic mechanism to address (loss- vs. gain-of-function), and the expression cassette used, rather than by the pathophysiological specifics of the disease (Brooks; Brooks et al., 2020). Incorporation of various therapeutic genes into a chosen AAV capsid facilitates development of simultaneous treatments for different diseases. This modular approach allows for swapping transgenes, selection of regulatory elements (enhancer, promoter, etc.), and modifying the capsid to target specific tissues and cell types. This platform approach has become a model for gene therapy for rare diseases, demonstrated by the NIH-backed PaVe-GT Initiative – a pilot project using the same AAV9 serotype to develop gene therapies for four rare diseases within organic acidemias and congenital myasthenia syndromes (Brooks et al., 2020).

Building on this concept, we propose adopting the AAV9 construct originally developed for SMA as a foundation for gene therapy in SRS with slight modifications. Zolgensma, an FDA-approved self-complementary AAV9, delivers a functional *SMN1* gene to SMA patients, crossing the blood-brain barrier and improving motor functions (Foust et al., 2010; Strauss et al., 2022).

This chapter outlines the methods and considerations for delivering a functional copy of the *SMS* gene using AAV9 in a mouse model for SRS that our group has recently characterized (Akinyele et al., 2024), along with molecular approaches to assess efficacy. We envision developing a future platform to evaluate the therapeutic potential of genetic interventions for SRS and other polyaminopathies.

2. Materials

2.1 Adeno-associated viral vector (AAV)

1. AAV9-SMS
2. Optional: AAV9-GFP (to assess gross biodistribution)
3. Optional: AAV-PHP.eB carrying SMS and/or GFP (to increase transduction in the brain)
4. Sterile Phosphate Buffer Saline (PBS)
5. Biosafety Cabinet (BSC) Class II; this may vary depending on institutional regulations.

2.2 Animals and genotyping

1. B6C3-$Sms^{em2Lutzy}$/J (B6C3F1. Sms^{G56S}; The Jackson Laboratory Strain # 033707).

   Hemizygous mice are generated by breeding female heterozygous (B6C3-$Sms^{em2Lutzy}$/J) with male B6C3F1/J (The Jackson Laboratory Strain # 100010).

   Wildtype male mice from the breeding colony are used as controls.
2. Metal ear tags
3. Surgical scissors
4. Tissue lysis buffer (25 mM NaOH/0.2 mM EDTA)
5. 40 mM Tris HCl (pH 5.5)

6. Heating block
7. Q5 High-Fidelity PCR (NEB, M0491)
8. PCR SuperMix (Life Technologies, 10572-014)
9. Primers to detect *Sms* mutation and determine sex:
   Sms-Forward: GTCTTCCAGCTGATGGTGAG
   Sms-Reverse: CAAGTGGTGGTGTTGCACAC
   Rbm31x/y-Forward: CACCTTAAGAACAAGCCAATACA
   Rbm31x/y-Reverse: GGCTTGTCCTGAAAACATTTGG
10. *Eco*RI restriction enzyme and the supplied CutSmart Buffer (NEB, R3101S)
11. Thermocycler PCR machine

2.3 Temporal vein injection in neonatal mice
1. Wet ice in a beaker/an ice bucket
2. Sterile paper absorbent
3. 0.5 mL 31 G insulin syringes
4. Heating pad

2.4 Tail vein injection in juvenile mice
1. 70% ethanol
2. Mouse restrainer
3. Overhead heat lamp
4. Warm water
5. 0.5 mL 31G insulin syringes

2.5 Tissue isolation and biodistribution evaluation
1. CO_2 euthanizing chamber
2. 50 mL conical tubes
3. IVIS imaging system
4. Parafilm
5. Liquid nitrogen

2.6 Expression analysis
1. RIPA buffer
   a. Buffer 1 (RIPA no detergent):
      i. 50 mM Tris-HCl pH 7.4
      ii. 150 nM NaCl
      iii. 1 mM EDTA
      iv. Protease inhibitor mini tablet (Thermo Scientific, A32953)
   b. Buffer 2 (RIPA double detergent):
      i. 50 mM Tris-HCl pH 7.4

ii. 150 nM NaCl
iii. 1 mM EDTA
iv. 2% deoxycholate
v. 2% NP40
vi. 2% Triton X-100
vii. Protease inhibitor mini tablet (Thermo Fisher, A32953)
2. 2 mL tubes prefilled with 1.4 mm ceramic beads (Fisher Sci, 15-340-153)
3. Tissue homogenizer (Bullet Blender Homogenizer, Next Advance)
4. BCA assay kit (Thermo Fisher, 23227)
5. Western blot reagents:
   4–12% Bis-Tris gradient gel (Thermo Fisher, NP0321BOX)
   MOPS SDS running buffer (Thermo Fisher, NP0001)
   LDS sample buffer (Thermo Fisher, NP0007)
   Sample reducing agent (Thermo Fisher, NP0009)
   iBlot 2 NC regular stacks (Thermo Fisher, iB23001)
   Ponceau stain (Sigma, P7170-1L Ponceau S solution 0.1% (w/v) in 5% acetic acid)
6. Protease inhibitor (Thermo Fisher, A32965)
7. Antibodies:
   Rabbit anti-SMS antibody [EPR9252(B)] (Abcam, ab156879)
   Rabbit anti-vinculin antibody [EPR8185] (Abcam, ab129002)
   Goat anti-rabbit IgG-HRP antibody (Bio-Rad, 1706515)
8. Tris-Buffered saline with Tween, pH 7.4 (Thermo Fisher, J77500-K2, dilute to 1x working solution)
9. Non-fat dry milk (Fisher Sci, NC9121673)
10. SuperSignal West Femto Maximum Sensitivity Substrate (Thermo Fisher, 34095)

3. Methods

3.1 AAV design

This section focuses on designing plasmids to overexpress the gene of interest (i.e., *SMS*) and the packaging into the appropriate AAV vector. The recombinant AAV is produced by triple transfection of a transfer plasmid carrying the gene of interest, a helper plasmid (encoding genes that mediate AAV replication), and a Rep/Cap plasmid (that determines the tissue serotypes) into a packaging cell line, such as human embryonic kidney cells (HEK293; Fig. 2).

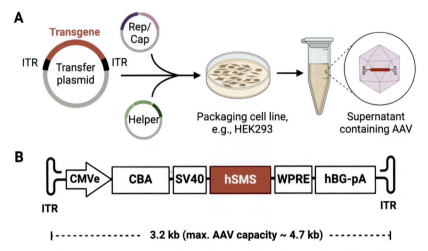

Fig. 2 Production and structure of recombinant AAV. (A) Overview of the triple transfection method for AAV production. The transfer plasmid with the transgene, helper plasmid for replication, and Rep/Cap plasmid for serotype specification are co-transfected into HEK293 cells. AAV is collected from the supernatant. (B) Vector map showing key component flanked by Inverted Terminal Repeats (ITRs): CMV enhancer (CMVe), chicken beta-actin (CBA) promoter, SV40 site, human *SMS* (hSMS) coding sequence as the transgene (red), woodchuck hepatitis post-transcriptional regulatory element (WPRE), and the human beta-globin polyadenylation (hBG-polyA) signal. The presented design of AAV9-SMS occupies 3.2 kb out of the ~4.7 kb genomic capacity of an AAV. *Figures are made using BioRender.*

3.1.1 Design of transfer plasmid

The typical transfer plasmid carries a gene expression cassette, which includes a promoter, a transgene, and a transcription termination signal flanked by two ends of inverted terminal repeats (ITRs). The ~4700 nucleotide space between the two ITRs determines the packaging load of an AAV.

A strong, ubiquitous promoter is used to achieve high transgene expression. Such promoters include the cytomegalovirus (CMV) promoter and a fusion of chicken β-actin promoter and a short variant of CMV early enhancer (CBA; Fig. 2). Alternatively, a tissue- or cell-specific promoter can be used to control gene expression in desired tissues or organs.

The full-length coding region of human *SMS* (NM_004595.5) is incorporated into the transfer plasmid (Fig. 2). It is advisable to apply codon optimization, while maintaining low CpG dinucleotide content, to improve the transgene expression (Wright, 2020; Xie et al., 2024). In parallel, a reporter gene such as *Luciferase, RFP,* or *GFP* can be packaged and tested side-by-side to enable gross biodistribution and/or transgene expression analyses (see Section 3.6 below).

Finally, we also incorporated several regulatory elements, including SV40 intron, the woodchuck hepatitis post-transcriptional regulatory element (WPRE), and the human beta-globin polyadenylation (hBG-polyA) signal sequence (Fig. 2). A publicly available webtool such as VectorBuilder (VectorBuilder) allows the swapping of transgenes and regulatory elements (promoter, enhancer, etc.) to construct a transfer plasmid in a straightforward manner.

Note: The ITR sequences of AAV vectors are highly unstable. They often accumulate deletions or undergo recombination during plasmid propagation in bacteria. Therefore, we recommend using recombinant-deficient strains, such as Stable Competent *E. coli,* and sequencing the entire plasmids to prevent spontaneous mutations and/or recombination.

3.1.2 Serotype selection and production

Depending on their serotype, AAVs can have specific tropism for specific organs and tissues in the body (Table 1). We opt for AAV serotype 9, which has broad tropism to most tissues in vivo, including the central nervous system (CNS) and skeletal muscles (Ganapathi et al., 2019; Rajasekaran et al., 2021), tissues that are affected in SRS. In parallel, we test AAV-PHP.eB, an engineered AAV9-derivative with vastly improved blood-brain barrier penetration (Chan et al., 2017; Mathiesen, Lock, Schoderboeck, Abraham, & Hughes, 2020). As illustrated in Fig. 2, the serotype of the AAV is determined by the Rep/Cap plasmids (commercially available at plasmid repositories). Several biotechnology companies or university-affiliated vector cores offer competitive packages for virus production, purification, titer determination, and quality control. Readers interested in in-house AAV production methods are referred to previously published protocols (Gray et al., 2011; Kimura et al., 2019; Shin, Yue, & Duan, 2012).

3.2 Preparation of AAV aliquots

This section outlines practical considerations in preparing virus aliquots before mouse injections. Multiple freeze-thaw cycles may reduce the viral titer and should be avoided.

1. Thaw viral stock on ice. Vortex occasionally to help remove ice crystals and ensure the virus is completely thawed.
2. In a biosafety cabinet, carefully open the tube. Depending on the viral titer, aliquot the virus into the practical volumes, *e.g.*, 20–100 µl, into 0.5 mL microfuge tube(s). Properly close and label the tubes.

Table 1 Tissue tropism of selected AAV serotypes.

AAV serotype	CNS	Lung	Liver	Skeletal muscle	Kidney	Heart	Pancreas	Retina	Bone
AAV1	✓	✓		✓		✓		✓	
AAV2	✓		✓		✓			✓	
AAV3		✓	✓			✓			
AAV4	✓					✓		✓	
AAV5	✓	✓	✓	✓				✓	
AAV6	✓	✓	✓	✓		✓			
AAV7			✓	✓					
AAV8			✓	✓			✓	✓	✓
AAV9	✓	✓	✓	✓	✓	✓		✓	✓
AAV-rh10	✓	✓	✓			✓			
AAV-PHP.B	✓								
AAV-PHP.eB	✓								
MyoAAV				✓		✓			
BI-hTFR1	✓		✓						

Note: Since new generations of AAVs are continuously being developed, Table 1 is not exhaustive. Readers are encouraged to tailor their selections based on scientific needs, such as cell types, genetic backgrounds, and preclinical animal models.

3. When all viruses have been aliquoted, dispose the virus tube and pipette tips into the biohazard waste container.
4. Aliquot equal volumes of PBS (similar to step 2).
5. Clean any equipment and work area based on institutional regulations.
6. Store aliquots in −80 °C freezer.
7. On injection day, take the required volume of viruses (depending on the titer) from each aliquot and bring it to the appropriate injection volume with PBS. Store any unused virus in a 4 °C fridge for up to a week.

3.3 Mouse genotyping

This section covers PCR-based method to determine the genotype and sex of the B6C3-$Sms^{em2Lutzy}$/J (also known as G56S) mice.

3.3.1 Tissue collection and DNA isolation

1. At the weaning age (3 weeks old), securely restrain the animal and apply metal ear tags. The tag should hang from the bottom of the ear.
2. Cut a small portion (2-3 mm) of the contralateral ear using sharp scissors. This procedure can be done on mice once their ears have developed (older than 8 days of age) and does not require anesthesia.
3. Place the ear snip in a 1.5 mL microfuge tube.
4. Add 75 μl of 25 mM NaOH /0.2 mM EDTA lysis buffer.
5. Incubate the samples at 98 °C for 1 h.
6. Add 75 μl of 40 mM Tris-HCl (pH 5.5).
7. Centrifuge at 4000 rpm for 3 min.
8. Measure DNA concentration using a NanoDrop and dilute to 40 ng/μl.

3.3.2 Detection of Sms mutation using PCR-Digest

1. Set up PCR reactions based on Table 2.

Table 2 Preparation of PCR reactions

Component	Volume (25 μl reaction)
DNA (40 ng)	1 μl
PCR SuperMix	22.5 μl
20 μM Sms-Forward	0.25 μl
20 μM Sms-Reverse	0.25 μl
DNA/RNA-free water	1 μl

2. Spin the tube briefly to collect the solutions at the bottom before running on a thermocycler using the following cycle conditions: 95 °C for 2 min, 35 cycles of (95 °C for 30 s, 60 °C for 30 s, 72 °C for 30 s), and final extension at 72 °C for 5 min.
3. Following the amplification, process the PCR products for restriction digest based on Table 3.
4. Spin the tube briefly and incubate at 37 °C for 15 min.
5. Run 10 μl of the reaction mixture on a 2.5 % agarose gel. Fig. 3A shows example of a genotyping result. The *Sms* mutation in the G56S mouse introduces an *Eco*RI cut site. Thus, the *Eco*RI digestion produces the following expected fragment sizes: WT = 330 bp, hemizygote (G56S) = 210 bp and 120 bp, and heterozygote = 330 bp, 210 bp and 120 bp.

3.3.3 PCR-based sex determination
1. Set up PCR reactions based on Table 4.
2. Briefly spin the tube to collect the solutions at the bottom before running it on a thermocycler with the following cycle conditions: 98 °C for 30 s, followed by 35 cycles of 98 °C for 10 s, 62 °C for 10 s, and 72 °C for 10 s, with a final extension at 72 °C for 2 min.

Table 3 Preparation of restriction digest reactions

Component	Volume (20 μl reaction)
PCR product	16 μl
10X CutSmart buffer	2 μl
*Eco*RI enzyme	2 μl

Table 4 Preparation of PCR reactions

Component	Volume (25 μl reaction)
DNA (40 ng)	1 μl
10 mM dNTPs	0.5 μl
20 μM Rbm31x/y-Forward	0.625 μl
20 μM Rbm31x/y-Reverse	0.625 μl
5X Q5 reaction buffer	5 μl
Q5 DNA polymerase	0.25 μl
DNA/RNA-free water	17 μl

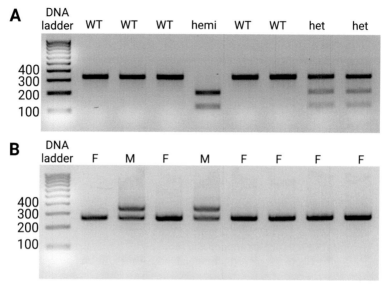

Fig. 3 Sms genotyping and sex determination. (A) A litter from heterozygous B6C3-Sms^{em2Lutzy}/J females and B6C3F1/J mice males is analyzed for the *Sms* mutation. PCR-digest produces fragments: WT = 330 bp, hemizygote (G56S) = 210 bp and 120 bp, and heterozygote = 330 bp, 210 bp, and 120 bp. (B) The same litter is analyzed for the sex-specific deletion in the *Rbm31x* gene, yielding a 269-bp product for XX (female; F) and two products, 269 bp and 353 bp, for XY (male; M) samples. *Figures are made using BioRender.*

3. Run 25 μl of the PCR product on a 2.5% agarose gel. The primers were designed to flank an 84-bp deletion in the X-linked *Rbm31x* gene relative to its Y-linked gametolog, *Rbm31y* (Tunster, 2017). This resulted in a single 269-bp product from XX (female) samples and two products, 269 bp and 353 bp, from XY (male) samples (Fig. 3B).

3.4 Temporal vein injection in neonatal mice

This section describes an efficient method of delivering AAV systemically into neonatal mice through the superficial temporal vein no later than postnatal day 2.

1. Retrieve virus aliquots (see Section 3.2) from −80 °C freezer and allow to thaw on ice.
2. Determine the virus dose to be injected into the mouse and calculate the amount of virus needed. Take the required volume of AAV and dilute in 1x sterile PBS if needed. Final volume of 20–50 μl is appropriate for neonatal mice.

Fig. 4 Temporal vein injection in neonatal mouse. (A) Neonatal mice in their crinkle bedding, separated from the dam. (B) Neonatal mouse anesthetized on wet ice. (C and D). Injection into the superficial temporal vein. The arrow indicates the visible vein (C) and needle positioning while injecting the AAV solution (D). (E) A visible black spot (arrow) at the site of injection, indicating successful injection. (F) The injected pup is covered with the dam's feces to mask handling odors and prevent rejection by the dam. *Figures are made using BioRender.*

3. Load the insulin syringe with the AAV vector from above.
4. Remove the dam from the neonates and carefully open the crinkle bedding (Fig. 4A).
5. To anesthetize the pups before injection, place a clean paper towel on an ice bucket. Working with one pup at a time, gently place the pup on the ice for 30–60 s or until it becomes immobile (Fig. 4B). Avoid leaving the animal on ice for too long to prevent hypothermia.
6. Gently hold the pup by the scruff, being careful not to apply excessive pressure to avoid injuring the animal. Identify the superficial temporal vein (Fig. 4C).
7. Once the vein is identified, hold the syringe with the needle beveled up. Secure the syringe between the index and middle fingers, with the ring finger providing additional support. Place the thumb on the plunger. Care must be taken to avoid dispensing AAV solution from the needle before the injection.
8. Position the needle tip at an angle of approximately 5–10 degrees above the temporal vein. Gently insert the needle into the vein and advance it about 1–2 mm along the vein's track (Fig. 4D). Proper insertion is typically indicated by blood filling the needle bevel. Once

the needle is inside the vein, slowly press the plunger to inject the virus solution into the mouse.
9. After the injection, wait 3–5 s before slowly withdrawing the needle. Gently apply pressure to the injection site with a clean, sterile absorbent to prevent bleeding. A successful injection is typically indicated by the appearance of a black blemish at the injection site (Fig. 4E).
10. Gently place the pup in a cage lined with crinkle bedding and placed the cage on a heating pad to help restore the pup's body temperature.
11. Before returning the dam to the cage, carefully rub the pups with the dam's feces or urine (Fig. 4F) to mask any foreign odors from handling, which could otherwise lead to rejection by the dam.

Note: For temporal vein injection, all pups are injected regardless of their genotypes. At the weaning age of 3-weeks, the mice are genotyped for sex and the *Sms* mutation.

3.5 Tail vein injection in juvenile mice

This section describes a method to administer AAV9 systemically via tail vein in the G56S mice of least 3 weeks of age. An alternative delivery method to tail- and temporal veins is retroorbital injection, which is suitable for all ages. This method is out of the scope of this paper; however, readers are referred to previously published protocols (Gessler, Tai, Li, & Gao, 2019; Yardeni, Eckhaus, Morris, Huizing, & Hoogstraten-Miller, 2011).

1. Weigh the mice prior to injection.
2. Retrieve virus aliquots (see Section 3.2) from $-80\,°C$ freezer and allow to thaw on ice.
3. Determine the virus dose to be injected into the mouse and calculate the amount of virus needed when corrected to the body weight. Take the required volume of AAV and dilute in 1x sterile PBS if needed. Final volume 100 μl is appropriate for tail vein injection in juvenile mice.
4. Load the insulin syringe with the AAV vector from above.
5. Place the mouse in a restrainer and secure its movement using the restrainer plug. Be careful not to over-restrain the mouse to prevent restricting its breathing (Fig. 5A).
6. Immerse the tip of the tail in warm water (slightly above $30\,°C$) (Fig. 5B) and position a heat lamp over the mouse, close to the tail, to dilate the veins and improve visibility.
7. Wipe the tail with tissue paper soaked in 70% ethanol.

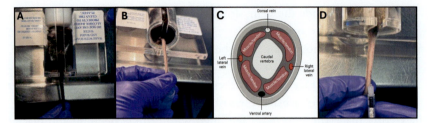

Fig. 5 Tail vein injection in juvenile mouse. (A) The mouse is positioned in a restrainer. (B) The tip of the tail is immersed in warm water to dilate the tail vein. (C) A schematic showing the four major blood vessels in the tail when the mouse is upright. Either left or right lateral veins (red) can be selected for injection. (D) Once the vein is visible, a marker is used to trace the vein (dotted red line) for easier identification during injection. The needle is then carefully inserted at a 20° angle along the marked line to inject the virus. *Figures are made using BioRender.*

8. Hold the tail straight and taut. In an upright position, the mouse's dorsal vein is at 12 o'clock, while the two lateral veins are located at the 3- and 9 o'clock positions (Fig. 5C). Choose one of the lateral veins, e.g., the right lateral vein, for the injection.
9. Using a marker, carefully trace the lateral vein line, starting from the base of the tail and moving toward the tip (Fig. 5D).
10. With the tail bent, insert the needle at a 20° angle into the vein, with the bevel facing up. Once the tip penetrates the vein halfway past the bevel, adjust the needle so that it aligns with the length of the vein (Fig. 5D).
11. Carefully dispense the 100 μl of the viral solution into the vein. If the needle is properly inserted into the vein, the plunger would move smoothly without resistance. Any resistance indicates that the needle has struck muscle, tendon, or vertebra (Fig. 5C) instead of the vein; in this case, withdraw the needle and re-inject at a site along the lateral vein line closer to the body.
12. After injection, carefully withdraw the needle and gently apply pressure with a clean, sterile paper towel to prevent any bleeding.
13. Return the mouse to the cage.

3.6 Tissue isolation and biodistribution evaluation

This section outlines the use of an in vivo imaging system (IVIS) to visualize fluorescent signals in organs isolated from animals injected with GFP-containing AAV. This methodology is particularly useful for ex vivo imaging and for comparing different AAV serotypes.

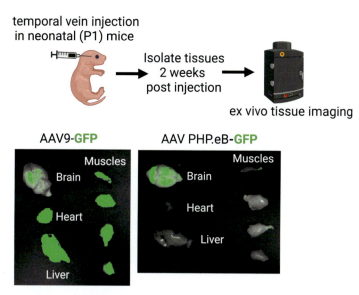

Fig. 6 Comparison of GFP expression in intact organs. Freshly isolated tissues from mice injected with either AAV9-GFP or AAVPHP.eB-GFP at equal doses. The observed GFP expression across different tissues highlights the different biodistributions/tropism between the two AAV serotypes. *Figures are made using BioRender.*

1. Euthanize the mouse in a CO_2 chamber.
2. Carefully dissect and isolate tissues of interest.
3. Place the fresh organs on parafilm and analyzed for fluorescence using the IVIS system. It is essential to calibrate the machine before use to reset the instrument back to zero to ensure consistent and reproducible results.
4. Set the IVIS fluorescent filter to 480–520 nm measurement range. The enhanced GFP has an excitation peak of 488 nm and emission peak of 510 nm. Fig. 6 presents an example of GFP expression in tissues isolated from mice injected with two different AAV serotypes, e.g., AAV9-GFP and AAV-PHP.eB-GFP at equal dose.
5. Following imaging, tissues can be snap frozen in liquid N2 for further molecular analyses.

3.7 Analysis of SMS protein expression using western blot

This section outlines the methods for tissue homogenization and analysis of SMS protein expression using western blot.

1. Place a small piece of freshly isolated tissues into a 2 mL microtube with ceramic beads.

2. Add 150 μl of RIPA buffer (without detergent) containing protease inhibitor.
3. Homogenize the tissue using a Bullet Blender homogenizer at maximum speed for 4 min. Afterward, incubate the tubes on wet ice for 1 min and centrifuge at 10,000 g at 4 °C for 1 min. If tissue remains visible, repeat the homogenization process (3-4 times) until the sample is fully homogenized.
4. Remove homogenate from the tube and add equal volume of RIPA buffer (with detergent and protease inhibitor). Incubate the samples on a roller at 4 °C for 1 h.
5. Following incubation, centrifuge samples at 12,000 g for 10 min at 4 °C.
6. Quantify the total protein concentration using BCA assay according to manufacturer's instructions.
7. Load 10–30 μg of protein onto a 4–12% Bis-Tris gel with MOPS running buffer.
8. Transfer the protein from the gel to a nitrocellulose membrane using the semi-dry transfer technique using the iBlot device at 20 V for 2 min, 23 V for 6 min, 25 V for 4 min, according to the manufacturer's instruction.
9. After transfer, stain the membrane with 2 mL of Ponceau stain to confirm successful protein transfer. Wash the membrane in TBST to remove Ponceau stain.
10. Block the membrane in 5% non-fat milk in TBST for 1 h at 4 °C.
11. Probe the membrane overnight at 4 °C with rabbit anti-SMS (1:1000 in 5% milk/TBST) and rabbit anti-vinculin (1:2500 in 5% milk/TBST) antibodies.
12. Wash the membrane 3 times for 5 min each in TBST, then incubate with goat anti-rabbit secondary antibody (1:5000 in 5% milk/TBST) for 1 h at room temperature.
13. Wash the membrane 3 times for 5 min with TBST.
14. Develop and visualize the membrane using enhanced chemiluminescence reagents per manufacturer's instructions. A representative image is shown in Fig. 7.

Note: Tissues may need to be divided into multiple pieces and/or processed accordingly to use in several downstream assays, including but not limited to protein analysis, enzymatic assays, histology, transcriptomic analysis, polyamine quantification. Furthermore, depending on the instruments, different homogenizing parameters may be required for different tissues. A pilot experiment may be necessary to achieve optimum tissue lysis.

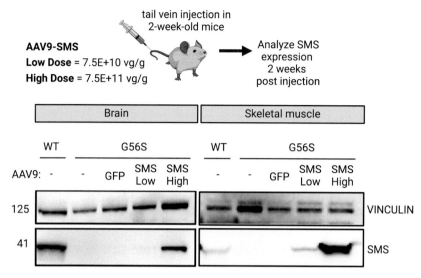

Fig. 7 **Comparison of SMS protein expression in mice injected with two different doses of AAV9-SMS via tail vein.** Two-week-old G56S mutant mice were injected via tail vein with AAV9-SMS at doses of 7.5×10^{10} and 7.5×10^{11} vg/g body weight, and AAV9-GFP at 7.5×10^{11} vg/g body weight. Two weeks post-injection, tissues were collected and SMS protein levels in the brain and skeletal muscle were determined by western blot. Vinculin serves as a loading control. *Figures are made using BioRender.*

4. Conclusion

AAV gene therapy has become the leading clinical candidate for several diseases, including rare neurological diseases. Ongoing efforts aim to optimize the use of the same viral delivery system, such as AAV9, to deliver genes to the correct tissues in the body. The methodology presented here represents the initial steps in developing and evaluating gene therapy to restore SMS expression in Snyder-Robinson syndrome mouse model. This approach could eventually be adapted for other conditions where gene supplementation may be beneficial, including but not limited to polyaminopathies (Bachmann, VanSickle, Michael, Vipond, & Bupp, 2024).

Acknowledgments

The University of Pittsburgh Institutional Animal Care and Use Committee (IACUC 22061237) approved the protocols. This work is supported by AFM-Telethon postdoctoral fellowship and RK Mellon trainee award (to O.A.), Development Fund from the Department of Pediatrics, University of Pittsburgh School of Medicine, NIH Director's New Innovator Award DP2-AR081047, and NIH R01-AR078872 (to D.U.K.). Members of the Kemaladewi Lab are acknowledged for their technical assistance and valuable inputs.

References

Akinyele, O., Munir, A., Johnson, M. A., Perez, M. S., Gao, Y., Foley, J. R., ... Kemaladewi, D. U. (2024). Impaired polyamine metabolism causes behavioral and neuroanatomical defects in a mouse model of Snyder-Robinson syndrome. *Disease Models & Mechanisms, 17*, 1–14.

Bachmann, A. S., VanSickle, E. A., Michael, J., Vipond, M., & Bupp, C. P. (2024). Bachmann-Bupp syndrome and treatment. *Developmental Medicine and Child Neurology, 66*, 445–455.

Brooks, P. J. https://ncats.nih.gov/research/research-activities/pave-gt.

Brooks, P. J., Ottinger, E. A., Portero, D., Lomash, R. M., Alimardanov, A., Terse, P., ... Lo, D. C. (2020). The platform vector gene therapies project: Increasing the efficiency of adeno-associated virus gene therapy clinical trial startup. *Human Gene Therapy, 31*, 1034–1042.

Bupp, C. P., Schultz, C. R., Uhl, K. L., Rajasekaran, S., & Bachmann, A. S. (2018). Novel de novo pathogenic variant in the ODC1 gene in a girl with developmental delay, alopecia, and dysmorphic features. *American Journal of Medical Genetics. Part A, 176*, 2548–2553.

Cason, A. L., Ikeguchi, Y., Skinner, C., Wood, T. C., Holden, K. R., Lubs, H. A., ... Schwartz, C. E. (2003). X-linked spermine synthase gene (SMS) defect: The first polyamine deficiency syndrome. *European Journal of Human Genetics: EJHG, 11*, 937–944.

Chan, K. Y., Jang, M. J., Yoo, B. B., Greenbaum, A., Ravi, N., Wu, W. L., ... Gradinaru, V. (2017). Engineered AAVs for efficient noninvasive gene delivery to the central and peripheral nervous systems. *Nature Neuroscience, 20*, 1172–1179.

Faundes, V., Jennings, M. D., Crilly, S., Legraie, S., Withers, S. E., Cuvertino, S., ... Banka, S. (2021). Impaired eIF5A function causes a Mendelian disorder that is partially rescued in model systems by spermidine. *Nature Communications, 12*, 833.

Foust, K. D., Wang, X., McGovern, V. L., Braun, L., Bevan, A. K., Haidet, A. M., ... Kaspar, B. K. (2010). Rescue of the spinal muscular atrophy phenotype in a mouse model by early postnatal delivery of SMN. *Nature Biotechnology, 28*, 271–274.

Ganapathi, M., Padgett, L. R., Yamada, K., Devinsky, O., Willaert, R., Person, R., ... Chung, W. K. (2019). Recessive rare variants in deoxyhypusine synthase, an enzyme involved in the synthesis of hypusine, are associated with a neurodevelopmental disorder. *American Journal of Human Genetics, 104*, 287–298.

Gessler, D. J., Tai, P. W. L., Li, J., & Gao, G. (2019). Intravenous infusion of AAV for widespread gene delivery to the nervous system. *Methods in Molecular Biology, 1950*, 143–163.

Gray, S. J., Choi, V. W., Asokan, A., Haberman, R. A., McCown, T. J., & Samulski, R. J. (2011). Production of recombinant adeno-associated viral vectors and use in in vitro and in vivo administration. *Current Protocols in Neuroscience* **Chapter 4**, Unit 4 17.

Kimura, T., Ferran, B., Tsukahara, Y., Shang, Q., Desai, S., Fedoce, A., ... Bachschmid, M. M. (2019). Production of adeno-associated virus vectors for in vitro and in vivo applications. *Scientific Reports. 9*, 13601.

Larcher, L., Norris, J. W., Lejeune, E., Buratti, J., Mignot, C., Garel, C., ... Whalen, S. (2020). The complete loss of function of the SMS gene results in a severe form of Snyder-Robinson syndrome. *European Journal of Medical Genetics, 63*, 103777.

Leung, M., Sanchez-Castillo, M., Belnap, N., Naymik, M., Bonfitto, A., Sloan, J., ... Ramsey, K. (2024). Snyder-Robinson syndrome presenting with learning disability, epilepsy, and osteoporosis: A novel SMS gene variant. *Rare, 2*.

Marhabaie, M., Hickey, S. E., Miller, K., Grischow, O., Schieffer, K. M., Franklin, S. J., ... Wilson, R. K. (2021). Maternal mosaicism for a missense variant in the SMS gene that causes Snyder-Robinson syndrome. *Cold Spring Harbor Molecular Case Studies. 7*.

Mathiesen, S. N., Lock, J. L., Schoderboeck, L., Abraham, W. C., & Hughes, S. M. (2020). CNS transduction benefits of AAV-PHP.eB over AAV9 are dependent on administration route and mouse strain. *Molecular Therapy. Methods & Clinical Development, 19*, 447–458.

Mouskou, S., Katerelos, A., Doulgeraki, A., Leka-Emiri, S., Manolakos, E., Papoulidis, I., … Voudris, K. (2021). Novel hemizygous missense variant of spermine synthase (SMS) gene causes Snyder-Robinson syndrome in a four-year-old boy. *Molecular Syndromology. 12*, 194–199.

Murray Stewart, T., Khomutov, M., Foley, J. R., Guo, X., Holbert, C. E., Dunston, T. T., … Casero, R. A., Jr. (2020). (R,R)-1,12-Dimethylspermine can mitigate abnormal spermidine accumulation in Snyder-Robinson syndrome. *The Journal of Biological Chemistry, 295*, 3247–3256.

Peron, A., Spaccini, L., Norris, J., Bova, S. M., Selicorni, A., Weber, G., … Mastrangelo, M. (2013). Snyder-Robinson syndrome: A novel nonsense mutation in spermine synthase and expansion of the phenotype. *American Journal of Medical Genetics. Part A, 161A*, 2316–2320.

Qazi, T. J., Wu, Q., Aierken, A., Lu, D., Bukhari, I., Hussain, H. M. J., … Qing, H. (2020). Whole-exome sequencing identifies a novel mutation in spermine synthase gene (SMS) associated with Snyder-Robinson syndrome. *BMC Medical Genetics, 21*, 168.

Rajasekaran, S., Bupp, C. P., Leimanis-Laurens, M., Shukla, A., Russell, C., Junewick, J., … Bachmann, A. S. (2021). Repurposing eflornithine to treat a patient with a rare ODC1 gain-of-function variant disease. *Elife, 10*.

Rodan, L. H., Anyane-Yeboa, K., Chong, K., Klein Wassink-Ruiter, J. S., Wilson, A., Smith, L., … Berry, G. T. (2018). Gain-of-function variants in the ODC1 gene cause a syndromic neurodevelopmental disorder associated with macrocephaly, alopecia, dysmorphic features, and neuroimaging abnormalities. *American Journal of Medical Genetics. Part A, 176*, 2554–2560.

Schultz, C. R., Bupp, C. P., Rajasekaran, S., & Bachmann, A. S. (2019). Biochemical features of primary cells from a pediatric patient with a gain-of-function ODC1 genetic mutation. *The Biochemical Journal, 476*, 2047–2057.

Shin, J. H., Yue, Y., & Duan, D. (2012). Recombinant adeno-associated viral vector production and purification. *Methods in Molecular Biology, 798*, 267–284.

Stewart, T. M., Foley, J. R., Holbert, C. E., Khomutov, M., Rastkari, N., Tao, X., … Casero, R. A., Jr. (2023). Difluoromethylornithine rebalances aberrant polyamine ratios in Snyder-Robinson syndrome. *EMBO Molecular Medicine*, e17833.

Strauss, K. A., Farrar, M. A., Muntoni, F., Saito, K., Mendell, J. R., Servais, L., … Macek, T. A. (2022). Onasemnogene abeparvovec for presymptomatic infants with three copies of SMN2 at risk for spinal muscular atrophy: The Phase III SPR1NT trial. *Nature Medicine, 28*, 1390–1397.

Tantak, M. P., Sekhar, V., Tao, X., Zhai, R. G., & Phanstiel, O. t (2021). Development of a redox-sensitive spermine prodrug for the potential treatment of Snyder-Robinson syndrome. *Journal of Medicinal Chemistry, 64*, 15593–15607.

Tao, X., Zhu, Y., Diaz-Perez, Z., Yu, S. H., Foley, J. R., Stewart, T. M., … Zhai, R. G. (2022). Phenylbutyrate modulates polyamine acetylase and ameliorates Snyder-Robinson syndrome in a Drosophila model and patient cells. *JCI Insight, 7*.

Tunster, S. J. (2017). Genetic sex determination of mice by simplex PCR. *Biology of Sex Differences, 8*, 31.

VectorBuilder. www.vectorbuilder.com.

Wang, D., Tai, P. W. L., & Gao, G. (2019). Adeno-associated virus vector as a platform for gene therapy delivery. *Nature Reviews. Drug Discovery, 18*, 358–378.

Wright, J. F. (2020). Codon modification and PAMPs in clinical AAV vectors: The tortoise or the hare? *Molecular Therapy: the Journal of the American Society of Gene Therapy, 28*, 701–703.

Xie, Q., Chen, X., Ma, H., Zhu, Y., Ma, Y., Jalinous, L., ... Xie, J. (2024). Improved gene therapy for spinal muscular atrophy in mice using codon-optimized hSMN1 transgene and hSMN1 gene-derived promotor. *EMBO Molecular Medicine, 16*, 945–965.

Yardeni, T., Eckhaus, M., Morris, H. D., Huizing, M., & Hoogstraten-Miller, S. (2011). Retro-orbital injections in mice. *Lab Animal, 40*, 155–160.

Zhang, Z., Norris, J., Kalscheuer, V., Wood, T., Wang, L., Schwartz, C., ... Van Esch, H. (2013). A Y328C missense mutation in spermine synthase causes a mild form of Snyder-Robinson syndrome. *Human Molecular Genetics, 22*, 3789–3797.

Ziegler, A., Steindl, K., Hanner, A. S., Kar, R. K., Prouteau, C., Boland, A., ... Park, M. H. (2022). Bi-allelic variants in DOHH, catalyzing the last step of hypusine biosynthesis, are associated with a neurodevelopmental disorder. *American Journal of Human Genetics, 109*, 1549–1558.

CHAPTER SIXTEEN

Methods to study polyamine metabolism during osteogenesis

Amin Cressman[a] and Fernando A. Fierro[a,b,*]

[a]Stem Cell Program, Institute for Regenerative Cures, University of California Davis, Sacramento, CA, United States
[b]Department of Cell Biology and Human Anatomy, University of California Davis, Davis, CA, United States
*Corresponding author. e-mail address: ffierro@ucdavis.edu

Contents

1. Introduction	294
2. Materials and reagents	296
2.1 Isolation and expansion of MSCs	296
3. Osteogenic differentiation and quantification	296
4. Polyamine modulation	297
5. Methods	297
5.1 Isolation and expansion of MSCs	297
6. Characterization of MSCs	299
7. Osteogenic differentiation and quantification methods	300
8. Osteogenic differentiation	301
9. Quantification of mineralization	301
10. Quantification of alkaline phosphatase activity	302
11. Quantification of gene expression	302
12. Polyamine modulation to investigate role in osteogenesis	304
13. Supplementation with polyamines and pathway inhibitors	304
14. Reproducibility and statistical considerations	305
15. Conclusions and future directions	305
References	306

Abstract

Mammalian polyamines, namely putrescine, spermidine, and spermine, have been implicated in many cellular homeostatic processes. Polyamines play a critical role in skeletal health as evidenced by recent studies and by skeletal disorders caused by polyamine imbalances, such as Snyder-Robinson Syndrome (SRS). However, very little is still known about the role of polyamines within bone development, homeostasis, and metabolism. Human bone marrow derived mesenchymal stromal cells (MSCs) provide a unique opportunity to study polyamines at a cellular and molecular level within the context of osteogenic differentiation and calcium deposition. Through in vitro work, mechanistic understanding of the role of polyamines within osteogenesis as well as the consequences of polyamine imbalance can provide new insights into potential therapeutics for those experiencing polyaminopathies. This chapter describes procedures to develop a human primary cell culture system and quantify osteoblastogenesis as a function of polyamine modulation.

1. Introduction

Polyamine regulation plays an important role in cellular homeostasis, although different cell types may exhibit diverse phenotypes following polyamine dysregulation. It has been suggested that levels of polyamines and their regulatory enzymes are dynamic during chondrogenesis and osteogenesis (Cressman, Morales, & Zhang, 2024; Rath & Reddi, 1981). Rath and Reddi (Rath & Reddi, 1981) measured polyamines during ectopic endochondral bone formation in rats. They found that putrescine, spermidine and spermine levels were low yet spiked by the end of chondrogenesis and dropped during ossification. Similarly, ornithine decarboxylase (ODC) activity was high early, but low at later stages. In mice, both spermine and spermidine inhibit bone loss caused by ovariectomy, primarily by disrupting osteoclast activity (Yamamoto, Hinoi, & Fujita, 2012). Similar observations were reported by Chevalier et al., who found that warm environs increase gut microbial-derived polyamine synthesis to mitigate ovariectomy-associated bone loss (Chevalier, Kieser, & Colakoglu, 2020). In humans, by measuring 135 metabolites in 1504 participants with 9-year follow up, Kong et al. found that only spermidine concentrations positively associated with fracture risk (Kong, Kim, & Shin, 2021), stressing the significance of studying polyamine metabolism in the context of skeletal health. Despite these observations, very little is known about how polyamines contribute to osteogenesis and skeletal homeostasis.

Various groups have investigated the effect of polyamines on osteogenic differentiation of multipotent mesenchymal stromal cells (MSCs) in culture, although with discordant results (Guidotti, Facchini, & Platano, 2013; Lee, Chen, & Huang, 2013; Tjabringa, Vezeridis, & Zandieh-Doulabi, 2006; Tjabringa, Zandieh-Doulabi, & Helder, 2008; Tsai, Lin, & Huang, 2015). Potential causes for these discrepancies include low number of biological replicates, tissue of origin of the cells (adipose vs. bone marrow), outcome measurements, and the concentrations used for both polyamines and inhibitors of the polyamine biosynthetic pathway.

A disease that epitomizes the relevance of polyamines for skeletal health is Snyder-Robinson Syndrome (SRS). SRS is a X-linked recessive disorder leading to deficient spermine synthase (SMS) activity in all tissues. Surprisingly, despite many tissues showing the pathognomic increased ratio of spermidine to spermine due to SMS deficiency, only specific tissues show functional defects (Schwartz, Peron, Kutler, & Snyder-Robinson Syndrome, 1993). One of the more debilitating and consistent phenotypes presented in SRS patients is severe osteoporosis, leading to common atraumatic fractures (Koerner et al., 2024). Recent mouse models of SRS also recapitulate the skeletal defects (Akinyele, Munir, & Johnson, 2024), although the underlying mechanism remains unknown.

To develop therapeutics for SRS and other polyaminopathies that affect the skeleton it is imperative to understand the underlying mechanisms of polyamine function within these cells. The multipotent differentiation capabilities of human bone marrow-derived MSCs provides a unique avenue for investigating polyamine metabolism in the context of osteogenesis and mineral deposition in vitro. Albert et al. showed that MSCs derived from SRS patients show impaired mineralization (Albert, Bhattacharyya, & Wolfe, 2015) and we have shown that MSCs with deficient SMS activity through shRNA knockdown also exhibited decreased mineral deposition, in addition to impaired proliferation, and altered mitochondrial metabolism (Ramsay, Alonso-Garcia, & Chaboya, 2019). These findings are recapitulated through chemical inhibition of SMS using N-cyclohexyl-1,3-propanediamine (CDAP) (Cressman et al., 2024). Developing these models has therefore contributed to further understand the etiology of SRS and provide a path to therapeutic exploration.

Supplementing polyamines into the osteogenic induction media (OIM) of MSCs have shown that both spermidine and spermine have a dose dependent inhibitory effect on mineralization, while not clearly affecting expression of osteogenic genes, or alkaline phosphatase activity (Cressman et al., 2024). Furthermore, upstream inhibition of ornithine decarboxylase (ODC1) utilizing difluoromethylornithine (DFMO) rescues polyamine imbalance (Stewart, Foley, & Holbert, 2023), and in our model system DFMO rescues the impaired mineral deposition and mitochondrial dysregulation displayed by MSCs treated with CDAP (Cressman et al., 2024). Although DFMO may have acute and toxic effect in the Gyro mouse, a model of SRS through a X-linked deletion, inactivating *Sms* and *Phex* (Wang, Levic, & Gratton, 2009), more studies utilizing DFMO as a therapeutic are needed in new models of SRS. Importantly, DFMO is approved by the FDA for the treatment of Trypanosomiasis and is under current evaluation in multiple active clinical trials for the treatment of diseases such as glioblastoma, neuroblastoma, and medulloblastoma, showing a positive safety profile. Also, the underlying mechanism still needs further exploration to potentially develop alternative therapeutic strategies. MSCs present a unique opportunity to interrogate osteogenesis at a cellular and molecular level within the context of polyamine modulation.

Here we describe methods for MSC isolation, expansion, characterization, osteogenic differentiation, and to quantify osteogenic differentiation. In addition, we describe methods for polyamine manipulation through direct supplementation or inhibition of enzymes involved in polyamine synthesis.

2. Materials and reagents
2.1 Isolation and expansion of MSCs

1. Bone marrow aspirate from healthy donor (CGT Global #CP001)
2. 90 μm pore cell strainers (Corning #431224)
3. Sterile tissue culture flasks (Corning)
4. Ficoll (Cytiva #17144003)
5. Phosphate buffered saline (PBS) (Gibco #10010-023)
6. Minimum essential media α (MEM-α) (Corning #10-022-CV)
7. Fetal bovine serum (FBS) (GeminiBio #100-500)
8. Trypsin (0.05 %) with EDTA (Corning #25-051-CI)
9. Trypan blue (Gibco #15250-061)
10. Hemocytometer (Hausser Scientific #3200)
11. (Optional) Fibroblast growth factor 2 (FGF2) (PeproTech #100-18B-100UG)

3. Osteogenic differentiation and quantification

1. Osteogenic induction medium (OIM): MEM-α + 10 % v/v FBS, 0.2 mM ascorbic acid (Sigma #A4544), 0.1 μM dexamethasone (USP #1176007), 10 mM β-glycerophosphate (Sigma #50020).
2. Formalin 10 % v/v (Fisher Scientific #F79P-4)
3. Alizarin Red S stain (Spectrum Chemical #AL200)
4. Phosphate buffered saline (PBS) (Gibco #10010-023)
5. Molecular grade water (Corning #46-000-CV)

6. 10 % v/v acetic acid (Millipore #AX0073-9)
7. Alkaline phosphatase lysis buffer: 1.5 mM Tris–HCl (BioRad #161-0798) +1 % v/v Triton-X 100 (Sigma #T8787)
8. P-nitrophenylphosphate (p-NPP) (Sigma #P7998)
9. Coomassie blue stain (BioRad #50002050)
10. TRI reagent (Zymo #R2050-2-200)
11. 100 % isopropanol (Thermo Scientific #389710025)
12. Quick-RNA Miniprep kit (Zymo #R2052)
13. High-Capacity cDNA Reverse Transcription Kit (Applied Biosystems #4368813)
14. Taqman Universal PCR Master Mix (Applied Biosystems #4369016)

4. Polyamine modulation

1. Aminoguanidine (Sigma #396494)
2. Putrescine (Sigma #P7505)
3. Spermidine (Sigma #S2626)
4. Spermine (Sigma #S3256)
5. N-cyclohexyl-1,3-propanediamine (CDAP) (spermine synthase inhibitor) (TCI #A0891)
6. Difluoromethylornithine (DFMO) (ornithine decarboxylase inhibitor) (Sigma #D193)
7. Decarboxylated S-adenosylhomocysteine (dcSAH) (spermidine synthase inhibitor (Seckute, McCloskey, & Thomas, 2011), not commercially available)
8. MDL72527 (Spermine/spermidine oxidase (SMOX) inhibitor) (Sigma #M2949)
9. SAM486A (S-adenosylmethionine decarboxylase (AMD1) inhibitor) (Cayman Chemical #34901)
10. MDL 72527 (polyamine oxidase (PAOX) inhibitor) (Sigma #M2949)

5. Methods
5.1 Isolation and expansion of MSCs

Mesenchymal stromal cells are a heterogeneous group of cells and can be isolated from various sources (Bianco, Cao, & Frenette, 2013). MSCs isolated from bone marrow aspirates have robust multipotent capabilities to differentiate into osteoblasts, adipocytes, and chondrocytes. The age and health of the bone marrow donor play a role in the number (and possibly quality) of the isolated and expanded cells (Stolzing, Jones, & McGonagle, 2008). It is also important to consider the passage, morphology, and proliferation rate when using MSCs for further experimentation. Potent MSCs plated on plastic culture dishes exhibit an elongated spindle shape (Fig. 1). When MSCs increase in passage, they become flattened, bigger, and exhibit stress fibers (actin bundles) which are visible using phase-contrast light microscopy. Supplementing the culture media with basic fibroblast growth factor (FGF-2) often helps to reverse this senescent phenotype (Ito, Sawada, & Fujiwara, 2007; Solchaga, Penick, & Porter, 2005). We recommend using MSCs within passages 2–8 for experiments, where each passage represents around 6 days in culture (3–4 population doublings).

1. Ensure that the bone marrow aspirate collection and use follow regulatory requirements, such as IRB approval.
2. Once in the laboratory, pass the bone marrow aspirate through a 90 μm pore strainer to collect potential bone spicules, and add an equal volume of PBS.

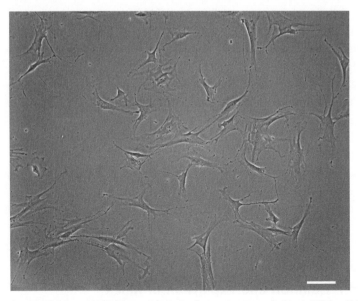

Fig. 1 Characteristic morphology of human bone marrow-derived MSCs. Although cells look "fibroblastic", the exact cell morphology can be quite variable. This picture was taken from cells in passage 6, seeded the day before at 4000 cells/cm^2. Scale bar is 100 μm.

3. In 50 mL conical tubes, gently aspirate 30 mL of the aspirate onto 20 mL of Ficoll, holding the tube at a 45° angle to avoid mixing.
4. Centrifuge at 700 x g for 30 min with the acceleration set to low and the break set to zero. This density gradient will separate mononuclear cells (often referred to as the buffy coat) from red blood cells and granulocytes.
5. Aspirate the buffy coat containing the mononuclear cells (white layer immediately above the Ficoll), fill the new tube with PBS to wash off any remnants of Ficoll from the cells and centrifuge at 700 x g for 5 min (using normal acceleration and brake modes).
6. Resuspend the cell pellet in MEM-α + 10 % FBS and plate into tissue culture flasks or plates at 2–5 × 10^5 cells/cm^2.
7. Allow MSCs to adhere for 2 days before washing 2–3 times with PBS, to remove non-adherent cells.
8. Follow with medium changes every 2–3 days for expansion. Here (passage 0) MSCs typically expand clonally in which cell morphology may vary among colonies.
9. To passage cells, use trypsin. First, wash cells once with PBS, then add trypsin (typically a volume that is half of the respective culture media volume) and incubate at 37 °C incubator for 7 min with a gentle agitation after 3 min.
10. Following this, add an equal volume of MEM-α + 10 % FBS to each flask/plate to stop the trypsin. Count cells in suspension using Trypan Blue exclusion dye and hemocytometer. For normal expansion of MSCs, we recommend plating cells at 1000 cells/cm^2 and harvest at ~70 % confluence (10,000 cells/cm^2).

6. Characterization of MSCs

MSCs are primarily isolated based on their ability to attach to plastics such as polystyrene and polyester, which are commonly used for tissue culture flasks and plates. In addition, MSCs possess three main characteristics: (1) a fibroblastic morphology when attached to a substrate, (2) a common immune phenotype (surface markers), and (3) trilineage differentiation potential (Dominici, Blanc, & Mueller, 2006). Although the exact morphology of MSCs varies even within a single culture, these typically resemble spindle-shape, flatten, fibroblasts. Fresh cultures of MSCs grow forming "colonies", which are frequently referred to as fibroblastic colony forming unites (CFU-F). Common surface markers used to determine the purity of human MSCs (determined by flow cytometry) include absence of hematopoietic markers CD11b (macrophages), CD14 (monocytes), CD34 (hematopoietic stem/progenitor cells), and CD45 (a pan-hematopoietic marker), and presence of mesenchymal markers CD73, CD90, and CD105 (Pittenger, Mackay, & Beck, 1999). By trilineage potential, we refer to the ability of MSCs to differentiate in vitro into osteoblasts, adipocytes, and chondrocytes (Pittenger et al., 1999). For the purpose of this book chapter, we will only focus on their osteogenic differentiation potential. Of note, although MSCs can robustly differentiate into calcium-precipitating osteoblasts, at least under conventional 2D culture conditions, these typically do not further mature to become matrix-embedded osteocytes (Kim & Adachi, 2021).

7. Osteogenic differentiation and quantification methods

Though there is increasing literature investigating polyamines in human disease, there is a dearth in understanding polyamines as a function of skeletal development and maintenance. MSCs present an opportunity to explore the role of polyamines in the context of osteogenic differentiation and mineralization. Osteogenic induction media (OIM) consists of standard culture media supplemented with ascorbic acid, dexamethasone, and β-glycerophosphate. Ascorbic acid (Vitamin C) was originally thought to be only necessary as a co-factor for collagen maturation, but it was recently shown to also be a critical epigenetic regulator of key osteogenic genes, such as *Bglap2* and *Dlx3* (Thaler, Khani, & Sturmlechner, 2022). Dexamethasone is a synthetic glucocorticoid that has been used as a supplemental factor for inducing osteogenic differentiation in MSCs for the past 30 years (Cheng, Yang, & Rifas, 1994). Dexamethasone induces osteoblastogenesis through activation of *Runx2,* a key transcription factor (Langenbach & Handschel, 2013). Activation of *Runx2* has been shown through multiple mechanisms such as MKP-1 (Phillips, Gersbach, & Wojtowicz, 2006), FHL2/b-catenin (Hamidouche, Hay, & Vaudin, 2008), and TAZ (Hong, Chen, & Xue, 2009). Lastly, β-glycerophosphate acts as the source of inorganic phosphate (Pi) needed to produce the apatite (calcium phosphate) crystals (Langenbach & Handschel, 2013).

8. Osteogenic differentiation

To induce osteogenic differentiation, seed cells at 10,000 cells/cm^2. The next day, change the growth media to osteogenic induction media (OIM). Change media every 3–4 days. The exact time in which cells are cultured in OIM is dictated by distinct stages of differentiation and mineralization, which are detailed below.

9. Quantification of mineralization

1. Alizarin Red S (ARS) stains calcium deposition as a marker for late-stage osteogenic differentiation (osteoblastogenesis) and we developed a method to quantify it based on a protocol previously developed by Gregory et al (Gregory, Gunn, & Peister, 2004). We recommend performing this assay in either 24-well or 12-well plates. For simplicity, we describe the protocol for 24-well plates only, but volumes can be simply doubled for 12-well plates.
2. Following 21 days in osteogenic medium, wash with 0.5 mL PBS and fix with 0.5 mL formalin, at room temperature for 15 min.
3. Following this, wash once with PBS, add 0.5 mL ARS, and incubate at room temperature for 20 min with gentle rocking.
4. Wash 3 times with PBS and once with molecular grade water (or until aspirate is clear) and take photos at low magnification for qualitative assessment.
5. For quantification, add 0.5 mL 10% v/v acetic acid and incubate at room temperature for 30 min with gentle rocking. Using a cell scraper, scrape off the monolayer of cells and collect cells and supernatant into microcentrifuge tubes.
6. Vortex the samples for 30 s and centrifuge at 10,000 x g for 5 min.
7. Transfer the supernatant, avoiding the pellet (cell debri), to a new microcentrifuge tube.
8. Pipette 100 μL in triplicate into a 96-well plate and read optical density (absorbance) at 405 nm on a spectrophotometer.

Notice that this method does not use an internal control to normalize values. It is therefore important to inspect the wells, to ensure that any specific treatment is not affecting cell viability. Differentiation is started with cells at high density (confluency), to maximize cell-to-cell signaling while minimizing further cell proliferation.

10. Quantification of alkaline phosphatase activity

1. Alkaline phosphatase (ALP) is a marker of pre-osteoblasts and therefore an early-stage osteogenic differentiation marker. It is best measured after 10 days of culture in OIM. We have found that at least 40,000 MSCs/well plated in 12-well plates are needed for optimal measurements.
2. Following 10 days of culture in OIM, lift cells using trypsin.
3. This is done by adding 500 µL/well and incubating at 37 °C incubator for 7 min with a gentle agitation after 3 min.
4. Following this, add an equal volume of MEM-α + 10 % FBS to each well to stop the trypsin. If cells have not fully detached, use a 1000 µL pipette tip to scrape the remaining cells and mix well prior to collection into microcentrifuge tubes.
5. Centrifuge at 700 x g for 5 min and remove the supernatant without disturbing the pellet.
6. Resuspend the cell pellet in 100 µL ALP lysis buffer, vortex for 30 s, and shake vigorously at 4 °C or in ice for 20 min.
7. Vortex again for 30 second and centrifuge at 10,000 x g for 5 min at 4 °C, to remove any cell debris.
8. After centrifugation, collect the supernatant and keep samples on ice or store at −80 °C.
9. To measure ALP activity, add 2 µL of each sample in triplicate into a 96-well plate. Add 200 µL of p-NPP, incubate at 37 °C in the dark for 10 min, and immediately read the optical density at 405 nm on a spectrophotometer.
10. To normalize samples to total protein content, add 10 µL sample in triplicate into a 96-well plate. Pipette 200 µL Coomassie blue reagent into each well and measure the optical density at 595 nm using a spectrophotometer.

11. Quantification of gene expression

1. Plating of cells and cell culture is as described above. The time of isolation is dictated by the gene of interest. We have shown that markers for osteogenesis (ALPL, IBSP, SPP1) as well as polyamine enzymes

(SAT1) are dynamic throughout osteogenesis (Cressman et al., 2024), so this must be planned accordingly. We recommend using at least 10^5 MSCs to yield sufficient high-quality mRNA.

2. To isolate RNA, remove culture media, add 400 µL of Trizol directly to each well, and incubate at room temperature for 5 min.
3. Collect the lysed cells into a microcentrifuge tube by scraping the bottom of the dish and triturating the cell suspension.
4. Add equal volume of 100 % isopropanol and mix prior to adding to the column provided in the Quick-RNA Miniprep Kit (Zymo).
5. Follow the protocol as described by the manufacturer.
6. We have found that during the elution step, we are able to get a better yield by eluting with 30 µL DNAse/RNase free water and allowing to soak into the column for 2–3 min prior to centrifugation.
7. Following this, keep the samples on ice and confirm the concentration using a nanodrop machine.
8. Concentration of at least 30 ng/µL is necessary to yield optimal results using RT-PCR.
9. To synthesize cDNA, we recommend using the High-Capacity cDNA Reverse Transcription Kit (Applied Biosystems).
10. 10 µL of RNA are needed for each reaction, mixing the sample with 2.0 µL 10X RT Buffer, 0.8 µL 25X dNTP mix (100 mM), 2.0 µL 10X RT Random Primers, 1.0 µL Multiscribe™ Reverse Transcriptase, 1.0 µL RNase inhibitor, and 3.2 µL RNase free water per reaction with a final volume of 20 µL.
11. The reaction protocol is 1. 95 °C for 3:00 min, 2. 95 °C for 0:30 s, 3. 85 °C for 0:30 ss, 4. 72 °C for 1:00 min, repeat steps 2-4 34 times, and 5. 72 °C for 5:00 min.
12. Use these samples immediately for RT-PCR, or store at 4 °C for up to 1 year.
13. For RT-PCR, we use Taqman Universal PCR Master Mix and Taqman probe/primer sets.
14. Dilute cDNA samples by adding DNase free water to obtain the required volume. The maximum dilution will depend on the RNA concentration obtained from the RNA extraction and the resultant concentration of cDNA, which should be even for all samples.
15. Each well will need 9.0 µL sample, 10 µL Taqman PCR Universal Master Mix, and 1.0 µL Taqman probe/primer set.
16. We recommend performing each reaction in duplicates.

17. GAPDH or B-Actin are common housekeeping genes used as loading controls for MSCs. The reaction protocol is as follows: 1. 50 °C for 2:00 min, 2. 95 °C for 10:00 min, 3. 95 °C for 0:15 s, 60 °C for 1:00 min, and repeat steps 3-4 40 times.

12. Polyamine modulation to investigate role in osteogenesis

Direct supplementation of polyamines and inhibition of enzymes within the polyamine pathway in vitro utilizing MSCs can provide valuable insights on their role during osteogenesis. To prevent toxicity when supplementing polyamines directly into cells, the diamine oxidase present in FBS can be inhibited with 1.0 mM aminoguanidine (AG), hence preventing polyamine oxidation into aminoaldehydes (Gahl & Pitot, 1978). We have found that concentrations of putrescine, spermidine, and spermine under 2.5 mM, 1.25 mM, and 0.31 mM, respectively, have little effect on cell viability (Cressman et al., 2024), while concentrations as little as 20 µM of spermidine and spermine have been shown to have an inhibitory effect on mineralization (Cressman et al., 2024).

To mimic the impaired SMS activity observed in SRS patients, our lab has either silenced SMS expression using lentiviral-delivered shRNA (Ramsay et al., 2019) or employed the inhibitor CDAP to deplete SMS function in MSCs. We observed toxicity induced by CDAP at concentrations of ≥ 0.625 mM and measured potent inhibition of mineralization and cytoplasmic vacuolization at 200 µM, as well as mitochondrial dysfunction 100 µM (Cressman et al., 2024). In contrast, DFMO showed no signs of toxicity at concentrations of up to 10 mM. However, DFMO was able to rescue phenotypes exhibited by CDAP with as little as 10–20 µM (Cressman et al., 2024). Though we have employed DFMO and CDAP to specifically study SRS, there are many polyamine pathway inhibitors that require investigation, further advancing the knowledge of polyamines in the context of bone homeostasis.

13. Supplementation with polyamines and pathway inhibitors

1. The polyamines and inhibitors are soluble in water, and we recommend making 1000X stock solutions for ease of use. These solutions can be prepared in non-sterile conditions but should be passed through a

0.2 μm filter within a biosafety cabinet for sterility purposes. The stock solutions can be stored at −20 °C for 12 months. To prevent degradation from freeze/thaw cycles, create smaller aliquots that can be stored and used for up to 3 months at 4 °C.
2. Prepare conical tubes with culture media beforehand and add chemicals to these at 1.0 μL per 1.0 mL of media, as opposed to spiking wells directly. This will reduce variability between replicates. Conduct medium changes every 3-4 days with OIM supplemented with the desired conditions. Utilize the methods described above to measure osteogenic outcomes.

14. Reproducibility and statistical considerations

As described above for ARS staining, ALP activity, and RT-PCR, each sample isolated from a single well is loaded in either duplicate or triplicate for reading. We do not consider this a replicate, but rather a method to confirm accuracy of the measurement. The technical replicates are the number of wells under the same condition. For example, for ARS staining in 24-well plates, it is easy to use 4 wells for each specific condition to be tested. In addition, we strongly encourage to use biological replicates, which are MSCs derived from different donors. For many laboratories working with human MSCs, the number of experiments is usually equivalent to the number of lots of MSCs tested. These biological replicates and independent experiments ultimately dictate the reproducibility of a specific finding. In our hands, to reach statistically conclusive results, using paired Student's t test (for 2 conditions) or 1-way ANOVA (for 3 conditions or more) requires using 4–8 lots of MSCs.

15. Conclusions and future directions

In this chapter we provide a framework for the exploration of polyamine participation in osteogenesis in vitro, though there is still much to be understood. As polyamines are highly conserved and are ubiquitous throughout cells, with expansive roles, mechanistic investigation has proven onerous. The techniques described above, and utilization of polyamine pathway enzyme inhibitors can begin to shed light onto the role polyamines play in skeletal development and turnover. Future research is also needed to develop specific chemical inhibition for other enzymes of the polyamine biosynthetic pathway and to assess and optimize exogenous polyamine uptake (Tantak, Sekhar, & Tao, 2021).

References

Akinyele, O., Munir, A., Johnson, M. A., et al. (2024). Impaired polyamine metabolism causes behavioral and neuroanatomical defects in a mouse model of Snyder-Robinson syndrome. *Disease Models & Mechanisms, 17*.

Albert, J. S., Bhattacharyya, N., Wolfe, L. A., et al. (2015). Impaired osteoblast and osteoclast function characterize the osteoporosis of Snyder-Robinson syndrome. *Orphanet Journal of Rare Diseases, 10*, 27.

Bianco, P., Cao, X., Frenette, P. S., et al. (2013). The meaning, the sense and the significance: Translating the science of mesenchymal stem cells into medicine. *Nature Medicine, 19*, 35–42.

Cheng, S. L., Yang, J. W., Rifas, L., et al. (1994). Differentiation of human bone marrow osteogenic stromal cells in vitro: Induction of the osteoblast phenotype by dexamethasone. *Endocrinology, 134*, 277–286.

Chevalier, C., Kieser, S., Colakoglu, M., et al. (2020). Warmth prevents bone loss through the gut microbiota. *Cell Metabolism, 32*, 575–590.e577.

Cressman, A., Morales, D., Zhang, Z., et al. (2024). Effects of spermine synthase deficiency in mesenchymal stromal cells are rescued by upstream inhibition of ornithine decarboxylase. *International Journal of Molecular Sciences, 25*.

Dominici, M., Le Blanc, K., Mueller, I., et al. (2006). Minimal criteria for defining multipotent mesenchymal stromal cells. The International Society for Cellular Therapy position statement. *Cytotherapy, 8*, 315–317.

Gahl, W. A., & Pitot, H. C. (1978). Reversal by aminoguanidine of the inhibition of proliferation of human fibroblasts by spermidine and spermine. *Chemico-Biological Interactions, 22*, 91–98.

Gregory, C. A., Gunn, W. G., Peister, A., et al. (2004). An Alizarin red-based assay of mineralization by adherent cells in culture: Comparison with cetylpyridinium chloride extraction. *Analytical Biochemistry, 329*, 77–84.

Guidotti, S., Facchini, A., Platano, D., et al. (2013). Enhanced osteoblastogenesis of adipose-derived stem cells on spermine delivery via beta-catenin activation. *Stem Cells and Development, 22*, 1588–1601.

Hamidouche, Z., Hay, E., Vaudin, P., et al. (2008). FHL2 mediates dexamethasone-induced mesenchymal cell differentiation into osteoblasts by activating Wnt/beta-catenin signaling-dependent Runx2 expression. *The FASEB Journal, 22*, 3813–3822.

Hong, D., Chen, H. X., Xue, Y., et al. (2009). Osteoblastogenic effects of dexamethasone through upregulation of TAZ expression in rat mesenchymal stem cells. *The Journal of Steroid Biochemistry and Molecular Biology, 116*, 86–92.

Ito, T., Sawada, R., Fujiwara, Y., et al. (2007). FGF-2 suppresses cellular senescence of human mesenchymal stem cells by down-regulation of TGF-beta2. *Biochemical and Biophysical Research Communications, 359*, 108–114.

Kim, J., & Adachi, T. (2021). Cell-fate decision of mesenchymal stem cells toward osteocyte differentiation is committed by spheroid culture. *Scientific Reports, 11*, 13204.

Koerner, T. L., Green, A. M., Pace-Farr, D. J., Zeitler, C. M., Schwartz, M. B., & Kutler, M. J. F. (2024). Bone manifestations in Snyder-Robinson syndrome. *Rare, 2*.

Kong, S. H., Kim, J. H., & Shin, C. S. (2021). Serum spermidine as a novel potential predictor for fragility fractures. *The Journal of Clinical Endocrinology and Metabolism, 106*, e582–e591.

Langenbach, F., & Handschel, J. (2013). Effects of dexamethasone, ascorbic acid and beta-glycerophosphate on the osteogenic differentiation of stem cells in vitro. *Stem Cell Research & Therapy, 4*, 117.

Lee, M. J., Chen, Y., Huang, Y. P., et al. (2013). Exogenous polyamines promote osteogenic differentiation by reciprocally regulating osteogenic and adipogenic gene expression. *Journal of Cellular Biochemistry, 114*, 2718–2728.

Phillips, J. E., Gersbach, C. A., Wojtowicz, A. M., et al. (2006). Glucocorticoid-induced osteogenesis is negatively regulated by Runx2/Cbfa1 serine phosphorylation. *Journal of Cell Science, 119*, 581–591.

Pittenger, M. F., Mackay, A. M., Beck, S. C., et al. (1999). Multilineage potential of adult human mesenchymal stem cells. *Science (New York, N. Y.), 284*, 143–147.

Ramsay, A. L., Alonso-Garcia, V., Chaboya, C., et al. (2019). Modeling Snyder-Robinson syndrome in multipotent stromal cells reveals impaired mitochondrial function as a potential cause for deficient osteogenesis. *Scientific Reports, 9*, 15395.

Rath, N. C., & Reddi, A. H. (1981). Changes in polyamines, RNA synthesis, and cell proliferation during matrix-induced cartilage, bone, and bone marrow development. *Developmental Biology, 82*, 211–216.

Schwartz, C. E., Peron, A., & Kutler, M. J. (1993). Snyder-Robinson syndrome. In M. P. Adam, H. H. Ardinger, & R. A. Pagon, (Eds.). *GeneReviews(R)*. Seattle (WA): University of Washington.

Seckute, J., McCloskey, D. E., Thomas, H. J., et al. (2011). Binding and inhibition of human spermidine synthase by decarboxylated S-adenosylhomocysteine. *Protein Science: A Publication of the Protein Society, 20*, 1836–1844.

Solchaga, L. A., Penick, K., Porter, J. D., et al. (2005). FGF-2 enhances the mitotic and chondrogenic potentials of human adult bone marrow-derived mesenchymal stem cells. *Journal of Cellular Physiology, 203*, 398–409.

Stewart, T. M., Foley, J. R., Holbert, C. E., et al. (2023). Difluoromethylornithine rebalances aberrant polyamine ratios in Snyder-Robinson syndrome. *EMBO Molecular Medicine*e17833.

Stolzing, A., Jones, E., McGonagle, D., et al. (2008). Age-related changes in human bone marrow-derived mesenchymal stem cells: Consequences for cell therapies. *Mechanisms of Ageing and Development, 129*, 163–173.

Tantak, M. P., Sekhar, V., Tao, X., et al. (2021). Development of a redox-sensitive spermine prodrug for the potential treatment of Snyder Robinson syndrome. *Journal of Medicinal Chemistry, 64*, 15593–15607.

Thaler, R., Khani, F., Sturmlechner, I., et al. (2022). Vitamin C epigenetically controls osteogenesis and bone mineralization. *Nature Communications, 13*, 5883.

Tjabringa, G. S., Vezeridis, P. S., Zandieh-Doulabi, B., et al. (2006). Polyamines modulate nitric oxide production and COX-2 gene expression in response to mechanical loading in human adipose tissue-derived mesenchymal stem cells. *Stem Cells, 24*, 2262–2269.

Tjabringa, G. S., Zandieh-Doulabi, B., Helder, M. N., et al. (2008). The polymine spermine regulates osteogenic differentiation in adipose stem cells. *Journal of Cellular and Molecular Medicine, 12*, 1710–1717.

Tsai, Y. H., Lin, K. L., Huang, Y. P., et al. (2015). Suppression of ornithine decarboxylase promotes osteogenic differentiation of human bone marrow-derived mesenchymal stem cells. *FEBS Letters, 589*, 2058–2065.

Wang, X., Levic, S., Gratton, M. A., et al. (2009). Spermine synthase deficiency leads to deafness and a profound sensitivity to alpha-difluoromethylornithine. *The Journal of Biological Chemistry, 284*, 930–937.

Yamamoto, T., Hinoi, E., Fujita, H., et al. (2012). The natural polyamines spermidine and spermine prevent bone loss through preferential disruption of osteoclastic activation in ovariectomized mice. *British Journal of Pharmacology, 166*, 1084–1096.

CHAPTER SEVENTEEN

Collection, preparation, and biobanking of clinical specimens for analysis in polyaminopathies

Elizabeth A. VanSickle[a,b], Chad R. Schultz[b,c], André S. Bachmann[b,c], and Caleb P. Bupp[a,b,c,*]

[a]Division of Medical Genetics, Corewell Health/Helen DeVos Children's Hospital, Grand Rapids, MI, United States
[b]International Center for Polyamine Disorders, Grand Rapids, MI, United States
[c]Department of Pediatrics and Human Development, College of Human Medicine, Michigan State University, Grand Rapids, MI, United States
*Corresponding author. e-mail address: caleb.bupp@corewellhealth.org

Contents

1. Introduction	310
2. Collection of biological samples	310
2.1 Identification of patients with polyaminopathies	310
2.2 Obtaining control samples	311
2.3 Informed consent	312
2.4 Remote sample collection, collection kits, and shipping	313
2.5 Collection timing, volume limitations, and other considerations	314
3. Preparation of biological samples	315
3.1 Blood	315
3.2 Urine	318
3.3 Skin	318
3.4 Tissue	319
4. Biobanking of biological samples	319
4.1 Storage considerations	319
4.2 Sample organization and tracking	320
Acknowledgments	320
Competing interests	320
References	321

Abstract

Polyaminopathies are a relatively new family of rare genetic syndromes recently described in the literature. These syndromes are involved in the biosynthesis of polyamines, which include putrescine, spermidine, and spermine. Polyamines are aliphatic molecular that are found in most life forms, including humans, and are essential for embryogenesis, organogenesis, and tumorigenesis. The five known polyaminopathies that have been described to date include Snyder-Robinson Syndrome (SRS), Bachmann-Bupp Syndrome (BABS),

Faundes-Banka Syndrome (FABAS), as well as neurodevelopmental disorders associated with variants in *DHPS* and *DOHH*. These syndromes share many overlapping clinical phenotypes, including developmental delay, hypotonia, and intellectual disability.

Here we describe details for identifying and obtaining high-quality biological samples from patients with polyaminopathies. This includes special considerations for the informed consent process and the collection and shipment of biological samples for patients with rare diseases, many of whom live in countries around the world. We also detail the technical protocols for the collection, processing, storage, and tracking of biological samples for downstream research analysis specific to research in polyaminopathies, as well as biobanking for future use.

1. Introduction

The discovery of genetic syndromes has its origin in pattern recognition through observation of dysmorphic features, distinctive medical issues, and familial inheritance. Many genetic conditions were noted, described, and named before the underlying causes were even identified. Early genetic testing like chromosome analysis helped link syndromes with mechanism of disease, like the report of three copies of a chromosome causing Down Syndrome in 1959 (Lejeune, Gautier, & Turpin, 1959) long after it was clinically described in 1866 (Down, 1866). The development of newer technologies has allowed for increased diagnosis of many genetic disorders that had been reported as well as the discovery of new families of disorders linked by molecular and biochemical mechanisms. One rare family of genetic syndromes called polyaminopathies has recently been defined, with four of the five known syndromes being first described in just the past five years. Polyamines are aliphatic molecules and are found in most living organisms, including humans. Polyamines include putrescine, spermidine, and spermine, and are involved in essential cellular functions including embryonic cell development, regulation of cell division, proliferation, and tumorigenesis.

2. Collection of biological samples
2.1 Identification of patients with polyaminopathies

The polyamine pathway shares this history with the first disorder described by Snyder and Robinson in 1969 (Snyder & Robinson, 1969) and the underlying genetic cause reported in 2003 (Cason et al., 2003). Though this syndrome does have some recognizable features such as intellectual disability, shorter stature, and osteoporosis, it is typically not diagnosed by only clinical observation. Most individuals with Snyder-Robinson Syndrome (SRS) will have a variant in the *SMS* gene identified on molecular testing, now usually through whole exome or whole genome sequencing. Prior to broad sequencing being clinically available and not cost prohibitive, some might have been found by gene panel testing, typically an intellectual disability gene panel or even a panel of X-linked disorders as that is the inheritance pattern of SRS.

With the discovery of Bachmann-Bupp Syndrome (BABS) in 2018 (Bupp, Schultz, Uhl, Rajasekaran, & Bachmann, 2018) the 2nd polyamine-related disorder, which is autosomal dominant, broad sequencing analysis is becoming the standard for patient identification. As other conditions caused by variants in *DHPS* (Ganapathi et al., 2019), *eIF5A* (Faundes-Banka Syndrome) (Faundes et al., 2021), and *DOHH* (Ziegler et al., 2022) have been described, a new class of disorders was coined in 2023 called polyaminopathies (Bachmann, VanSickle, Michael, Vipond, & Bupp, 2024). All these disorders remain ultra-rare, and molecular results can identify variants of uncertain significance which can provide a clinical challenge. The spectrum of phenotype is still evolving and somewhat unknown, and the main features of most polyaminopathies (intellectual disability, developmental delay, hypotonia, seizures) are nonspecific enough to not clearly delineate uncertainty into pathogenicity. In these situations, additional methodologies like metabolomics or epigenetics are emerging as ways to clarify uncertainty. Unfortunately, those do not illuminate every case. Thus, there is a need for sample collection to allow for performance of analysis for patients suspected of and confirmed to have polyaminopathies.

2.2 Obtaining control samples

Biological samples collected from unaffected control patients are useful tools when performing clinical research. These samples allow researchers to establish a baseline comparison for the patients with polyaminopathies. Age- and sex-matched control samples from healthy patients can be difficult to obtain in a pediatric population but are especially useful for metabolomics analysis to ensure that changes seen in the metabolomic profile are not due to naturally-occurring changes throughout childhood. Adding the ability to screen and consent healthy controls into an institutional review board (IRB)-approved research study can aid in the collection of control samples. It can still be difficult to identify and obtain samples from healthy pediatric patients, however, as pediatric patients admitted to the hospital rarely meet inclusion criteria for control participants. If obtaining control samples from healthy pediatric patients is not possible, certain biological samples may be able to be purchased through commercial biobanks. Additionally, pooled pediatric control samples specifically for use in metabolomics analysis may be considered, such as NIST SRM 1950 Metabolites in Human Plasma from Millipore Sigma.

2.3 Informed consent

As part of an IRB-approved research study, informed consent documents must also be reviewed and approved by each institution's own IRB. Ongoing review typically occurs annually after initial approval. Separate consent documents should be written for proband patients as well as biological family members and control patients. Special consideration for informed consent into clinical research studies should be given for publishing deidentified clinical and genetic information into open access databases, which is often required when publishing findings into peer-reviewed academic journals. This may often require a separate section for the patient or parent and/or guardian to opt in or out, but typically does not limit their participation in the study as a whole. The consent form should be provided to the patient and family ahead of the consent discussion for them to have time to read and review. Patients with hearing or visual impairments must be provided with accessible options during the consent discussion to ensure that all parts of the consent and discussion are understood prior to the researcher obtaining consent. To accommodate patients who are unable to travel to your site location, seeking IRB-approval for virtual consent through Microsoft Teams or Zoom is also beneficial.

Due to the rare nature of polyaminopathies, patients are being identified all over the world and many are non-English speaking. In this case, the consent discussion may be led by two certified medical interpreters in the patient's primary language, using the original English informed consent document. This means that the interpreters must read each word verbatim to the patient and parent(s)/guardian(s) in their primary language after the study team reads it in English. While this option does save money up front for the study team by paying for hourly interpretation services, it does require the family to attend a very lengthy consent discussion, usually lasting 2 h or more. Researchers can also choose to have their informed consent documents translated by a certified medical translator. While this does typically mean a larger investment up front, the benefit of having fully translated informed consent documents, especially in more common languages, means ease of understanding for the patient and parent(s)/guardian(s), as well as a more seamless consent discussion. These translated informed consent documents mean that only one certified medical interpreter is required and would be there to provide only basic interpretation between the researcher and the family.

2.4 Remote sample collection, collection kits, and shipping

Similar to the considerations given for informed consent due to the rare nature of polyaminopathy research are the considerations that must be given for biospecimen collection, processing, storage, and shipping. As there are less than 200 patients with polyaminopathies known worldwide, this means that researchers must be nimble when it comes to collecting biospecimens from patients. Since many patients identified with one of these rare disorders are unable to travel for sample collection, it is often best for the researcher to coordinate the collection, processing, storage, and shipping of these samples with a laboratory that's close to and convenient for the patient.

To meet the needs of the biological samples to be collected for analysis in polyaminopathies, the remote laboratory must have access to all materials and equipment, as detailed later in Section 3, as well as dry ice for shipping. Once a remote laboratory has been identified, the study team must ensure that the lab understands the standard operating protocol (SOP) for the collection, processing, storage, and shipment of the biological samples to be collected. Samples should be collected ideally on a Monday or Tuesday to ensure that the ambient sample collected for lymphoblastoid cell line (LCL) establishment remains viable, as this sample needs to be shipped ambient same-day. The study team will then prepare prelabeled collection kits with the patient's deidentified study ID, sample type, and collection date. These kits will include all necessary blood tubes and cryotubes. The study team will ensure that the remote laboratory has received the kit and that all materials are intact prior to confirming the collection with the family.

Once the samples have been collected, processed, and stored, the study team will communicate with the remote laboratory to arrange for shipment of the samples back to the study team. Domestic shipments within the United States are generally straightforward, but teams should still ensure that ample dry ice is included in each shipment. Each shipment should contain approximately 6–7 kg (13–17 lbs) of dry ice in a polystyrene box, inside of a cardboard outer package with a UN1845 sticker. The dry ice pellets should be in a 4–6 cm (2–3 in) layer on the bottom, with frozen specimen bags laid flat on top of the layer, and additional dry ice layered on top of the specimens.

It may be prudent to split the samples into two shipments, in the event that a package is lost or delayed. Samples should always be shipped on a Monday or Tuesday to ensure receipt by the study team by the end of the week. Ambient samples may be shipped in a cushioned package and may include a cold or cool pack, especially if shipping during warm temperatures (Gordy, Tashjian, Lee, Movassaghi, & Yong, 2019).

International shipments require additional measures to ensure that samples arrive in a timely manner and are viable upon arrival. Similar to domestic shipments within the United States, shipments from other countries to the United States should only be sent on a Monday to prevent a delay if the shipment is held at customs. More dry ice should also be used for international shipments, with the recommendation being at least 12–14 kg (26–34 lbs). A UN1845 dry ice sticker is still required for international shipments of biospecimens. The study team should identify a shipping company that can refill dry ice at each destination, if needed. Additionally, some shipping companies monitor the temperature of shipments for the researchers to have a record of any temperature excursions that the samples may have taken during transit. For most countries, multiple copies of a customs invoice will be required, as well as an Airway Bill and Dangerous Goods declaration. Three copies of these documents are required to be printed, signed, and affixed in the document pouch on the outside of the package. Both the remote laboratory and the study team should be mindful of holidays in both countries that may delay transit (Gordy et al., 2019).

2.5 Collection timing, volume limitations, and other considerations

Putrescine, spermidine, and spermine can be found in many foods and are important sources of polyamines. It's not known if or how dietary polyamine intake may impact downstream analysis of biological samples, so therefore, there are no standard fasting or dietary recommendations for patients with polyaminopathies (Muñoz-Esparza et al., 2019). Additionally, patients with polyaminopathies often present with phenotypes that would make fasting or specific dietary considerations difficult. Therefore, families may keep a food diary of food and drinks consumed for 24 h prior to sample collection, but this is not a requirement for collection.

For pediatric patients, maximum daily blood draw volumes must be considered when prioritizing and collecting biological samples. The maximum amount of blood that may be drawn in a pediatric patient per

24 h period is equal to 2.5 mL/kg/day. For example, this means that a baby weighing for 4 kg (8.8 lbs) may only have 10 mL blood collected in total per day, including both clinical and research blood draws (Howie, 2011). For young children weighing less than 10 kg, this will often mean adjusting volumes for blood samples collected, as detailed in Section 3.1. When collecting blood from young children, priority is typically given to blood for LCL establishment, as well as RBCs for enzyme activity and polyamine analysis.

3. Preparation of biological samples
3.1 Blood

3.1.1 Materials and equipment
1. Eppendorf™ Research™ Plus Pipettors
2. Eppendorf™ Research™ Plus Pipet Tips
3. Revco® Ultima PLUS −80 °C freezer or comparable (Thermo Scientific)
4. Sorvall™ LYNX 6000 Superspeed Centrifuge or comparable (Thermo Fisher #75006590)
5. BD Vacutainer™ Glass Blood Collection Tubes with Acid Citrate Dextrose (ACD) Solution A. (BD# 364816).
6. BD Vacutainer™ Glass Blood Collection Tubes with Acid Citrate Dextrose (ACD) Solution B. (BD# 364816-1).
7. PAXgene Blood RNA Tube (BD# 762165).
8. BD Vacutainer™ K2EDTA Collection Tubes, 6 mL volume. (BD# 367835).
9. Fisherbrand™ Externally and Internally Threaded Cryogenic Storage Vials (Thermo Fisher #10-500-26)
10. Falcon™ 15 mL Conical Centrifuge Tubes or comparable (Thermo Fisher #14-959-53 A)

3.1.2 ACD A tube collection, processing, and storage
Whole blood collected into ACD A tubes can be used for downstream ODC enzyme activity and polyamine analysis via HPLC or biobanked for future research use.
1. Draw minimum 1 mL blood into BD Vacutainer™ ACD A Tube.
2. Gently invert 8−10 times.
3. Process sample within 2 h of collection. Store upright at room temperature.

4. Gently invert 8-10 times immediately before processing.
5. Using Eppendorf™ Research™ Plus Pipettors and pipet tips, aliquot 500 mL whole blood into Fisherbrand™ Externally and Internally Threaded Cryogenic Storage Vials.
6. Transfer to −80 °C for long-term storage or for at least 24 h before shipping, if applicable.

3.1.3 ACD B tube collection and processing

Whole blood collected into ACD B tubes can be used for LCL establishment. Blood should be collected on a Monday or Tuesday and shipped same day to appropriate lab for LCL establishment to ensure maximum cell viability.
1. Draw minimum 2 mL blood into BD Vacutainer™ ACD B Tube.
2. Gently invert 8-10 times.
3. Store upright at room temperature and away from light until ready to ship.

3.1.4 PAXGene blood RNA tube collection, processing, and storage

Whole blood collected into PAXGene RNA tubes can be used for downstream RNA extraction and RNA sequencing.
1. Draw minimum 2.5 mL blood into PAXGene RNA Tube. If there are volume considerations for neonates, volumes as low as 1.2 mL may be utilized, depending on the lab performing the RNA extraction. Confirm with your lab before collecting less than the minimum recommended 2.5 mL.
2. Gently invert 8−10 times.
3. Store upright at room temperature away from light for 2−4 h to ensure complete lysis of blood cells.
4. Transfer tube to −20 °C for at least 2 h.
5. Transfer tube to −80 °C for long-term storage, if RNA will not be extracted immediately, or for at least 24 h before shipping, if applicable.
6. If the PAXgene Blood RNA Tube was stored at 2–8 °C, −20 °C, or −70 °C after blood collection, first equilibrate it to room temperature and then store it at room temperature for 2 h before starting the procedure.

3.1.5 BD Vacutainer™ K2EDTA tubes collection, processing, and storage

Whole blood collected into K2EDTA tubes may be used for DNA extraction. Plasma isolated from K2EDTA tubes may be used for downstream metabolomics/lipodomics analysis via LC/MS, as described in a separate chapter of this book (see Sheldon et al.) and pharmacokinetic analysis for patients on DFMO treatment as described in a separate chapter of this book (see Dowling et al.). Buffy Coat/RBC isolated from K2EDTA tubes may be used for downstream ODC activity analysis and polyamine analysis via HPLC as described in a separate chapter of this book (see Schultz, Bupp, Rajasekaran, & Bachmann, 2019).

3.1.5.1 DNA extraction
1. Draw minimum 3 mL blood into BD Vacutainer K2EDTA Tube.
2. Gently invert 8–10 times.
3. Store upright at room temperature and transfer to lab same-day for DNA extraction.
4. Transfer tube to −80 °C for long-term storage, if DNA will not be extracted immediately, or for at least 24 h before shipping, if applicable.

3.1.5.2 Plasma/buffy coat isolation
1. Pre-cool centrifuge to 4 °C.
2. Draw minimum 3 mL blood into tube.
3. Gently invert 8–10 times.
4. Process sample within 2 h of collection. Store upright at room temperature.
5. Centrifuge tubes at 800–1000 × g for 10 min with the brake off to avoid disrupting the density gradient during deceleration.
6. The following layers should be visible after centrifugation
    a. Yellowish plasma layer (supernatant)
    b. Grayish-white buffy coat layer containing leukocytes
    c. Deep red layer of erythrocytes (red blood cells, RBCs)
7. Using a pipette, remove plasma and place into Falcon™ 15 mL Conical Centrifuge Tubes.
    a. Note: Do not disturb the buffy coat layer while collecting the plasma.
8. Centrifuge plasma at 2000 × g for 10 min with brake on to remove platelets and debris.
9. While plasma is spinning, using a pipetter, transfer 200 uL aliquots of buffy coat/RBC layers to Fisherbrand™ Externally and Internally Threaded Cryogenic Storage Vials.
10. Immediately place specimens into a −80 °C freezer.
11. Once plasma has finished spinning, using a pipetter, transfer 250 uL aliquots of plasma to Fisherbrand™ Externally and Internally Threaded Cryogenic Storage Vials.
12. Immediately place specimens into a −80 °C freezer.

3.2 Urine

Urine samples may be used for downstream metabolomic analysis via LC/MS as described in another chapter of this book (see Sheldon et al.).

3.2.1 Materials and equipment
1. Thermo Scientific™ Samco™ Clicktainer™ Vials and Specimen Containers (Thermo# 13-712-55)
2. Eppendorf™ Research™ Plus Pipettors
3. Eppendorf™ Research™ Plus Pipet Tips
4. Revco® Ultima PLUS −80 °C freezer or comparable (Thermo Scientific)
5. Fisherbrand™ Externally and Internally Threaded Cryogenic Storage Vials (Thermo Fisher #10-500-26)

3.2.1.1 Urine collection and storage
1. Using clean catch method, instruct patient to collect urine in Thermo Scientific™ Samco™ Clicktainer™ Vials and Specimen Containers.
2. Store at 4 °C if not processing immediately. Process within 24 h of collection.
3. Using pipettor and pipet tips, aliquot 250 uL urine into pre-labeled Fisherbrand™ Externally and Internally Threaded Cryogenic Storage Vials.
4. Immediately place specimens into a −80 °C freezer.

3.3 Skin

Skin specimens collected from skin punch biopsies may be used for the establishment of primary human dermal fibroblasts.

3.3.1 Materials and equipment
1. Integra™ Miltex™ Standard Biopsy Punches or comparable (Fisher# 12-460-399)
2. 70% Alcohol
3. 90% fetal bovine serum (FBS)/10% dimethyl sulfoxide (DMSO) solution
4. Fisherbrand™ Externally and Internally Threaded Cryogenic Storage Vials (Thermo Fisher #10-500-26)
5. Mr. Frosty™ Freezing Container (Thermo Fisher# 5100-0001)
6. Revco® Ultima PLUS −80 °C freezer or comparable (Thermo Scientific)

3.3.1.1 Skin collection and storage
1. Prepare skin to be biopsied with 70% alcohol solution and betadine in standard sterile fashion.

2. Perform 4 mm punch biopsy using Integra™ Miltex™ Standard Biopsy Punch or comparable
3. Place skin specimen in 90% FBS/10% DMSO solution in Fisherbrand™ Externally and Internally Threaded Cryogenic Storage Vial.
4. Freeze overnight at −80 °C in Mr. Frosty™ Freezing Container.
5. Move skin specimen to liquid nitrogen (vapor phase) until ready to establish primary dermal fibroblast cells (Schultz et al., 2019).

3.4 Tissue
1. 90% fetal bovine serum (FBS)/10% dimethyl sulfoxide (DMSO) solution
2. Fisherbrand™ Externally and Internally Threaded Cryogenic Storage Vials (Thermo Fisher #10-500-26)
3. Mr. Frosty™ Freezing Container (Thermo Fisher# 5100-0001)
4. Revco® Ultima PLUS −80 °C freezer or comparable (Thermo Scientific)

3.4.1 Tissue collection and storage
1. Place tissue specimen in 90% FBS/10% DMSO solution in Fisherbrand™ Externally and Internally Threaded Cryogenic Storage Vial.
2. Freeze overnight at −80 °C in Mr. Frosty™ Freezing Container.
3. Move tissue specimen to liquid nitrogen (vapor phase) until ready to process.

4. Biobanking of biological samples
4.1 Storage considerations

There are many considerations when deciding on storage conditions for biospecimens. Aside from initial analyses to be performed on clinical samples collected from patients with polyaminopathies, future research questions and specific utility of biospecimens may be uncertain at the time of collection and storage. Storage conditions should be standardized to ensure quality of samples and downstream research studies. Freezers should have remote temperature monitoring in place that can notify multiple study personnel in the event of a temperature deviation or freezer failure. Study teams should also have back up freezers that can be readily available to use for sample storage in these scenarios. Similarly, liquid nitrogen dewars should be equipped with remote monitoring systems that can alert study teams in the event of low liquid nitrogen volume.

As described in Section 3, initial activities performed on biological samples collected from patients with polyaminopathies often include enzyme activity analysis, polyamine detection, LCL establishment, RNA sequencing, and metabolomics/lipodomics analysis. The protocols detailed in Section 3 include recommendations for vial size and aliquot volume that are suitable for these anticipated analyses, while minimizing unnecessary freeze/thaw cycles. Optimal preservation of LCL after establishment is storage in the vapor phase of liquid nitrogen, while optimal preservation of blood sample derivatives is at −80 °C (Perry, Jasim, Hojat, & Yong, 2019).

4.2 Sample organization and tracking

Study personnel who access biospecimens should log storage conditions at the time of collection, as well as any protocol deviations impacting biological samples, including freezer failure, freeze/thaw incidents, and other temperature changes. Biological specimens should have a unique identifier with all pertinent details of the sample. Identifiers should be deidentified and should never include PHI. Information on labels should be legible and ideally printed instead of handwritten. Labels that can withstand extreme cold temperatures should always be utilized. A laboratory information management system (LIMS) should be utilized to ensure that all biospecimens are logged and tracked throughout their entire life cycle. Each sample should be tracked and logged with its specific location including freezer, shelf, rack, drawer, box, and space number. Any protocol deviations noted during sample collection, processing, and storage should be noted in the LIMS, as these details can be essential for troubleshooting downstream research assays.

Acknowledgments

National Institutes of Health grant award R01 HD110500 (A.S.B., C.P.B.) for the provided funds to study BABS and SRS and we thank the MSU-Spectrum Health Alliance Corporation (SH-MSU-ACF RG101298) for the provided support to generate the data shown in this book chapter. We would like to thank each of the patients and families impacted by polyaminopathies for trusting us with this important work.

Competing interests

A.S.B. and C.P.B. are listed inventors of two U.S. patents (US 11,273,137 B2 and US 12,194,010 B2) issued on March 15, 2022 and January 14, 2025, respectively, entitled "Methods and compositions to prevent and treat disorders associated with mutations in the *ODC1* gene" and Michigan State University and Corewell Health have an exclusive licensing agreement with Orbus Therapeutics, Inc. C.P.B. provides consulting services for Orbus Therapeutics. A.S.B. is sole inventor of a U.S. patent (US 9,072,778) issued on July 7, 2015, entitled "Treatment regimen for N-Myc, C-Myc, and L-Myc amplified and overexpressed tumors". No potential conflicts of interest were disclosed by the other authors.

References

Bachmann, A. S., VanSickle, E. A., Michael, J., Vipond, M., & Bupp, C. P. (2024). Bachmann-Bupp syndrome and treatment. *Developmental Medicine and Child Neurology, 66*(4), 445–455. https://doi.org/10.1111/dmcn.15687.

Bupp, C. P., Schultz, C. R., Uhl, K. L., Rajasekaran, S., & Bachmann, A. S. (2018). Novel de novo pathogenic variant in the ODC1 gene in a girl with developmental delay, alopecia, and dysmorphic features. *American Journal of Medical Genetics. Part A, 176*(12), 2548–2553. https://doi.org/10.1002/ajmg.a.40523.

Cason, A. L., Ikeguchi, Y., Skinner, C., Wood, T. C., Holden, K. R., Lubs, H. A., ... Schwartz, C. E. (2003). X-linked spermine synthase gene (SMS) defect: The first polyamine deficiency syndrome. *European Journal of Human Genetics: EJHG, 11*(12), 937–944. https://doi.org/10.1038/sj.ejhg.5201072.

Down, J. L. (1866). Observations on an ethnic classification of idiots. *Mental Retardation, 33*(1), 54–56.

Faundes, V., Jennings, M. D., Crilly, S., Legraie, S., Withers, S. E., Cuvertino, S., et al. (2021). Impaired eIF5A function causes a mendelian disorder that is partially rescued in model systems by spermidine. *Nature Communications, 12*(1), 833. https://doi.org/10.1038/s41467-021-21053-2.

Ganapathi, M., Padgett, L. R., Yamada, K., Devinsky, O., Willaert, R., Person, R., et al. (2019). Recessive rare variants in deoxyhypusine synthase, an enzyme involved in the synthesis of hypusine, are associated with a neurodevelopmental disorder. *American Journal of Human Genetics, 104*(2), 287–298. https://doi.org/10.1016/j.ajhg.2018.12.017.

Gordy, D., Tashjian, R. S., Lee, H., Movassaghi, M., & Yong, W. H. (2019). Domestic and international shipping of biospecimens. *Methods in Molecular Biology (Clifton, N. J.), 1897*, 433–443. https://doi.org/10.1007/978-1-4939-8935-5_35.

Howie, S. R. (2011). Blood sample volumes in child health research: Review of safe limits. *Bulletin of the World Health Organization, 89*(1), 46–53. https://doi.org/10.2471/BLT.10.080010.

Lejeune, J., Gautier, M., & Turpin, R. (1959). Etude des chromosomes somatiques de neuf enfants mongoliens [Study of somatic chromosoms from 9 mongoloid children]. *Comptes Rendus Hebdomadaires des Seances de l'Academie des Sciences, 248*(11), 1721–1722.

Muñoz-Esparza, N. C., Latorre-Moratalla, M. L., Comas-Basté, O., Toro-Funes, N., Veciana-Nogués, M. T., & Vidal-Carou, M. C. (2019). Polyamines in food. *Frontiers in Nutrition, 6*, 108. https://doi.org/10.3389/fnut.2019.00108.

Perry, J. N., Jasim, A., Hojat, A., & Yong, W. H. (2019). Procurement, storage, and use of blood in biobanks. In W. Yong (Vol. Ed.), *Biobanking. Methods in molecular biology: Vol. 1897*, New York, NY: Humana Press. https://doi.org/10.1007/978-1-4939-8935-5_9.

Schultz, C. R., Bupp, C. P., Rajasekaran, S., & Bachmann, A. S. (2019). Biochemical features of primary cells from a pediatric patient with a gain-of-function ODC1 genetic mutation. *The Biochemical Journal, 476*(14), 2047–2057. https://doi.org/10.1042/BCJ20190294.

Snyder, R. D., & Robinson, A. (1969). Recessive sex-linked mental retardation in the absence of other recognizable abnormalities. Report of a family. *Clinical Pediatrics, 8*(11), 669–674. https://doi.org/10.1177/000992286900801114.

Ziegler, A., Steindl, K., Hanner, A. S., Kar, R. K., Prouteau, C., Boland, A., et al. (2022). Bi-allelic variants in DOHH, catalyzing the last step of hypusine biosynthesis, are associated with a neurodevelopmental disorder. *American Journal of Human Genetics, 109*(8), 1549–1558. https://doi.org/10.1016/j.ajhg.2022.06.010.

CHAPTER EIGHTEEN

An enhanced method for the determination of polyamine content in biological samples by dansyl chloride derivatization and HPLC

Ashley Nwafor, Tracy Murray Stewart, and Robert A. Casero, Jr.[*]

Department of Oncology, Johns Hopkins University School of Medicine, Baltimore, MD, United States
*Corresponding author. e-mail address: rcasero@jhmi.edu

Contents

1. Introduction	324
2. Rationale	326
3. Polyamine standards preparation	327
3.1 Equipment	327
3.2 Reagents	327
3.3 Procedure	328
4. Cell, media, and urine sample lysate preparation	328
4.1 Equipment	328
4.2 Reagents	328
4.3 Procedure	328
4.4 Notes	328
5. Tissue sample lysate preparation	329
5.1 Equipment	329
5.2 Reagents	329
5.3 Procedure	329
5.4 Notes	330
6. Dansyl chloride labeling of polyamines for HPLC	330
6.1 Equipment	330
6.2 Reagents	330
6.3 Procedure	330
6.4 Notes	331
7. HPLC analysis of polyamine content	331
7.1 Equipment	331
7.2 Reagents	332
7.3 Procedure	332
7.4 Notes	333

Methods in Enzymology, Volume 715
ISSN 0076-6879, https://doi.org/10.1016/bs.mie.2025.01.066
Copyright © 2025 Elsevier Inc. All rights are reserved, including those for text and data mining, AI training, and similar technologies.

8. Results	334
Acknowledgments	335
References	335

Abstract

The polycationic alkylamines known as polyamines are ubiquitous organic cations found in all forms of life and are involved in multiple biological processes including RNA and DNA replications and conformational changes, transcription, translation, regulation of ion channels, free radical scavenging, and cell proliferation. Therefore, polyamine homeostasis is critical to normal cellular function and health. In cancer and other hyperproliferative diseases, polyamines and their metabolism are frequently dysregulated due to the increased requirement for polyamines under conditions of rapid, or continued, uncontrolled growth. Consequently, polyamine function and metabolism are of considerable interest as possible targets for antiproliferative therapies. As the awareness of the importance of polyamines in health and disease has increased, so has the need for rapid and precise measurements of polyamines in biological samples, including cells, blood, serum, urine, and other biological fluids. Here we present an improvement over one of the standard high performance liquid chromatography methods that is faster, uses fewer reagents, and is at least as precise as previously published methods.

1. Introduction

Putrescine (Put), spermidine (Spd), and spermine (Spm) are naturally occurring polycationic alkylamines that are essential for the normal growth and function of essentially all eukaryotic cells (Fig. 1). Prokaryotes also have a polyamine requirement for growth and survival, although most do not synthesize Spm, and several have additional members of the class that can include branched structures and/or different numbers of methylene spacers between amine/imine moieties (Casero, Murray Stewart, & Pegg, 2018). The cellular content of polyamines in normal cells is precisely controlled by a highly regulated polyamine metabolic pathway, coupled with active transport systems. In most mammalian cells, polyamines are derived from the urea cycle amino acid ornithine, itself a product of the activity of arginase I on arginine (Fig. 1).

Ornithine is converted to Put by the activity of ornithine decarboxylase (ODC1), and Put is subsequently converted to Spd and, ultimately, Spm by the successive addition of aminopropyl groups to the opposite amino-terminal groups of Put by the enzymes Spd synthase (SRM) and Spm synthase (SMS), respectively. All three polyamines can also be transported

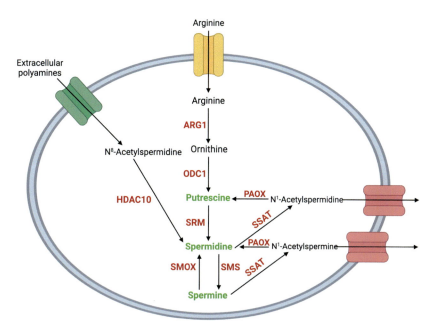

Fig. 1 The polyamine pathway through which the native polyamines putrescine, spermidine, and spermine are derived from ornithine, which is derived from arginine via arginase 1. Spermidine and spermine can be acetylated by Spermidine/spermine N^1-acetyltransferase (SSAT) for export, or these acetylated polyamines can be catabolized by polyamine oxidase (PAOX). *Fig. 1 was made using BioRender.*

from the extracellular environment by a poorly characterized, energy-dependent transport system (Holbert, Cullen, Casero, & Stewart, 2022). Polyamine content can also be modulated through the polyamine catabolic pathway through oxidation and/or acetylation and export (Fig. 1). Although the importance of polyamines in normal cell function and proliferation, and the dysregulation of polyamine content and metabolism in hyperproliferative diseases has long been known, only recently have the molecular tools been available to examine the molecular roles of polyamine more specifically in living systems.

Recent studies have implicated normal polyamine function to be critical in normal aging (Arthur, Jamwal, & Kumar, 2024; Ni & Liu, 2021). Multiple ongoing and completed clinical trials have demonstrated the utility of inhibiting polyamine biosynthesis in combination with other therapeutic agents to treat specific forms of neoplastic diseases, including pancreatic cancer, neuroblastoma, lung and prostate cancers (NCT05254171, NCT06219174, NCT06059118) (Yang, 2024). Additionally, there is growing recognition that genetic abnormalities in the genes responsible for the control of polyamine metabolism contribute to an expanding list of genetic diseases and syndromes, including Bachmann/Bupp and Snyder-Robinson syndromes (Afrin et al., 2023; Bachmann, VanSickle, Michael, Vipond, & Bupp, 2024; Murray-Stewart, Dunworth, Foley, Schwartz, & Casero, 2018; Stewart et al., 2023). It is also important to note that several recent studies have demonstrated that polyamine metabolism plays a critical role in the normal development of the human immune system and that abnormal polyamine metabolism likely contributes to the establishment of a tumor-permissive immune microenvironment (Chin, Bieberich, Stewart, & Casero, 2022; Holbert, Cullen, Casero, & Stewart, 2022).

Taken together, the recent advances in the study of the roles of polyamines in human health and disease necessitate a rapid, cost-effective analytic method that allows precise determination of the polyamine content in biological specimens. Here we present a significant improvement of a standard high performance liquid chromatography (HPLC) method that is faster, uses less reagents, and is at least as precise as previously published methods (Kabra, Lee, Lubich, & Marton, 1986). Applicable to a variety of biological samples, this method saves both time and resources, allowing for a greater number of samples to be analyzed more efficiently.

2. Rationale

While new studies continue to demonstrate that polyamines play a crucial role in cell growth and function, one main challenge has been establishing a fast, consistent method that not only aids in native polyamine concentration determination, but also relates associated enzyme activity to non-native polyamines in several different sample types. One such association is cytosolic histone deacetylase-10 (HDAC10), which respectively deacetylases N^8-acetylspermidine (N^8-AcSpd) (Stewart et al., 2022). Previous difficulties in detecting multiple polyamine derivatives in one biological sample, such as with N^8-AcSpd and N^1-acetylspermidine (N^1-AcSpd), included isomers eluting as a single peak, the requirement for separate standards for different derivates, and lengthy run times. This improved HPLC method includes gradients for the effective and rapid detection of native polyamines, acetylated polyamines, and polyamine analogs in one standard, including the partial separation of N^8-AcSpd and N^1-AcSpd isomers (Fig. 2), significant for further study of related enzymatic activity.

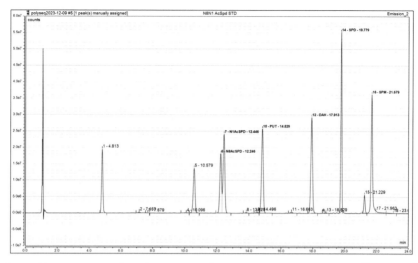

Fig. 2 Chromatogram of a polyamine standard containing 0.02 nmol/mL N^8-AcSpd, N^1-AcSpd, putrescine (PUT), spermidine (SPD), and spermine (SPM) obtained using the modified Gradient III for the partial separation of N^8AcSpd and N^1-AcSpd isomers (12.246 min and 12.446 min, respectively) with 1,7-diaminoheptane (DAH) as the internal standard.

3. Polyamine standards preparation

3.1 Equipment

1. Vacuum filter/storage bottle system (Corning #431097).

3.2 Reagents

1. Hydrochloric Acid (J.T. Baker #9535-33).
2. HPLC-grade water (Fisher #W5-4).
3. 1, 7-Diaminoheptane (Sigma #D17408-5G; MW: 130.23 g/mol).
4. N^8-Acetylspermidine dihydrochloride (Sigma #A-3658; MW: 260.2 g/mol).
5. N^1-Acetylspermidine dihydrochloride (Cayman #9001535; MW: 260.2 g/mol).
6. Putrescine dihydrochloride (Sigma #P-7505; MW: 161.1 g/mol).
7. N^1-Acetylspermine trihydrochloride (Fluka #01467; MW: 353.76 g/mol).
8. Spermidine trihydrochloride (Sigma #S-2501; FW: 254.6 g/mol).
9. Spermine tetrahydrochloride (Sigma #S2876-5G; MW: 348.18 g/mol).
10. Di(ethyl)dihydroxyhomospermine (SBP101; Panbela Therapeutics, Inc., Waconia, MN, USA).
11. N^1,N^{11}-bis(ethyl)norspermine (BENSpm; Patrick Woster, Medical University of South Carolina, USA).

3.3 Procedure

1. Dilute hydrochloric acid with HPLC-grade water to yield a 0.1 N stock solution; filter with a vacuum filter bottle system.
2. Dissolve DAH in 0.1 N HCl to yield a 40 μM stock solution; aliquot in 1.5 mL tubes and store at −20 °C.
3. Dissolve native polyamines (putrescine, spermidine, and spermine) in 0.1 N HCl to yield 10 mM stock solutions; aliquot in 1.5 mL tubes and store at −20 °C.
4. Dissolve acetylated polyamines and polyamine analogs in 0.1 N HCl to yield 1 mM stock solutions; aliquot in 1.5 mL tubes and store at −20 °C.
5. Add 4 μL each of 10 mM putrescine, spermidine, and spermine, plus 40 μL of 1 mM acetylated polyamines or analogs as desired to be detected for 1 mL of standard. Bring volume up to 1 mL with 0.1 N HCl. Aliquot and store at −20 °C.

4. Cell, media, and urine sample lysate preparation

4.1 Equipment
1. Mini vortex mixer (Fisher, model #02215365).
2. Microcentrifuge (Thermo Scientific, model #75002447).

4.2 Reagents
1. Perchloric Acid (J.T. Baker #9652-01).

4.3 Procedure
1. Dilute perchloric acid with HPLC-grade water to yield a 1.2 N stock solution; aliquot in 15 mL conical tubes and store at 4 °C.
2. Add 50 μL of cell lysate, media (see Note 1), or urine sample to 50 μL 1.2 N perchloric acid.
3. Vortex briefly.
4. Centrifuge samples for 10–15 min at 12,000–15,000 g, 4 °C.
5. Collect 50 μL of sample supernatant for dansyl chloride labeling (see Note 2).

4.4 Notes
1. Concentrated media samples will be dried down. As a result, instead add 100 μL 0.6 N perchloric acid directly to each sample before spinning.
2. This perchloric acid lysate extraction procedure may be performed ahead of time and lysates may be stored at −20 °C.

5. Tissue sample lysate preparation

5.1 Equipment
1. Homogenizer (modified Dremel 4000).
2. Mini vortex mixer (Fisher, model #02215365).
3. Microcentrifuge (Thermo Scientific, model #75002447).

5.2 Reagents
1. Ornithine decarboxylase (ODC) breaking buffer:
   - EDTA (Invitrogen #15575-038).
   - Dithiothreitol (DTT; Fermentas #R0861).
   - HPLC-grade water (Fisher #W5-4).
   - Hydrochloric Acid (J.T. Baker #9535-33).
   - Trizma base (Sigma #T1503-5KG).
2. Perchloric Acid (J.T. Baker #9652-01).

5.3 Procedure
1. Prepare Tris-HCl by dissolving Trizma base in HPLC-grade water to yield a 0.5 M stock solution, then pH to 7.5 with HCl. Store at 4 °C.
2. Make up 50 mL ODC breaking buffer by combining 2.5 mL 0.5 M Tris-HCl (pH 7.5), 10 mL 0.5 M EDTA, and 0.02 g DTT, and bringing up to volume with HPLC H_2O.
3. Place tissue samples on ice to thaw (see Note 1).
4. Place homogenizer tip in 15 mL conical and place on ice.
5. Make up two small beakers of HPLC-grade water: one with, one without ice.
6. Place a 15-mL aliquot of ODC breaking buffer on ice and add 200 µL buffer to each sample tube (see Notes 2-3).
7. Transfer the homogenizer tip to the conical containing the remaining aliquot of breaking buffer, keeping the conical on ice.
8. When everything is set up, begin by lifting the homogenizer from the buffer conical, making sure to knock off excess buffer from the tip before placing it in a sample tube.
9. Turn on the homogenizer for 5 s, moving the tip up and down to homogenize the tissue sample.
10. Turn off the homogenizer, then back on to repeat step 8, ensuring the sample remains on ice whilst doing so to prevent heat buildup from the homogenizer.
11. Turn off the homogenizer then place the tip in the room temperature HPLC-grade water beaker, turning it on for 5 s to rinse.

12. Repeat step 10 for the iced HPLC-grade water beaker, then place the homogenizer tip back in the ODC breaking buffer conical on ice.
13. Repeat steps 7–11 for each sample.
14. After every sample is homogenized, transfer 50 μL of each to pre-labeled tubes.
15. Perform perchloric acid extraction (from Section 4.3) on homogenized samples immediately, or store for later at −80 °C.

5.4 Notes
1. Samples may first have to be transferred to flat bottom tubes so the homogenizer can fit to the bottom.
2. Depending on the number of samples and tissue size, more than 200 μL breaking buffer may be needed.
3. When adding breaking buffer, samples must be kept on ice the entire time, and for the duration of the procedure.

6. Dansyl chloride labeling of polyamines for HPLC

6.1 Equipment
1. Light-shielded (1.5 and 15 mL) and standard microcentrifuge tubes (1.5 and 2 mL).
2. Water bath (Precision Scientific model #182, #66643).
3. Sep-Pak C18 cartridges (Waters #WAT023590).
4. Vac-Elut (Analytichem International model #AI 6000).
5. 0.2-μm nylon syringe filters (Thermo Scientific #42213-NN).

6.2 Reagents
1. Sodium carbonate monohydrate (Sigma #S4132-1KG).
2. Acetone (Sigma #270725-1 L).
3. Dansyl chloride (Sigma #03641-100MG).
4. Proline (Sigma #P0380-100G).
5. HPLC-grade acetonitrile (Fisher #A998-4).
6. HPLC-grade water (Fisher #W5-4).

6.3 Procedure
1. Turn on and set water bath to 70 °C.
2. Place 50 μL of sample supernatant from perchloric acid extraction (from Section 4.3) or polyamine/analog standard (from Section 3.3; 40 pmol/μL; see Note 1) in a 1.5 mL tube with 50 μL of 40 μM DAH (from Section 3.3), the internal standard.

Determination of polyamine content in biological samples 331

3. Add 200 μL saturated sodium carbonate (Na$_2$CO$_3$) to each tube.
4. Make up a fresh aliquot of 100 mg/mL dansyl chloride by adding 1 mL acetone directly to a 100-mg bottle; transfer to a light-shielded 1.5 mL tube.
5. Dilute the 100 mg/mL aliquot of dansyl chloride 1:10 with acetone for the respective number of samples in a light-shielded 15 mL conical. Add 200 μL to each sample tube and cap quickly to avoid evaporation.
6. Vortex tubes for 10–15 s
7. Place racked tubes in 70 °C water bath for 10 min (see Note 1).
8. Remove the tubes from the water bath and store them in the dark until cooled to room temperature (about 10 min).
9. Prepare fresh proline solution (250 mg/mL) for the respective number of samples in HPLC-grade water. Add 100 μL to each sample tube, vortex briefly, then incubate in the dark for 10 min.
10. Set up Vac-Elut vacuum apparatus with Sep-Pak C18 column cartridges.
11. Turn on vacuum and wash each column with 2 column volumes of 100% HPLC-grade acetonitrile, followed by 2 column volumes of 35% acetonitrile (diluted with HPLC-grade water).
12. Turn off vacuum and pipette entire contents of dansyl chloride-labeled sample or standard onto column. Turn on vacuum and draw samples through.
13. Wash each column twice with 35% acetonitrile.
14. Turn off vacuum and insert rack with 2 mL collection tubes into apparatus. Place nylon syringe filters at the base of each column.
15. Add 1 mL 100% acetonitrile to each column then turn on vacuum to draw samples through into collection tubes.
16. Store samples at −20 °C.

6.4 Notes

1. Polyamine/analog standards do not get perchloric acid extracted. As such, just add 50 μL DAH to 50 μL standard directly.
2. Heat block inserts may be placed on top of the tubes once in the water bath to ensure the caps do not pop open.

7. HPLC analysis of polyamine content
7.1 Equipment

1. Vanquish HPLC system (Thermo Scientific, Vanquish quaternary pump model #VC-P20-A, autosampler model #VC-A12-A, column compartment model #VC-C10-A, and fluorescence detector model #VF-D50-A).

2. Universal uniguard holder for 4.0/4.6 mm ID3 (Thermo Scientific #850-00).
3. Syncronis C18 10 × 4.0 mm 5-μm drop-in guards (Thermo Scientific #97105-014001).
4. Syncronis C18 5-μm, 150 × 4.6 mm reversed-phase column (Thermo Scientific #97105-154630).
5. SureSTART 2-mL amber glass screw top HPLC vials (Thermo Scientific #6ASV9-2P).
6. SureSTART GOLD-grade conical glass inserts for HPLC vials (Thermo Scientific #6PME02CG).
7. SureSTART 9-mm screw caps for HPLC vials (Thermo Scientific #6ASC9ST1).
8. Chromeleon Chromatography Data System Software (CDS) 7 (Thermo Scientific).
9. 96-well microplates (Greiner #655101).

7.2 Reagents
1. HPLC-grade acetonitrile (Fisher #A998-4).
2. HPLC-grade water (Fisher #W5-4).
3. Protein assay dye reagent concentrate (Bio-Rad #5000006).

7.3 Procedure
1. Place conical inserts in HPLC vials.
2. Load 150 μL standard or sample in each vial, leaving an empty vial at the front of the rack for a blank injection, followed by the standard vial(s), then sample vials.
3. Create an injection sequence in the Chromeleon software by inputting the number of vials to run (see Note 1) and dilution factor (see Note 2), as well as selecting the appropriate gradient within the instrument method (see Table 1A-2 and Note 3), processing method, and report template. Table 2
4. Run HPLC with a constant flow rate of 1 mL/min, 10-μL injection volumes, and monitoring of the column effluent at an excitation wavelength of 345 nm and an emission wavelength of 505 nm at a sensitivity of 7 (see Note 4).
5. Perform a Bradford assay (Bradford, 1976) on the original sample lysates to input the protein concentrations under "Weight" within the Chromeleon software.

Table 1 Composition of gradients IIA & B for native polyamines and polyamine analogs.

(A) Native and acetylated polyamines

Elution Time (min)	Percentage of solvent A	B
0	45	55
4	80	20
7.5	90	10
10	95	5
13.5	95	5

(B) Native, acetylated, and polyamine analogs

Elution time (min)	Percentage of solvent A	B
0	45	55
4	80	20
7.5	90	10
10	95	5
18	95	5

(A) Gradient IIA: for native polyamines only and/or acetylated polyamines (DASpm, Put, N¹AcSpm, Spd, and Spm). Solvent A, HPLC-grade acetonitrile; solvent B, HPLC-grade water. Equilibration time, 6 min; total time 19.5 min (B) Gradient IIB: for native, acetylated, and/or polyamine analogs SBP101 & BENS. Solvent A, HPLC-grade acetonitrile; solvent B, HPLC-grade water. Equilibration time, 6 min; total time 24 min.

Table 2 Composition of gradient III for N^8- N^1AcSpd partial separation.

Gradient III elution time (min)	Percentage of solvent A	B
0	45	55
16	80	20
16.5	90	10
18	95	5
24	95	5

Gradient III: for the partial separation of N^8– N^1AcSpd as depicted in Fig. 2. Solvent A, HPLC-grade acetonitrile; solvent B, HPLC-grade water. Equilibration time, 6 min; total time 30 min.

7.4 Notes

1. Within the injection sequence, the number of injections per vial may be set. This protocol allows for standard and sample injections to be run in duplicate; to run more than two injections per vial, more eluate would have to be added to the vials after two 10-μL injections.

2. The dilution factor should also be set within the injection sequence under "Dilution." After acid extracting 50 µL of original sample, eluting with 1000 µL ACN⁻, and injecting 10 µL onto the column, the dilution factor for this protocol as written is 4000.
3. Within the instrument method, the column temperature should be set to 50 °C. The needle height should be set to 13,500 µm to avoid breaking the bottom of the conical inserts.
4. The flow rate, injection volume, excitation and emission wavelengths, and FLD sensitivity can all be set within the instrument method. This method can then be copy and pasted or set as the default to use for future sequences, or multiple instrument methods can be created.
5. Final polyamine concentrations are reported relative to total cellular protein in the lysate as nmol polyamine/mg protein, determined by the method of Bradford (Bradford, 1976).

8. Results

When the sequence is complete, perform all peak processing within the "Data Processing" tab in Chromeleon. If the instrument method included different FLD sensitivities, a higher or lower sensitivity may instead be selected to report the data within this tab as well. Within the sequence report (generated in the "Report Designer" tab), the "Summary Report" displays a summarized review of every injection. Edit the layout and orientation of the summary report table columns as desired, being sure to include injection numbers, names, component columns for all present polyamines, area and retention time columns for DAH, and a protein column. Both DAH columns allow for review of the consistency of the entire sequence. To add or edit a column and its properties, including header and units, select the desired column, then "Summary Table." Here, one emission can be set for the entire report as well, or different emissions can be analyzed before selecting one by setting each column channel to "< Selected Channel >." Ensure all sample protein concentrations are input under "Weight" in the injection sequence, then properly adjust all component units to nmol/mgP where "P" stands for "protein," and the protein column units to mg/mL. Note that Chromeleon will automatically set "1.000" as the blank and standard injection protein concentrations. This can be removed after exporting the report. The report can be exported as a whole, just one section, and as several different file types. The report, or a section, of just one injection may also be exported.

Acknowledgments

Work in the Stewart/Casero laboratory is supported by grants from the US National Institutes of Health (HD110500, CA204345 and CA235863), the Samuel Waxman Cancer Research Foundation, the University of Pennsylvania Orphan Disease Center Million Dollar Bike Ride (MDBR-20–135 SRS), the Patrick C. Walsh Prostate Cancer Research Fund, the Chan Zuckerberg Initiative and a research contract with Panbela Therapeutics, Inc. The authors also wish to thank all the laboratory members for their ongoing support.

References

Afrin, A., Afshan, T. S., VanSickle, E. A., Michael, J., Laarman, R. L., & Bupp, C. P. (2023). Improvement of dermatological symptoms in patients with Bachmann-Bupp syndrome using difluoromethylornithine treatment. *Pediatric Dermatology, 40*(3), 528–531.

Arthur, R., Jamwal, S., & Kumar, P. (2024). A review on polyamines as promising next-generation neuroprotective and anti-aging therapy. *European Journal of Pharmacology, 978*, 176804.

Bachmann, A. S., VanSickle, E. A., Michael, J., Vipond, M., & Bupp, C. P. (2024). Bachmann-Bupp syndrome and treatment. *Developmental Medicine and Child Neurology, 66*(4), 445–455.

Bradford, M. M. (1976). A rapid and sensitive method for the quantitation of microgram quantities of protein utilizing the principle of protein-dye binding. *Analytical Biochemistry, 72*, 248–254.

Casero, R. A., Jr., Murray Stewart, T., & Pegg, A. E. (2018). Polyamine metabolism and cancer: Treatments, challenges and opportunities. *Nature Reviews. Cancer, 18*(11), 681–695.

Chin, A., Bieberich, C. J., Stewart, T. M., & Casero, R. A., Jr. (2022). Polyamine depletion strategies in cancer: Remodeling the tumor immune microenvironment to enhance anti-tumor responses. *Medical Sciences (Basel), 10*(2).

Holbert, C. E., Cullen, M. T., Casero, R. A., & Stewart, T. M. (2022). Polyamines in cancer: Integrating organismal metabolism and antitumour immunity. *Nature Reviews Cancer, 22*(8), 467–480.

Kabra, P. M., Lee, H. K., Lubich, W. P., & Marton, L. J. (1986). Solid-phase extraction and determination of dansyl derivatives of unconjugated and acetylated polyamines by reversed-phase liquid chromatography: Improved separation systems for polyamines in cerebrospinal fluid, urine and tissue. *Journal of Chromatography, 380*(1), 19–32.

Murray-Stewart, T., Dunworth, M., Foley, J. R., Schwartz, C. E., & Casero, R. A., Jr. (2018). Polyamine homeostasis in snyder-robinson syndrome. *Medical Sciences (Basel), 6*(4), 112.

Ni, Y. Q., & Liu, Y. S. (2021). New insights into the roles and mechanisms of spermidine in aging and age-related diseases. *Aging and Disease, 12*(8), 1948–1963.

Stewart, T. M., Foley, J. R., Holbert, C. E., Khomutov, M., Rastkari, N., Tao, X., & Casero, R. A., Jr. (2023). Difluoromethylornithine rebalances aberrant polyamine ratios in Snyder-Robinson syndrome. *EMBO Molecular Medicine, 15*(11), e17833.

Stewart, T. M., Foley, J. R., Holbert, C. E., Klinke, G., Poschet, G., Steimbach, R. R., & Casero, R. A., Jr. (2022). Histone deacetylase-10 liberates spermidine to support polyamine homeostasis and tumor cell growth. *The Journal of Biological Chemistry, 298*(10), 102407.

Yang, J. (2024). Approval of DFMO for high-risk neuroblastoma patients demonstrates a step of success to target MYC pathway. *British Journal of Cancer, 130*(4), 513–516.

CHAPTER NINETEEN

Cell-based determination of HDAC10-mediated polyamine deacetylase activity

Ishika Gupta, Ashley Nwafor, Robert A. Casero, Jr., and Tracy Murray Stewart*

Sidney Kimmel Comprehensive Cancer Center, Johns Hopkins School of Medicine, Baltimore, MD, United States
*Corresponding author. e-mail address: Tracy.Murray.Stewart@jhmi.edu

Contents

1. Introduction	338
2. Rationale	339
3. HDAC10-dependent growth rescue	340
3.1 Equipment and materials	340
3.2 Reagents	341
3.3 Procedure	341
3.4 Expected outcomes	346
3.5 Optimization and troubleshooting	346
4. Summary	347
References	348

Abstract

Among histone deacetylases, HDAC10 is unique in its substrate preference for a specific acetylated polyamine, N^8-acetylspermidine (N^8-AcSpd), over other acetylated polyamines and peptides. As a polyamine deacetylase, HDAC10 catalyzes the conversion of N^8-AcSpd into spermidine, thereby enabling the cell to utilize this acetylated derivative to support polyamine homeostasis. Therefore, the level of HDAC10-mediated PDAC activity in a particular tissue and its exposure to extracellular N^8-AcSpd, a byproduct of certain intestinal microbes, may directly contribute to the maintenance of intracellular polyamine concentrations. This chapter provides detailed methods for determining relative levels of HDAC10-mediated polyamine deacetylase activity using cell-based assays. These cost-efficient methods are useful for identifying tissue-specific differences in PDAC activity and may also be adapted to enable high-throughput screening of effectors of HDAC10 function, such as HDAC inhibitors.

1. Introduction

The histone deacetylases (HDACs) comprise a family of enzymes whose substrate preferences include acetylated histones as well as non-histone proteins. While histone-modifying Class I HDACs generally localize to the nucleus, members of Class II HDACs can be exclusively cytoplasmic or can shuttle between the two compartments. Recently, substrate specificities for acetylated molecules such as polyamines and fatty acids have been identified for certain HDAC isozymes (Chen, Xie, Chen, & Zhuang, 2022; Hai, Shinsky, Porter, & Christianson, 2017). Originally cloned and characterized as a Class II deacetylase in 2002, HDAC10 is found in both the nucleus and cytoplasm (Fischer et al., 2002; Guardiola & Yao, 2002; Kao, Lee, Komarov, Han, & Evans, 2002; Tong, Liu, Bertos, & Yang, 2002). However, only recently was it reported that HDAC10 is unique in its substrate preference for a specific acetylated polyamine, N^8-acetylspermidine (N^8-AcSpd), over other polyamines, histones and peptides (Hai et al., 2017). This has allowed for the development of highly selective inhibitors of HDAC10 that are based the polyamine moiety (Steimbach et al., 2022). We previously demonstrated that loss of HDAC10 gene expression in tumor cells prohibits the conversion of N^8-AcSpd into spermidine (Spd), indicating that HDAC10 is solely responsible for this polyamine deacetylase (PDAC) activity (Steimbach et al., 2022; Stewart et al., 2022). As the only mammalian enzyme known to deacetylate N^8-AcSpd, the conversion of this substrate into spermidine can be used as an indicator of HDAC10 enzymatic activity.

The cell-based growth rescue assay described in this chapter relies on the requirement of proliferating cells for polyamines to sustain renewal. Inhibition of *de novo* polyamine biosynthesis in vitro can inhibit cellular proliferation. However, cells can overcome this limitation by invoking a compensatory increase in the uptake of polyamines from the extracellular environment. The supplementation of exogenous polyamines into culture medium is a commonly used "rescue" strategy in assigning polyamine-dependent mechanisms and phenotypes. 2-Difluoromethylornithine (DFMO), a commonly used inhibitor of polyamine biosynthesis, inactivates ornithine decarboxylase (ODC) – the first rate-limiting step of the pathway (Fig. 1). Early studies demonstrated the ability of the most common polyamines and their derivatives to support proliferation in the presence of DFMO when supplied exogenously. Since the discovery of N^8-AcSpd as the preferred substrate of HDAC10, we have adapted this rescue method as a means with which to

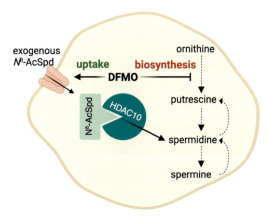

Fig. 1 Schematic of HDAC10 assay rationale. Inhibiting polyamine biosynthesis with DFMO stimulates uptake of exogenous N^8-AcSpd, which is deacetylated by HDAC10 to rescue polyamine homeostasis and growth inhibition. *Created in BioRender (2024) https://BioRender.com/q88g395.*

confirm (1) the substrate specificity of HDAC10 in situ; (2) if exogenous N^8-AcSpd could support polyamine homeostasis under polyamine-limiting conditions; and (3) the requirement for HDAC10 in deacetylating N^8-AcSpd (Stewart et al., 2022).

2. Rationale

The ability of N^8-AcSpd to rescue DFMO-mediated growth inhibition specifically through the polyamine deacetylase activity of HDAC10 allows the use of standard, cost-effective cellular proliferation assays to investigate potential differences in activity among samples as well as the effects of potential activators or repressors of HDAC10 PDAC activity in a cellular context. For straightforward interpretation of growth assay results, treatment conditions modulating HDAC10 activity (i.e., genetic knockout or inhibition) should have minimal effects on growth rate. Cells are assayed in 96-well plates, enabling multiple determinations with a variety of treatment conditions. The example assay provided (Fig. 2) is performed using the parental LNCaP prostate cancer cell line in comparison with a subclone in which the *HDAC10* gene has been disrupted using CRISPR/Cas9 gene editing. We have also used this method to investigate novel inhibitors of HDAC10, one of which is now commercially available and can be used as a positive control (DKFZ-748) (Steimbach et al., 2022). Importantly, by

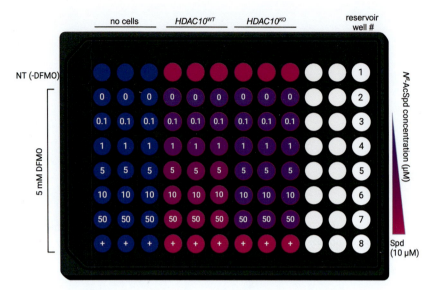

Fig. 2 Sample layout of a microtiter plate for HDAC10 assay. Each condition for each cell line is performed in triplicate, all supplemented with 1 mM AG (with or without DFMO), and combinations of DFMO with increasing concentrations of N^8-AcSpd (0, 0.1, 1, 5, 10, and 50 μM). The plate contains blanks (blue, due to presence of resazurin) and untreated cells of both populations (purple, conversion of resazurin to resorufin due to cellular metabolism). Spermidine (Spd) at 10 μM is included as a positive control for rescue (+). *Created with BioRender.com/j48n087.*

reducing the number of initial test concentrations, this method is also conducive to high-throughput screening of effector compounds, followed by more detailed dose responses on potential hits. Results can be subsequently validated by scaling up the treatment conditions of interest to provide sufficient lysate for analysis of intracellular polyamine levels (i.e., the accumulation of N^8-AcSpd or its conversion into Spd) by HPLC (see Chapter 18) (Nwafor, Stewart, & Casero, 2025).

3. HDAC10-dependent growth rescue
3.1 Equipment and materials
- SpectraMax M5 microplate reader or similar (Molecular Devices)
- 96-well flat-bottom tissue culture plates, black with clear bottom (Corning #7491)
- T-75 flask (Sarstedt #83.3911.002)
- 5% CO_2 incubator at 37 °C

- Pipettes (0.001–1 mL, single-channel; 0.01–0.2 mL, multichannel)
- Multichannel reservoir
- Refrigerated benchtop centrifuge
- Bio-Rad TC10 cell counter and counting chambers

3.2 Reagents
- Complete growth medium as appropriate for cells of choice (including fetal bovine serum or bovine calf serum)
- Sterile phosphate-buffered saline (PBS), pH 7.4 (#10010-049; Gibco)
- Trypan Blue Solution, 0.4% (#15250-061; Gibco)
- HDAC10 inhibitor (DKFZ-748; MedChemExpress #HY-151590) or *HDAC10*-knockout cell line
- Aminoguanidine hemisulfate (AG; #A7009; Sigma Chemical Co.)
- α-Difluoromethyl-DL-ornithine monohydrochloride monohydrate (DFMO; CAS# 96020-91-6; provided by Dr. Patrick Woster)
- N^8-Acetylspermidine dihydrochloride (N^8-AcSpd; #A3658; Sigma Chemical Co.)
- Spermidine trihydrochloride (Spd; #S2501; Sigma)
- CellTiter-Blue Cell Viability reagent (#G8088; Promega)

3.3 Procedure
3.3.1 Cell culture and plate setup
1. Cultivate cells in a T-75 flask with complete media containing 10% FBS and 1% penicillin/streptomycin until they reach the required confluency.
2. Remove and discard the spent medium.
3. Add 3 mL of pre-warmed trypsin/EDTA solution to the flask and transfer it to a 37 °C incubator.
4. Check for cell detachment under a microscope.
5. Quench trypsinization by adding >3 mL of complete culture medium.
6. Transfer the cell suspensions into sterile 15 mL conical tubes.
7. Centrifuge the cells at 500 x g for 5 min to pellet them.
8. Carefully discard the supernatant.
9. Resuspend the cell pellet in 1 mL of sterile PBS.
10. To count the cells, mix 10 μL of the cell suspension with 10 μL of trypan blue and count using a TC10 cell counter.
11. Seed the cells at 2000 cells per well in a 96-well plate (**Note 1**) by diluting them in a volume of complete culture medium sufficient to seed 100 μL per well (2×10^4 cells/mL), in triplicate for each condition to be assayed (see Fig. 2 for an example plate setup).

12. Dispense 100 μL of the cell suspension into each well of a 96-well plate using a multi-channel pipette.
13. Incubate the plate overnight at 37 °C with 5% CO_2 to allow cell attachment.

Note:

Seeding density should be determined for individual cell lines to elicit the full polyamine-dependent rescue curve in 96 h. For the LNCaP cells used here, 2000 cells per well was determined to be optimal, while 1000 cells/well is appropriate for HeLa and HCT116 cells.

3.3.2 Cell treatment and proliferation measurement

1. Prepare respective stock solutions of AG, DFMO, N^8-AcSpd, and Spd in ddH$_2$O, sterilize each using 0.45-micron syringe filters, aliquot, and store at −20°C.

Compound	FW	Stock concentration
AG	123.1	1 M
DFMO	236.6	0.5 M
N^8-AcSpd	260.2	10 mM
Spd	254.6	10 mM

2. Prepare fresh treatment media containing 1 mM AG (see **Note 1**). Media volume should be sufficient to provide 100 uL/well, including "no cells" wells to be used for background readings (10 mL is appropriate for the plate setup in Fig. 2). Transfer 1 mL into well #1 of a sterile 12-well reagent reservoir to be used for untreated control wells in row "A". Add DFMO to the remaining 9 mLs to yield a final concentration of 5 mM (**Note 2**). Aliquot 1-mL volumes into wells 2 through 8 of the reagent reservoir.
3. For the DFMO-containing wells, supplement with increasing concentrations of N^8-AcSpd (0, 0.1, 1, 5, 10, and 50 μM) and mix by pipetting.
4. As a positive control, add 10 μM spermidine to well #8 and mix thoroughly by pipetting.
5. Remove plated cells from incubator and carefully aspirate the original media using a Pasteur pipette. Gently transfer 100 μL of treatment media from wells 1–8 of the reagent reservoir into rows A–H, respectively, using a multi-channel pipette to fill by column.

6. Return plate to 37 °C and incubate for 96 h.
7. After incubation, observe cells under microscope and note obvious differences in cell density.
8. Add 20 μL of CellTiter-Blue reagent per well using a multi-channel pipette. Avoiding bubble formation, briefly swirl the plate to ensure homogenous distribution of the reagent.
9. Return the plate to 37°C and continue incubation for an additional 2-4 h (see **Note 3**).
10. Measure the fluorescence ($560_{Ex}/590_{Em}$) using a plate reader (SpectraMax M5 or similar) to determine effects on cell growth. The lid should be removed, and the plate shaken for 5 s prior to reading (see **Note 4**). The conversion of resazurin (dark blue and non-fluorescent) to resorufin (pink and highly fluorescent) indicates cellular metabolic activity.

Notes:
1. AG is used in this assay as an inhibitor of bovine serum amine oxidase (BSAO) (Murray Stewart, Dunston, Woster, & Casero, 2018). Therefore, it should be added to culture medium prior to the addition of N^8-AcSpd, a known substrate of BSAO (Gahl & Pitot, 1982; Sharmin et al., 2001).
2. The acidity of DFMO alters the color of culture media containing pH indicators. This is expected and should adjust upon incubation.
3. With our assay conditions, 2 h of incubation with CellTiter-Blue is sufficient for commonly used cancer cell lines, including HCT116, HeLa, and LNCaP. The conversion from dark blue to pink is easily visible and can be monitored throughout; however, seeding density and incubation time should be optimized for new cell lines of interest.
4. Absorbance measurements can be used as an alternative to fluorescence. However, absorbance is less sensitive and requires additional calculations due to overlapping spectra.

3.3.3 Quantification, visualization, and statistical analysis
1. Transfer fluorescence values into Excel or similar software, including data from treated, control (untreated), and blank wells.
2. Calculate cell viability percentage using the formula:

$$Cell\ viability\ (\%) = \left(\frac{RFU\ of\ treated\ cells\ -\ RFU\ of\ blank}{RFU\ of\ control\ -\ RFU\ of\ blank}\right) \times 100$$

3. Create a dose-response curve by plotting percentage of live cells against the concentration of N^8-AcSpd and using statistical software (e.g., GraphPad Prism) to calculate standard deviation among triplicate wells (Fig. 3).
4. Once conditions are optimized, perform at least 2 additional independent biological experiments, each with technical triplicates. Calculate cell viability as % of untreated control (as above) and plot the mean values of each experiment versus the concentration of added N^8-AcSpd. Fit nonlinear regression curves and analyze the data using statistical software (e.g., GraphPad Prism) to calculate standard error of the means, identify differences between the curves, and perform significance tests (e.g., 2-way ANOVA with multiple comparisons).

3.3.4 Sample preparation for HPLC analysis of N^8-AcSpd deacetylation

Upon identification of apparent differences in HDAC10 activity based on growth rescue by N^8-AcSpd, appropriate treatment conditions can be chosen for verification of the conversion of N^8-AcSpd into Spd, which is

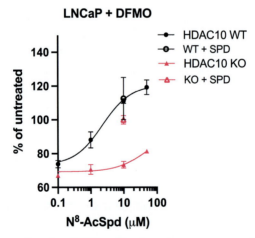

Fig. 3 Example results of a CellTiter Blue-based rescue assay. CRISPR/Cas9 gene editing was used to create an HDAC10 knockout version of LNCaP prostate cancer cells. The loss of HDAC10 polyamine deacetylase activity in the KO cells was demonstrated by a lack of rescue by the substrate N^8-AcSpd when polyamine biosynthesis was blocked by DFMO. Spermidine (Spd) at 10 μM is included as a positive control and effectively rescued growth in both WT and KO cell lines (open symbol). Graph was created using GraphPad Prism.

redistributed to spermine and putrescine in the re-establishment of homeostatic polyamine pools (Fig. 4) (Stewart et al., 2022).

1. Scale up chosen treatment conditions from the plate-based assay appropriately for T-25 flasks (5-mL total volume per flask).
2. Incubate treated cells for 96 h at 37 °C.
3. Assess and annotate differences in cell density visible under the microscope.
4. Collect cells with trypsin-EDTA and pellet at 500 x g, 4 °C, 5 min.
5. Resuspend each in a known volume of cold PBS, on ice, and aliquot 10 µL for cell counts using Trypan blue.
6. Repeat centrifugation to pellet cells; aspirate supernatant
7. Calculate amount of lysis buffer (**Note 1**) needed to resuspend each pellet at a final concentration of 2×10^7 cells/mL:

$$\frac{total\ cell\ \#\ in\ pellet}{2\ \times\ 10^7 \frac{cells}{mL}} = mL\ lysis\ buffer$$

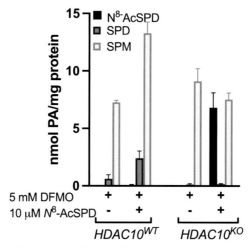

Fig. 4 Example results of intracellular polyamine quantitation by HPLC of HeLa S3 cells with genetic knockout of HDAC10. In WT cells, DFMO treatment reduces SPD and SPM levels, which are replenished by coincubation with N^8-AcSpd. The loss of HDAC10 polyamine deacetylase activity in the KO cells prevents the conversion of N^8-AcSpd into SPD, and subsequently SPM, accompanied by abundant accumulation of the substrate. *Graph was created using GraphPad Prism and data previously reported (Stewart et al., 2022).*

8. Resuspend each pellet by pipetting 2-3 times and quick-freeze immediately in an ethanol/dry ice bath. Store at −80 °C for subsequent processing.
9. In preparation for HPLC analysis, acid-extract 50 μL of each lysate and label amine groups with dansyl chloride according to the methods described in Chapter 18 of this volume (Nwafor et al.).
10. Centrifuge the remaining lysate (not acid-extracted) at ~14,000 x g, 4 °C, 10 min, and remove the supernatant for analysis of total protein content based on interpolation on a bovine serum albumin standard curve using Bio-Rad protein assay dye reagent (#50000006).
11. Perform HPLC as described in Chapter 18 using gradient III for resolution of N^8-AcSpd (Nwafor, Stewart, & Casero, 2025). Quantified polyamines will be reported relative to total cellular protein as determined in step 10.

Notes:
1. Cells can be resuspended in any of the buffers commonly used for assay of polyamine metabolic enzymes, including ODC, SSAT, and SMOX. SSAT breaking buffer can be prepared as a 10x solution (50 mM HEPES, pH 7.2; 10 mM DTT), aliquoted, and frozen at −20°C. Before use, dilute to 1x with ice-cold HPLC water.

3.4 Expected outcomes

In cells expressing HDAC10, the addition of N^8-AcSpd is expected to rescue DFMO-mediated growth inhibition by restoring polyamine levels. Conversely, in HDAC10-knockout cells or those treated with HDAC10 inhibitor, the lack of viable cell recovery indicates that HDAC10 activity is crucial for maintaining polyamine metabolism and cellular survival in the absence of biosynthesis through ODC. This correlates with an accumulation of intracellular N^8-AcSpd without restoration of the main polyamine pools, indicative of the lack of PDAC activity (Fig. 4). Note that we consistently observe a small increase in cell growth in cells lacking HDAC10 activity when using high concentrations of N^8-AcSpd (Fig. 3). We attribute this to a small amount of spermidine present in the procured powder, which is detectable by HPLC.

3.5 Optimization and troubleshooting
1. Conditions including seeding density, DFMO concentration, and incubation time should be empirically optimized upon first usage with alternative cell lines to produce a rescue curve resembling that in Fig. 3 for WT cells.

2. For comparison of HDAC10-inhibitor treated cells with HDAC10-KO cells, pretreatment with the HDAC10 inhibitor is recommended prior to the addition of DFMO. This can be accomplished by seeding the cells in 50 μL volumes per well, allowing ~4 h for attachment, then adding 50 μL of complete medium containing the inhibitor at 2X for a 24 h pretreatment. The medium is removed and replaced with 100 μL containing AG, fresh HDAC10 inhibitor, DFMO, and N^8-AcSpd for the final 96-h incubation.
3. Note that this proliferation assay performs best when used to compare cell lines or conditions with equal growth rates. Growth inhibition or cytotoxicity beyond that induced by DFMO may confound results and should be interpreted carefully.
4. Pretreatment with DFMO in the absence of exogenous polyamines may improve curve separation.
5. Avoid bubble formation during fluorescence reading.
6. Prolonged exposure to light can increase background fluorescence, leading to decreased assay sensitivity.
7. Background fluorescence may increase in wells without cells after extended incubation periods.
8. Insufficient cell numbers or inadequate incubation duration may result in low fluorescence.
9. If resorufin is converted to non-fluorescent hydro-resorufin, decrease the incubation time with CellTiter-Blue Reagent to maintain fluorescence levels.

4. Summary

Aberrant HDAC10 expression is associated with disease, and its expression in tumor cells has been associated with prognosis and outcomes in several cancer types, including neuroblastoma, melanoma, gastric, colorectal, and renal cancers (Cheng et al., 2021; Ling, Li, Peng, Yang, & Seto, 2024; Zhang et al., 2022). Evidence also indicates roles for HDAC10 in inflammatory and immune processes that may impact the course of cancer and other diseases (Bugide, Gupta, Green, & Wajapeyee, 2021; Dahiya et al., 2020; Yang et al., 2024). HDAC10 has been reported to contribute to critical cell functions including autophagy, apoptosis, cell cycle control, and epigenetic regulation (Cheng et al., 2021; Lambona, Zwergel, Fioravanti, Valente, & Mai, 2023). However, the mechanisms regulating HDAC10 expression and activity have been largely unexplored. The methods described in this chapter enable the selective analysis of HDAC10 activity based on its unique substrate preference for N^8-acetylspermidine as the sole human enzyme capable of this deacetylation reaction. These flexible and low-cost cell-based assays allow the researcher to investigate potential modulators of HDAC10 activity, expanding our knowledge of this enzyme and its impact on disease.

References

Bugide, S., Gupta, R., Green, M. R., & Wajapeyee, N. (2021). EZH2 inhibits NK cell-mediated antitumor immunity by suppressing CXCL10 expression in an HDAC10-dependent manner. *Proceedings of the National Academy of Sciences of the United States of America, 118*(30).

Chen, H., Xie, C., Chen, Q., & Zhuang, S. (2022). HDAC11, an emerging therapeutic target for metabolic disorders. *Frontiers in Endocrinology (Lausanne), 13*, 989305.

Cheng, F., Zheng, B., Wang, J., Zhao, G., Yao, Z., Niu, Z., & He, W. (2021). Histone deacetylase 10, a potential epigenetic target for therapy. *Bioscience Reports, 41*(6).

Dahiya, S., Beier, U. H., Wang, L. Q., Han, R. X., Jiao, J., Akimova, T., ... Hancock, W. W. (2020). HDAC10 deletion promotes Foxp3(+) T-regulatory cell function. *Scientific Reports, 10*(1), 424.

Fischer, D. D., Cai, R., Bhatia, U., Asselbergs, F. A., Song, C., Terry, R., ... Cohen, D. (2002). Isolation and characterization of a novel class II histone deacetylase, HDAC10. *The Journal of Biological Chemistry, 277*(8), 6656–6666.

Gahl, W. A., & Pitot, H. C. (1982). Polyamine degradation in foetal and adult bovine serum. *The Biochemical Journal, 202*(3), 603–611.

Guardiola, A. R., & Yao, T.-P. (2002). Molecular cloning and characterization of a novel histone deacetylase HDAC10. *Journal of Biological Chemistry, 277*(5), 3350–3356.

Hai, Y., Shinsky, S. A., Porter, N. J., & Christianson, D. W. (2017). Histone deacetylase 10 structure and molecular function as a polyamine deacetylase. *Nature Communications, 8*, 15368.

Kao, H. Y., Lee, C. H., Komarov, A., Han, C. C., & Evans, R. M. (2002). Isolation and characterization of mammalian HDAC10, a novel histone deacetylase. *The Journal of Biological Chemistry, 277*(1), 187–193.

Lambona, C., Zwergel, C., Fioravanti, R., Valente, S., & Mai, A. (2023). Histone deacetylase 10: A polyamine deacetylase from the crystal structure to the first inhibitors. *Current Opinion in Structural Biology, 82*, 102668.

Ling, H., Li, Y., Peng, C., Yang, S., & Seto, E. (2024). HDAC10 inhibition represses melanoma cell growth and BRAF inhibitor resistance via upregulating SPARC expression. *NAR Cancer, 6*(2), zcae018.

Murray Stewart, T., Dunston, T. T., Woster, P. M., & Casero, R. A., Jr. (2018). Polyamine catabolism and oxidative damage. *The Journal of Biological Chemistry, 293*(48), 18736–18745.

Nwafor, A., Stewart, T. M., & Casero, R. A., Jr. (2025). *An enhanced method for the determination of polyamine content in biological samples by dansyl chloride derivatization and HPLC, Methods in Enzymology*. Academic Press. https://doi.org/10.1016/bs.mie.2025.01.066.

Sharmin, S., Sakata, K., Kashiwagi, K., Ueda, S., Iwasaki, S., Shirahata, A., & Igarashi, K. (2001). Polyamine cytotoxicity in the presence of bovine serum amine oxidase. *Biochemical and Biophysical Research Communications, 282*(1), 228–235.

Steimbach, R. R., Herbst-Gervasoni, C. J., Lechner, S., Stewart, T. M., Klinke, G., Ridinger, J., ... Miller, A. K. (2022). Aza-SAHA derivatives are selective histone deacetylase 10 chemical probes that inhibit polyamine deacetylation and phenocopy HDAC10 knockout. *Journal of the American Chemical Society, 144*(41), 18861–18875.

Stewart, T. M., Foley, J. R., Holbert, C. E., Klinke, G., Poschet, G., Steimbach, R. R., ... Casero, R. A., Jr. (2022). Histone deacetylase-10 liberates spermidine to support polyamine homeostasis and tumor cell growth. *The Journal of Biological Chemistry, 298*(10), 102407.

Tong, J. J., Liu, J., Bertos, N. R., & Yang, X. J. (2002). Identification of HDAC10, a novel class II human histone deacetylase containing a leucine-rich domain. *Nucleic Acids Research, 30*(5), 1114–1123.

Yang, M., Qin, Z., Lin, Y., Ma, D., Sun, C., Xuan, H., ... Han, L. (2024). HDAC10 switches NLRP3 modification from acetylation to ubiquitination and attenuates acute inflammatory diseases. *Cell Communication and Signaling: CCS, 22*(1), 615.

Zhang, T., Wang, B., Gu, B., Su, F., Xiang, L., Liu, L., ... Chen, H. (2022). Genetic and molecular characterization revealed the prognosis efficiency of histone acetylation in pan-digestive cancers. *Journal of Oncology, 2022*, 3938652.

CHAPTER TWENTY

Drosophila melanogaster imaginal disc assays to study the polyamine transport system

Shannon L. Nowotarski* and Justin R. DiAngelo
Division of Science, Penn State Berks, Reading, PA, United States
*Corresponding author. e-mail address: sln167@psu.edu

Contents

1. Introduction	352
2. Rationale	353
3. Preparation of *Drosophila* imaginal discs	354
3.1 Equipment	354
3.2 Reagents	354
3.3 Procedure	355
3.4 Notes	356
4. Assessing disc development in the presence or absence of 20-hydroxyecdysone	357
4.1 Equipment	357
4.2 Reagents	357
4.3 Procedure	357
4.4 Notes	357
5. Imaginal disc assay to study the polyamine transport system	358
5.1 Equipment	358
5.2 Reagents	358
5.3 Procedure	358
5.4 Notes	359
6. Results	359
Acknowledgments	359
References	360

Abstract

Polyamine metabolism in higher eukaryotes is well studied; however, the mechanism of how the polyamines putrescine, spermidine and spermine enter the cell remains unclear. An effective approach to investigate potential players that function in the uptake of polyamines involves using the *Drosophila melanogaster* imaginal disc assay. Leg imaginal discs dissected from *Drosophila melanogaster* wandering third star larvae can be assessed for leg development after 18 h of treatment with hormones to induce this process. The protocol described here details how to use genetically manipulated *Drosophila melanogaster* to test candidate genes involved in the polyamine transport system, how to dissect leg imaginal discs and how to assess the entry of polyamines into the cells of the imaginal disc.

1. Introduction

Polyamines are small, organic cations that are essential for normal cell growth and development (Miller-Fleming, Olin-Sandoval, Campbell, & Ralser, 2015). Due to their charge, polyamines can bind to DNA, RNA, and phospholipids (Casero, Murray Stewart, & Pegg, 2018; Feuerstein, Williams, Basu, & Marton, 1991; Matthews, 1993; Pegg, 2009; Schuber, 1989). Under homeostatic conditions, the polyamines, putrescine, spermidine and spermine, are maintained in the millimolar range in the cell via highly regulated biosynthetic, catabolic, and transport mechanisms (Pegg, 2009).

While the biosynthesis and catabolism of polyamines have been well characterized, the transport of polyamines has only been well studied in prokaryotes and simple eukaryotes (Igarashi & Kashiwagi, 1999; Nowotarski, Woster, & Casero, 2013). In contrast, the transport of polyamines in higher eukaryotes remains enigmatic. Polyamines are unable to passively diffuse through the cell membrane due to their charge. In higher eukaryotes it has been shown that polyamine uptake is both energy and membrane potential dependent. Additionally, Ca^{2+} or Mg^{2+} are required for the activity of the polyamine transport system (PTS) (Casero et al., 2018; Poulin, Casero, & Soulet, 2012). There are three hypothesized models for the import of polyamines in higher eukaryotes and all three models share the idea that polyamines are sequestered in vesicles upon entering the cell (Belting et al., 2003; Soulet, Gagnon, Rivest, Audette, & Poulin, 2004; Uemura, Stringer, Blohm-Mangone, & Gerner, 2010). However, the precise mechanism of how polyamines are taken up by cells remains unknown.

An experimental model that is increasingly being used to study polyamines and the polyamine pathway is *Drosophila melanogaster* (*Drosophila* or fruit fly). This model organism is emerging as a powerful tool due to its genetics (polyamine enzymes share high homology between humans and fruit flies and the fruit fly genome is easier to manipulate), short generation time, and large offspring number (Burnette & Zartman, 2015; Leon, Fruin, Nowotarski, & DiAngelo, 2020). For example, *Drosophila* have recently been used to model Snyder Robinson Syndrome, an X-linked recessive syndrome characterized by loss of function spermine synthase mutations

(Li et al., 2017). *Drosophila* imaginal discs have also been used to study the PTS (Stump, Casero, Phanstiel, DiAngelo, & Nowotarski, 2022; Tsen et al., 2008; Wang, Phanstiel, & von Kalm, 2017).

Imaginal discs are epithelial tissues that during development can form the eyes, wings, legs and other discernible structures in the adult fly. After embryogenesis, a hatched *Drosophila* larvae undergoes further development via three larval instar stages. At the end of the larval stages to the beginning of the pupal stage ecdysteroids begin to increase in the animals. The hormone ecdysone regulates these developmental transitions where increased ecdysone stimulates animal molting and advances the larvae to the pupal stage of development. Moreover, ecdysone promotes the differentiation of the imaginal discs (Jaszczak & Halme, 2016). During the third larval instar stage of *Drosophila* development, imaginal leg discs can be dissected and placed into culture conditions with 20-hydroxyecdysone (Jaszczak & Halme, 2016; Tsen et al., 2008). The progression of leg development can still occur under such in vitro conditions and can thus be utilized as a great system for studying development. *Drosophila* imaginal discs are prepared from animals that are genetically manipulated to study a candidate gene involved in the PTS and from wild-type flies. To test the role of a gene in regulating the PTS, we used the Gal4/UAS system. This system can be used to combine transgenic flies that overexpress or decrease the expression of a candidate PTS gene using the leg imaginal disc-specific driver MJ33a-Gal4 (Hrdlicka et al., 2002; Stump et al., 2022).

2. Rationale

The imaginal leg disc assay in *Drosophila* has been utilized to better characterize the PTS by dissecting the third instar larval imaginal disc in genetically modified animals and determining whether the discs develop normally when in the presence of cytotoxic compounds known to enter the cell via the PTS (Tsen et al., 2008; Wang et al., 2017). Moreover, specificity of polyamine import can be accessed via polyamine add back experiments.

In polyamine add back experiments, the polyamines putrescine, spermidine, and spermine alone or in combination can be added to the in vitro media along with difluoromethylornithine (DFMO). DFMO is a pharmacological inhibitor of ornithine decarboxylase (ODC), the metabolic enzyme that converts ornithine to putrescine. Thus, DFMO can inhibit polyamine production and deplete the polyamines, namely putrescine and spermidine, making the cells rely on the import of exogenous polyamines present in the media. If the exogenous polyamine is able to enter the cell via the PTS then the imaginal disc should develop into a leg if 20-hydroxyecdysone is also present. Legs cultured with DFMO but without polyamines should not develop; this is due to the cytostatic effect of DFMO in vitro (Mamont, Duchesne, Grove, & Bey, 1978; Stump et al., 2022; Wang et al., 2017). The imaginal disc assay described provides an excellent tool for studying potential players involved in the PTS by utilizing the power and ease of *Drosophila* genetics and development to manipulate the expression of putative genes important for polyamine transport.

3. Preparation of *Drosophila* imaginal discs

3.1 Equipment
1. Fly incubator (Genesee Scientific #59-400L).
2. Dissecting microscope (Olympus #SZ61).
3. Sylgard 184 Silicone Elastomer Kit to make a silicon-lined dissection dish (Dow Chemical #2646340).
4. Size 5 Dumont forceps, Biology tip (Fine Science Tools #11252-20).

3.2 Reagents
1. Drosophila agar (Genesee Scientific #66-103).
2. Sucrose (Genesee Scientific #62-112).
3. Whole yeast (Genesee Scientific #62-103).
4. Yellow cornmeal (Genesee Scientific #62-101).
5. Narrow *Drosophila* vials (Genesee Scientific #32-116).
6. Large cotton balls (Genesee Scientific #51-101).
7. Potassium chloride (Amresco #0395-500G; MW: 74.5 g/mol).
8. Potassium dihydrogen phosphate (Sigma #1048730250; MW: 136.09 g/mol).
9. Sodium chloride (Amresco #X190-1KG; MW: 58.44 g/mol).
10. Sodium dihydrogen phosphate (Sigma #S9390; MW: 141.96 g/mol).
11. Magnesium sulfate (Sigma #M-5921; MW: 246.47 g/mol).
12. Magnesium chloride (Amresco #J364-100G; MW: 95.21 g/mol).
13. Calcium chloride (Sigma #C1016-100G; MW: 110.98 g/mol).
14. glucose (Geyer Instructional Products #213104; MW: 180.16 g/mol).
15. l-glutamine (Sigma #G-3126; MW: 146.14 g/mol).
16. glycine (Fisher #G46-1; MW: 75.07 g/mol).

17. l-leucine (Sigma #L8000-25G; MW: 131.17 g/mol).
18. l-proline (Sigma #P0380-100G; MW: 115.13 g/mol).
19. l-serine (Sigma #S4500-1G: MW: 105.09 g/mol).
20. l-valine (Sigma #V0500-1G: MW: 117.15 g/mol).
21. Bovine serum albumin (Sigma #05470-25 G).

3.3 Procedure

1. Grow *Drosophila* on cornmeal-sugar-yeast medium (9 g Drosophila agar, 100 mL Karo Lite Corn Syrup, 65 g cornmeal, 40 g sucrose, and 25 g whole yeast in 1.25 L of water) in fly vials with additional dry yeast added to the top of the food (see Note 1).
2. Set up fly matings using the appropriate genotypes for the candidate polyamine transport gene.
3. Allow fertilized eggs to lay on the food for 2-3 days at 25°C (see Note 2).
4. Make 2x Minimal Robb's media consisting of: 80 mM KCl, 0.8 mM KH_2PO_4, 80 mM NaCl, 0.8 mM $NaH_2PO_4 \cdot 7H_2O$, 2.4 mM $MgSO_4 \cdot 7H_2O$, 2.4 mM $MgCl_2 \cdot 6H_2O$, 2 mM $CaCl_2 \cdot 2H_2O$, 20 mM glucose, 8.0 mM l-glutamine, 0.32 mM glycine, 1.28 mM l-leucine, 0.64 mM l-proline, 0.32 mM l-serine and 1.28 mM l-valine, pH 7.2 (see Note 3).
5. Immediately prior to use, add 20 μL of 10 % bovine serum albumin (BSA) (w/v) per mL of Robb's media (Fristrom, Logan, & Murphy, 1973; Wang et al., 2017).
6. Make Ringer's solution consisting of: 130 mM NaCl, 5 mM KCl, 15 mM $CaCl_2 \cdot 2H_2O$ (see Note 4).
7. Immediately prior to use, add BSA to a final concentration of 0.1 % (w/v) to the Ringer's solution.
8. When fly embryos have reached the wandering third instar larvae stage of development, remove a larvae and dissect leg imaginal discs (Fig. 1) in Ringer's solution using a silicone-lined petri dissection dish that allows for pinning down the larva during the dissection (see Note 5).
9. Place all dissected imaginal discs into a well of a 6-well plate that has 2 mL of Ringer's solution in the well while additional dissections are occurring (see Note 6).
10. To prepare for culture conditions, add 1 mL of 1x Minimal Robb's media to each well of a 12-well plate (see Note 7).
11. Rinse the leg imaginal discs in Ringer's solution before moving the discs to the 12-well plate as described in step 10. One leg imaginal disc will go into each well (see Note 8).

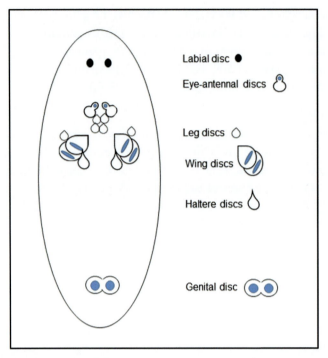

Fig. 1 *Drosophila melanogaster* larvae have 6 leg imaginal discs. The imaginal discs of the *Drosophila melanogaster* larvae develop into adult tissues such as: eyes, wings, legs and genitals. *Drosophila* larvae contain 6 leg imaginal discs that can be dissected during the third larvae instar stage of development for the assay described in this protocol.

3.4 Notes

1. Maintain and propagate the flies on a 12 h:12 h light:dark cycle at 25°C.
2. The leg imaginal discs are dissected from the wandering third instar larvae.
3. Use deionized water to make this media. The media can be stored at −20 °C.
4. Use deionized water and store at 4°C.
5. Dissections need to occur quickly to avoid prolonged storage in the Ringer's solution. We recommend dissecting and obtaining leg imaginal discs for no more than an hour before introducing them to the culture conditions.

6. To transfer discs to each well, use a p-1000 barrier tip with the tip cut so that the discs do not get stuck in the tip.
7. Please note that this requires that the 2x stock solution (from step 4) needs to be diluted to a 1x working solution. Use deionized water to dilute the stock to a working solution.
8. To transfer discs to each well, use a p-1000 barrier tip with the tip cut.

4. Assessing disc development in the presence or absence of 20-hydroxyecdysone

4.1 Equipment
1. Fly incubator (Genesee Scientific #59-400L).
2. Dissecting microscope (Olympus #SZ61).

4.2 Reagents
1. 20-hydroxyecdysone (Sigma #H5142-10mg).
2. DMSO (Sigma #D8418-50 mL).

4.3 Procedure
1. Dissect and prepare imaginal discs as described in Section 3.
2. Dilute the 20-hydroxyecdysone to a working concentration of 1 mg/mL (see Note 4.1).
3. To wells that are positive controls, add 20-hydroxyecdysone to a final concentration of 1 μg/mL in the wells with the imaginal disc and 1X Minimal Robb's media. Swirl to mix.
4. To wells that are negative controls, add 1 μL DMSO in the wells with the imaginal discs and 1X Minimal Robb's media. Swirl to mix.
5. Imaginal discs are incubated for 18 h at 25°C and then scored as developed or non-developed discs (see Note 4.2).
6. For each experiment, the percent development was determined by [(number of developed discs)/(total number of discs)] × 100.

4.4 Notes
1. This concentration is required for the leg imaginal discs to fully develop in vitro (Jaszczak & Halme, 2016).
2. Fully developed discs in which the leg is fully extended and partially developed discs in which the leg is not fully extended are scored as developed. Non-developed discs show no sign of development (Fig. 2).

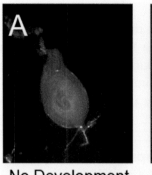
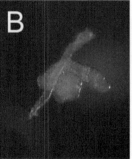

No Development   Development

Fig. 2 Leg imaginal discs dissected from *Drosophila melanogaster* can either develop or not develop. (A) Image of a leg imaginal disc from *Drosophila melanogaster* wandering third instar larvae that did not develop. (B) Image of leg imaginal disc from *Drosophila melanogaster* wandering third instar larvae that developed after 18 h in the presence of 20-hydroxyecdysone (1 μg/mL).

5. Imaginal disc assay to study the polyamine transport system

5.1 Equipment
1. Fly incubator (Genesee Scientific #59-400L).
2. Dissecting microscope (Olympus #SZ61).

5.2 Reagents
1. 20-hydroxyecdysone (Sigma #H5142-10mg).
2. Putrescine (Sigma #P-7505; MW: 161.1 g/mol).
3. Spermidine (Sigma #S2626-1G; MW: 145.25 g/mol).
4. Spermine trihydrochloride (Sigma #S2501; MW: 254.63 g/mol).
5. Difluoromethylornithine (Patrick Woster, Medical University of South Carolina, USA).
6. Ant 44 (Otto Phanstiel, University of Central Florida, USA).

5.3 Procedure
1. Prepare fly leg imaginal discs as outlined in Section 3. For each dissection session remember to include discs that will be positive controls, negative controls and experimental discs. These conditions need to be assessed in wild-type animals and in animals that have a PTS candidate gene manipulated during each experiment performed.

2. To test whether a specific gene is involved in the transport of polyamines, the leg imaginal discs that have been dissected from the wandering third instar larvae can be incubated in 1x Minimal Robb's media containing 20-hydroxyecdysone (1 μg/mL final concentration) and a known inhibitor of the PTS. In our experiments Ant44 was used at a final concentration of 50 μM (Stump et al., 2022; Wang et al., 2017).
3. Imaginal discs are incubated for 18 h at 25°C and then scored as developed or non-developed discs (Fig. 2). For each experiment, the percent development was determined by [(number of developed discs)/(total number of discs)] × 100 (see Note 5.1).
4. To further characterize if the genetic manipulation impacted the uptake of a specific polyamine(s), leg imaginal discs in 1x Minimal Robb's media can be incubated with 200 μM putrescine, 200 μM spermidine and/or 200 μM spermine along with 20-hydroxyecdysone (1 μg/mL), and 10 mM DFMO (see Note 5.2).
5. Imaginal discs are incubated for 18 h at 25°C and then scored as developed or non-developed discs (Fig. 2). For each experiment, the percent development was determined by [(number of developed discs)/(total number of discs)] × 100.

5.4 Notes
1. Wild-type animals' discs should not develop in the presence of a PTS inhibitor.
2. These concentrations may need to be optimized for individual experiments.

6. Results

Once all experiments are completed, the data can be accumulated and expressed as means ± standard error. Depending on the comparisons being made, unpaired Student's t-tests or ANOVAs comparing the mean percent development will need to be performed to determine whether the gene of interest is important in regulating polyamine transport.

Acknowledgments
Work in the Nowotarski and DiAngelo laboratories is supported by Penn State Berks. The authors wish to thank Coryn Stump for her technical assistance.

References

Belting, M., Mani, K., Jonsson, M., Cheng, F., Sandgren, S., Jonsson, S., & Fransson, L. A. (2003). Glypican-1 is a vehicle for polyamine uptake in mammalian cells: A pivital role for nitrosothiol-derived nitric oxide. *The Journal of Biological Chemistry, 278*(47), 47181–47189. https://doi.org/10.1074/jbc.M308325200.

Burnette, M., & Zartman, J. J. (2015). Spatiotemporal patterning of polyamines in Drosophila development. *Amino Acids, 47*(12), 2665–2670. https://doi.org/10.1007/s00726-015-2093-z.

Casero, R. A., Jr., Murray Stewart, T., & Pegg, A. E. (2018). Polyamine metabolism and cancer: Treatments, challenges and opportunities. *Nature Reviews. Cancer, 18*(11), 681–695. https://doi.org/10.1038/s41568-018-0050-3.

Feuerstein, B. G., Williams, L. D., Basu, H. S., & Marton, L. J. (1991). Implications and concepts of polyamine-nucleic acid interactions. *Journal of Cellular Biochemistry, 46*(1), 37–47. https://doi.org/10.1002/jcb.240460107.

Fristrom, J. W., Logan, W. R., & Murphy, C. (1973). The synthetic and minimal culture requirements for evagination of imaginal discs of Drosophila melanogaster in vitro. *Developmental Biology, 33*(2), 441–456. https://doi.org/10.1016/0012-1606(73)90149-8.

Hrdlicka, L., Gibson, M., Kiger, A., Micchelli, C., Schober, M., Schock, F., & Perrimon, N. (2002). Analysis of twenty-four Gal4 lines in Drosophila melanogaster. *Genesis (New York, N. Y.: 2000), 34*(1-2), 51–57. https://doi.org/10.1002/gene.10125.

Igarashi, K., & Kashiwagi, K. (1999). Polyamine transport in bacteria and yeast. *The Biochemical Journal, 344 Pt*(3), 633–642. Retrieved from. https://www.ncbi.nlm.nih.gov/pubmed/10585849.

Jaszczak, J. S., & Halme, A. (2016). Arrested development: Coordinating regeneration with development and growth in Drosophila melanogaster. *Current Opinion in Genetics & Development, 40*, 87–94. https://doi.org/10.1016/j.gde.2016.06.008.

Leon, K. E., Fruin, A. M., Nowotarski, S. L., & DiAngelo, J. R. (2020). The regulation of triglyceride storage by ornithine decarboxylase (Odc1) in Drosophila. *Biochemical and Biophysical Research Communications, 523*(2), 429–433. https://doi.org/10.1016/j.bbrc.2019.12.078.

Li, C., Brazill, J. M., Liu, S., Bello, C., Zhu, Y., Morimoto, M., & Zhai, R. G. (2017). Spermine synthase deficiency causes lysosomal dysfunction and oxidative stress in models of Snyder-Robinson syndrome. *Nature communications, 8*(1), 1257. https://doi.org/10.1038/s41467-017-01289-7.

Mamont, P. S., Duchesne, M. C., Grove, J., & Bey, P. (1978). Anti-proliferative properties of DL-alpha-difluoromethyl ornithine in cultured cells. A consequence of the irreversible inhibition of ornithine decarboxylase. *Biochemical and Biophysical Research Communications, 81*(1), 58–66. Retrieved from. http://www.ncbi.nlm.nih.gov/pubmed/656104.

Matthews, H. R. (1993). Polyamines, chromatin structure and transcription. *Bioessays: News and Reviews in Molecular, Cellular and Developmental Biology, 15*(8), 561–566. https://doi.org/10.1002/bies.950150811.

Miller-Fleming, L., Olin-Sandoval, V., Campbell, K., & Ralser, M. (2015). Remaining mysteries of molecular biology: The role of polyamines in the cell. *Journal of Molecular Biology, 427*(21), 3389–3406. https://doi.org/10.1016/j.jmb.2015.06.020.

Nowotarski, S. L., Woster, P. M., & Casero, R. A., Jr. (2013). Polyamines and cancer: Implications for chemotherapy and chemoprevention. *Expert Reviews in Molecular Medicine, 15*, e3. https://doi.org/10.1017/erm.2013.3.

Pegg, A. E. (2009). Mammalian polyamine metabolism and function. *IUBMB Life, 61*(9), 880–894. https://doi.org/10.1002/iub.230.

Poulin, R., Casero, R. A., & Soulet, D. (2012). Recent advances in the molecular biology of metazoan polyamine transport. *Amino Acids, 42*(2-3), 711–723. https://doi.org/10.1007/s00726-011-0987-y.

Schuber, F. (1989). Influence of polyamines on membrane functions. *The Biochemical Journal, 260*(1), 1–10. Retrieved from. http://www.ncbi.nlm.nih.gov/entrez/query.fcgi?cmd=Retrieve&db=PubMed&dopt=Citation&list_uids=2673211.

Soulet, D., Gagnon, B., Rivest, S., Audette, M., & Poulin, R. (2004). A fluorescent probe of polyamine transport accumulates into intracellular acidic vesicles via a two-step mechanism. *The Journal of Biological Chemistry, 279*(47), 49355–49366. https://doi.org/10.1074/jbc.M401287200.

Stump, C. L., Casero, R. A., Jr, Phanstiel, O. T., DiAngelo, J. R., & Nowotarski, S. L. (2022). Elucidating the role of Chmp1 overexpression in the transport of polyamines in Drosophila melanogaster. *Medical Sciences (Basel), 10*(3), https://doi.org/10.3390/medsci10030045.

Tsen, C., Iltis, M., Kaur, N., Bayer, C., Delcros, J. G., von Kalm, L., & Phanstiel, O. T. (2008). A Drosophila model to identify polyamine-drug conjugates that target the polyamine transporter in an intact epithelium. *Journal of Medicinal Chemistry, 51*(2), 324–330. https://doi.org/10.1021/jm701198s.

Uemura, T., Stringer, D. E., Blohm-Mangone, K. A., & Gerner, E. W. (2010). Polyamine transport is mediated by both endocytic and solute carrier transport mechanisms in the gastrointestinal tract. *American Journal of Physiology. Gastrointestinal and Liver Physiology, 299*(2), G517–G522. https://doi.org/10.1152/ajpgi.00169.2010.

Wang, M., Phanstiel, O., & von Kalm, L. (2017). Evaluation of polyamine transport inhibitors in a drosophila epithelial model suggests the existence of multiple transport systems. *Medical Sciences (Basel), 5*(4), https://doi.org/10.3390/medsci5040027.

CHAPTER TWENTY ONE

A fluorescence-based assay for measuring aminopropyltransferase activity

Pallavi Singh, Jae-Yeon Choi, and Choukri Ben Mamoun[*]

Department of Internal Medicine, Section of Infectious Diseases, Yale School of Medicine, New Haven, CT, United States
*Corresponding author. e-mail address: choukri.benmamoun@yale.edu

Contents

1. Introduction	364
2. Assay design	366
3. Things to keep ready beforehand	368
4. Key resources table	368
5. Preparation of reagents and standard curves for APT reactions	371
6. Setting up spermidine and spermine synthase reactions and detection using the DAB-APT assay	376
7. Validation of APT reactions using thin-layer chromatography (TLC) and mass spectrometry	380
8. Quantification and statistical analysis	382
9. Summary	384
References	386

Abstract

Polyamines (PAs) are small polycationic alkylamines that are essential for numerous cellular processes and found in all living cells. The three principal polyamines, putrescine (PUT), spermidine (SPD), and spermine (SPM), have been shown to play crucial roles in cellular function and implicated in several diseases including infectious diseases, cancer and neurodegenerative disorders. As such, the enzymes involved in polyamine biosynthesis are promising targets for developing antimicrobial, antineoplastic and neuroprotective therapies. Aminopropyl transferases (APTs) are key enzymes in this pathway, catalyzing the formation of spermidine from putrescine and spermine from spermidine. While in most eukaryotes and prokaryotes, the spermidine synthase and spermine synthase activities are catalyzed by distinct enzymes, some organisms such as *Plasmodium falciparum* have a single enzyme, which catalyzes both reactions with varying efficiency. To date, efforts to inhibit APTs have focused primarily on substrate analogs, often with limited selectivity. A major challenge in discovering novel inhibitors has been the lack of an assay suitable for high-throughput chemical screening. We have recently developed DAB-APT, the first fluorescence-based assay for measuring APT activity, using 1,2-diacetyl benzene (DAB) which reacts with putrescine, spermidine, and spermine to form fluorescent conjugates, with fluorescence intensity correlating to carbon chain length. The DAB-APT assay has been validated using APT enzymes from *Saccharomyces cerevisiae*, and *P. falciparum*, and has been found to be suitable for high-throughput screening of large chemical libraries. This assay represents a significant advancement, offering a valuable tool for identifying potential inhibitors of APT enzymes and accelerating drug discovery efforts in cancer, neurobiology, and infectious diseases.

1. Introduction

Polyamines are a class of small organic molecules distinguished by their polycationic nature, possessing multiple amino groups which impart unique chemical properties crucial for cellular function (Schibalski, Shulha, Tsao, Palygin, & Ilatovskaya, 2024). Evolutionarily, polyamines are conserved across a wide range of organisms, from bacteria to humans, underscoring their fundamental biological importance (Michael, 2018). The three most studied polyamines are putrescine, spermidine, and spermine; each exhibit specific structural features that influence their biological functions. Putrescine, a di-amine, consists of a four-carbon chain with two amino groups. Spermidine, synthesized from putrescine, features an additional aminopropyl group, while spermine, the most complex, includes two aminobutyl groups, resulting in a highly branched, positively charged structure (Bowman, Tabor, & Tabor, 1973; Dudley, Rosenheim, & Starling, 1926). Polyamines are essential molecules that regulate diverse functions across different organisms. In bacteria and archaea, polyamines stabilize nucleic acids and cellular structures, facilitating growth and stress adaptation (Michael, 2018; Terui, Ohnuma, Hiraga, Kawashima, & Oshima, 2005). In eukaryotes, including plants, fungi, and animals, polyamines are crucial for regulating cell proliferation, differentiation, and apoptosis by stabilizing chromatin, influencing gene expression, and modulating ribosomal function. They also play significant roles in stress responses, helping organisms manage environmental challenges (Rhee, Kim, & Lee, 2007; Solmi et al., 2023). During development, polyamines contribute to cellular differentiation and tissue formation, while in animals, they offer neuroprotection and regulate immune responses (Blazquez, 2024; Brooks, 2013; Pegg, 2016; Seiler & Raul, 2005; Tang, Xia, Liang, Ma, & Liu, 2021; Valdes-Santiago, 2013). Additionally, polyamines are involved in reproduction, influencing processes such as spermatogenesis in animals and seed germination in plants (Lefevre, Palin, & Murphy, 2011). This broad spectrum of functions highlights the evolutionary conservation and critical importance of polyamines in maintaining cellular and organismal homeostasis.

Polyamine biosynthesis involves a series of enzymatic reactions that convert the amino acids into polyamines, with ornithine and methionine as primary precursors. The pathway initiates with the decarboxylation of ornithine by ornithine decarboxylase (ODC), producing putrescine (Morris & Pardee, 1965; Raina & Janne, 1968). Putrescine serves as the substrate for the synthesis of spermidine and spermine. Spermidine is generated through the transfer of an aminopropyl group from decarboxylated S-adenosylmethionine (dcSAM) to putrescine, a reaction catalyzed by spermidine synthase, an aminopropyl transferase (APT) enzyme (Ikeguchi, Bewley, & Pegg, 2006). The subsequent formation of spermine involves the addition of an aminopropyl group, also derived from dcSAM, facilitated by another APT enzyme, spermine synthase (Pegg & Michael, 2010) (Scheme 1).

The polyamine biosynthetic pathway represents an attractive target for therapeutic intervention due to its crucial role in regulating cell growth, differentiation, and apoptosis. Dysregulation of this pathway is linked to various diseases, including cancer (Casero, Stewart, & Pegg, 2018; Han et al., 2024; Murray-Stewart, Woster, & Casero, 2016), neurodegenerative disorders (Brooks, 2024; Cason et al., 2003; Cervelli et al., 2022; Morrison & Kish, 1995), and infectious diseases (Firpo et al., 2021; Huang, Zhang, Chen, & Zeng, 2020; Kaiser, 2023), making it an attractive candidate for drug development. Specifically, APT enzymes, involved in spermidine and spermine formation, represent promising targets for antimicrobial therapies due to the essential role of polyamines in the survival and virulence of many pathogens. In *P. falciparum*, the causative agent of malaria, polyamine

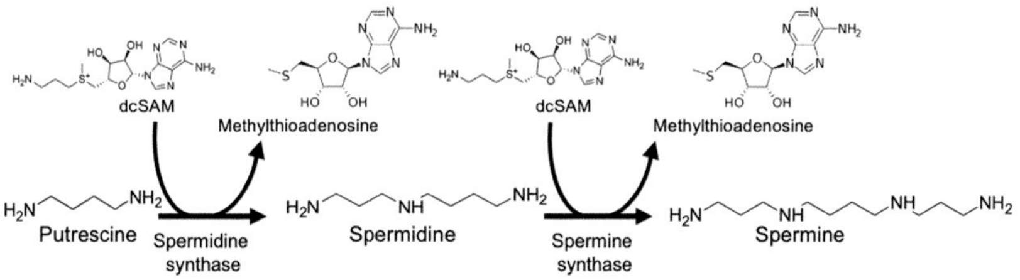

Scheme 1 Reactions catalyzed by APT enzymes, spermidine synthase and spermine synthase.

biosynthesis is critical for the parasite's rapid growth and replication within human erythrocytes. Inhibiting the APT enzyme PfSPDS could disrupt polyamine production, leading to impaired parasite development and survival, making it an attractive target for antimalarial therapy. Similarly, polyamine biosynthesis is crucial for bacterial pathogens like *Escherichia coli*, *Helicobacter pylori*, *Pseudomonas aeruginosa*, *Staphylococcus aureus* and *Salmonella typhimurium* where polyamines contribute to bacterial colonization, biofilm formation, microbial carcinogenesis and protection from oxidative and acid stress (Mendez et al., 2020; Shah & Swiatlo, 2008; Thongbhubate, Nakafuji, Matsuoka, Kakegawa, & Suzuki, 2021; Tofalo, Cocchi, & Suzzi, 2019; Zhang & Au, 2017). Targeting APT enzymes in these bacteria could pave the way for the development of novel antibiotics.

The search for novel small molecules or natural products targeting APT enzymes has been hindered by the lack of assays suitable for high-throughput screening (HTS) of chemical libraries. To date, four primary methods have been employed to measure APT enzyme activity in vitro. The classical method involves radioisotope-based assays, using radioactive substrates such as [^{14}C]-Putrescine and [^{14}C]-Spermidine to measure APT activity (Haider et al., 2005; Zappia, Cacciapuoti, Pontoni, & Oliva, 1980). A second approach utilizes dansylation of polyamines synthesized during APT reactions, followed by quantification using high-performance liquid chromatography (HPLC) (Saeki, Uehara, & Shirakawa, 1978). A third approach employs capillary electrophoresis with laser-induced fluorescence (CE-LIF), where spermidine is derivatized with 7-fluoro-4-nitrobenzo-2-oxa-1,3-diazole (NBD-F) (Sano & Nishino, 2007). The fourth approach uses a specific monoclonal antibody against the byproduct, 5'-methyl-thioadenosine (MTA), coupled with a homogeneous time-resolved fluorescence technique (Enomoto et al., 2006). Although these assays are sensitive and specific for measuring APT activity, they are low-throughput and involve labor-intensive procedures. The DAB-APT assay described here presents a breakthrough, offering a HTS platform for large chemical libraries, facilitating the search for novel APT enzyme inhibitors for therapeutic applications.

2. Assay design

Here, we present a simple and user-friendly fluorescence assay for measuring APT enzyme activity, using the aromatic hydrocarbon 1,

2-diacetyl benzene (1,2-DAB). Previous studies by Médici, de Maria, Otten, and Straathof (2011), and Choi et al (Choi et al., 2020) demonstrated that 1,2-DAB reacts specifically with the primary amines of various molecules such as gamma-aminobutyric acid (GABA), tyramine, and ethanolamine, but does not react with the primary amines that are attached to α-carboxylated compounds such as glutamate, tyrosine and serine. These reactions form fluorescent adducts that can be easily detected using a plate reader, with excitation and emission wavelengths of 364 nm, and 425 nm, respectively. In our recent study, we showed that DAB-mediated adduct formation with polyamines - putrescine, spermidine and spermine - can be employed to measure the activity of APT enzymes from various organisms (Singh et al., 2024). The DAB-APT assay generates differential fluorescence between the substrate and product of the reaction. In spermidine synthase-catalyzed reactions, the fluorescence signal for the product spermidine is approximately 4-fold higher than for the substrate putrescine. Similarly, in spermine synthase-catalyzed reactions, the product spermidine shows about a 1.5-fold increase in fluorescence compared to the substrate spermidine (Scheme 2).

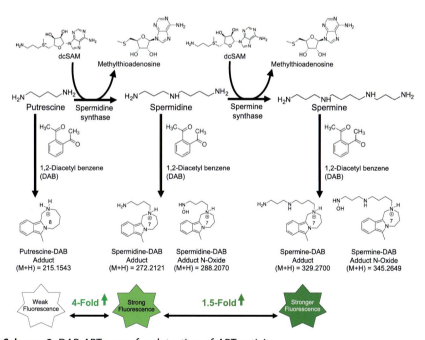

Scheme 2 DAB-APT assay for detection of APT activity.

3. Things to keep ready beforehand

To carry out the DAB-APT assay, recombinant APT enzymes must first be expressed and purified. In our studies, we have primarily used spermidine synthase and spermine synthase enzymes expressed as recombinant fusion proteins with an N-terminal maltose-binding protein (MBP) tag. (A detailed description of the procedure to purify MBP-tagged *Saccharomyces cerevisiae* Spe3, Spe4, and *Plasmodium falciparum* spermidine synthase (PfSPDS) enzymes is found in Section 6c)). To generate a standard curve for the APT reactions, you will need the appropriate substrates (putrescine or spermidine for spermidine synthase and spermine synthase reactions, respectively) and the co-substrate dcSAM. Additionally, you will need the reaction products, spermidine and spermine, as well as the byproduct methylthioadenosine (MTA), and 1,2-diacetyl benzene (1,2-DAB) for fluorescence detection. All required components, including putrescine, spermidine, spermine, dcSAM, MTA, 1,2-DAB, and buffer reagents, are commercially available and listed in the key resources table.

4. Key resources table

Reagent or resource or equipment	Source	Identifier
Chemicals		
Putrescine dihydrochloride	Millipore Sigma	P7505-25G
Spermidine	ThermoFisher Scientific	A19096.03
Spermine	Millipore Sigma	S4264-1G
Decarboxylated S-adenosyl Methionine	BOC Sciences	S4264-1G (custom quote)
5'-deoxy-5'-methylthioadenosine	Millipore Sigma	D5011
Boric acid	Millipore Sigma	B0394-100G
Sodium tetraborate	Millipore Sigma	221732-100G

Sodium chloride	Millipore Sigma	S6546-1L
Potassium phosphate monobasic	Millipore Sigma	P0662-25G
Potassium phosphate dibasic	Millipore Sigma	P3786-100G
Triton X-100	Millipore Sigma	X100-500ML
Ethylenediaminetetraacetic acid	Millipore Sigma	324506-100ML
Dithiothreitol (DTT)	Millipore Sigma	10197777001
Bovine Serum Albumin (BSA)	Millipore Sigma	A9647-100G
1,2-diacetyl benzene (1,2-DAB)	Millipore Sigma	242039-100MG
2-mercaptoethanol	Millipore Sigma	M6250-100ML
96 Well Black/Clear Bottom Plate, Non-Treated Surface, No Lid, Non-Sterile	ThermoFisher Scientific	265301
LB Broth (Miller)	Millipore Sigma	L3522-1KG
SOC medium	ThermoFisher Scientific	15544034
LB Agar Plates	ThermoFisher Scientific	J63197.EQF
Ampicillin sodium salt	Millipore Sigma	A0166-25G
Isopropyl-β-D-thiogalactopyranoside (IPTG)	Research Products International (RPI)	I56000-25.0
Amylose resin	New England Biolabs (NEB)	E8021S
Poly-Prep® Chromatography Columns	Bio-Rad	7311550
QIArack	Qiagen	19015
Tris-HCl	American Bio	AB14043-01000
Benzonase	Santa Cruz	sc-391121B
CHAPS Detergent (3-((3-cholamidopropyl) dimethylammonio)-1-propanesulfonate)	ThermoFisher Scientific	28300
Glycerol	Millipore Sigma	C988U16
Protease Inhibitor Cocktail	Millipore Sigma	11836170001
Maltose	Fisher Scientific	M75-100
10x Tris/Glycine/SDS (running buffer)	Bio-Rad	1610732

4–20% Mini-PROTEAN® TGX Stain-Free™ Protein Gels, 15 well, 15 µl	Bio-Rad	4568096
4x Laemmli Sample Buffer, Bio Rad	Bio-Rad	1610747
Bio-Safe™ Coomassie Stain	Bio-Rad	1610786
n-butanol	MP Biochemicals	C822D82
Acetic acid	J.T. Baker	9508-00
Pyridine	Millipore Sigma	P57506
Methanol	J.T.Baker	9070-05
TLC plates (500 µm), PK6F; 20 x 20 cm	Cytiva	4861830
Ninhydrin 99%	ThermoFisher Scientific	A10409-09
Bacterial Strains		
Rosetta 2 (DE3) comp. cells, Fisher Scientific	Millipore Sigma	713973
Enzymes		
MBP-*S. cerevisiae* Spe3	Genscript USA Inc	Custom synthesized
MBP-*S. cerevisiae* Spe4	Genscript USA Inc	Custom synthesized
MBP-*P. falciparum* Spermidine synthase	Genscript USA Inc	Custom synthesized
Equipments		
Sorval Centrifuge	Beckman Coulter	Avanti J-20 XP
Microcentrifuge	Eppendorf	5415D
Fisherbrand™ Variable Speed Nutator	Fisher Scientific	S06622
Sonicator	Fisher Scientific	FB432A
SDS-PAGE electrophoresis	Bio-Rad	552BR 238184
Amicon® Ultra-4 Centrifugal Filter Unit	Millipore Sigma	UFC801024D
Plate reader	Agilent	Biotek Synergy H1
Transilluminator box	Bane Bio	BL101/5000 K

The plasmids (pMAL) carrying MBP-tagged *S. cerevisiae* Spe3, Spe4, and *P. falciparum* spermidine synthase (PfSPDS) were used to transform Rosetta 2 (DE3) *E. coli* competent cells and recombinant proteins were purified as described in detail in Section 5.c.

5. Preparation of reagents and standard curves for APT reactions

a) Transformation of pMAL-Spe3, pMAL-Spe4, and pMAL-PfSPDS plasmids

Note: Sterile techniques should be followed when **handling** SOC medium and *E. coli* used for heterologous protein expression.

i. The plasmids (pMAL-Spe3, pMAL-Spe4, and pMAL-PfSPDS) were purified and resuspended in sterile distilled water to a final concentration of 0.1 µg/µL.

ii. Rosetta 2 (DE3) competent *E. coli* cells were thawed on ice for 15-20 minutes. Approximately 40 µl of cells were aliquoted into three pre-chilled microcentrifuge tubes.

iii. To each tube, 100 ng of either pMAL-Spe3, pMAL-Spe4, or pMAL-PfSPDS plasmid was added, and the cells were incubated on ice for 20 minutes.

iv. The cells underwent a heat shock at 42°C for exactly 30 seconds, followed by immediate placement on ice for 2 minutes. Afterward, 1 mL of pre-warmed SOC medium was added to each tube.

v. The tubes were incubated on a shaker at 37°C for 1 hour.

vi. After incubation, 100 µL of the transformed cell suspension from each tube was spread onto LB-ampicillin agar plates. The plates were incubated overnight at 37°C.

vii. The next morning, the plates were removed from the incubator and stored at 4°C for further use.

b) Making buffers for protein purification

I) Bacterial lysis buffer
  i. 25 mM Tris-HCl (pH 8.0)
  ii. 500 mM NaCl
  iii. 0.5% Glycerol
  iv. 1X Protease inhibitor cocktail
  v. 0.002% CHAPS
  vi. 250 U/mL Benzonase

II) Binding buffer
  i. 25 mM Tris-HCl (pH 8.0)
  ii. 500 mM NaCl
  iii. 0.5% Glycerol
  iv. 1 mM DTT
  v. 0.5 mM EDTA

III) Elution buffer
  i. 25 mM Tris-HCl (pH 8.0)
  ii. 500 mM NaCl
  iii. 0.5% Glycerol
  iv. 1 mM DTT
  v. 0.5 mM EDTA
  vi. 10 mM Maltose

c) Purification of MBP-Spe3, MBP-Spe4, and MBP-PfSPDS proteins

Note: LB broth used for growing bacteria must be autoclaved before use, and ampicillin should be freshly prepared from powder and filter-sterilized prior to adding it to the LB broth.

1. **Inoculation and Primary Culture**: Pick a single bacterial colony from the LB plates (one each for MBP-Spe3, MBP-Spe4, and MBP-PfSPDS as per step 5.a.vii.) and inoculate 20 mL LB broth containing 200 µg/mL ampicillin. Grow these primary cultures overnight (12–16 hours) at 37°C in an incubator shaker at 180–200 rpm.
2. **Secondary Culture**: The next morning, set up 1-liter secondary cultures for each protein by inoculating 10 mL of the primary culture from step 5.c.i. into fresh LB broth.
3. **Growth and Induction**: Grow the secondary cultures at 37°C and 200 rpm until the optical density at 600 nm (OD600) reaches approximately 0.6. Remove 1 mL from each culture as uninduced control samples. To the remaining cultures, add isopropyl β-d-1-thiogalactopyranoside (IPTG) to a final concentration of 1 mM. Continue incubating the cultures, along with the uninduced controls, overnight at 16°C with 200 rpm shaking.
4. **Harvesting Cells**: Harvest the cells from both the induced cultures and uninduced controls by centrifugation at 5000 × g for 30 minutes at 4°C. Discard the supernatant.
5. **Validation of Protein Induction**: To verify protein induction, take 1 mL from the induced cultures and uninduced controls, centrifuge at 6000 × g for 10 minutes, and discard the supernatant. Resuspend the cell pellets in 1X Laemmli buffer, mix, and boil at 95°C for 10 minutes. Run 5 µL of each sample on SDS-PAGE gels, and check protein induction using Coomassie staining.
6. **Cell Lysis**: Resuspend the remaining induced cell pellets in 20 mL of lysis buffer (Section 6.b.i.) and incubate on ice for 30 minutes.
7. **Sonication**: Sonicate the resuspended cells at 10% amplitude, using a 5-second ON and 5-second OFF cycle for a total of 10 minutes, with breaks every 2.5 minutes.

8. **Supernatant Collection**: After sonication, centrifuge the cell lysates at 12,000 × g for 45 minutes at 4°C. Collect the supernatant, discarding the cell debris. For protein expression analysis, take 50 μL of the supernatant from each protein, add 15 μL of 4X Laemmli buffer, boil at 95°C for 10 minutes, and run on SDS-PAGE gels to confirm protein expression using Coomassie staining.
9. **Amylose Resin Preparation**: Pack six poly-prep chromatography columns with 2 mL of amylose resin (50% slurry), two for each of MBP-Spe3, MBP-Spe4, and MBP-PfSPDS. Wash the columns with 10 column volumes (CV) of chilled binding buffer.
10. **Protein Binding to the amylose resin**: After washing, close the bottom of the columns with snap-off tips. Add the supernatants containing the proteins to their respective columns. Cap the tops of the columns and place them on a nutator shaker for 3 hours at 4°C to allow the proteins to bind to the amylose resin.
11. **Protein Elution**: After the binding step, place the columns on a QIArack, remove the top caps and bottom snap-off tips, and allow the unbound supernatant to flow through by gravity. Wash the columns three times with 20 C.V. of chilled binding buffer. Elute the MBP-Spe3, MBP-Spe4, and MBP-PfSPDS proteins by adding 10 mL of elution buffer to each column and collecting 10 fractions of 1 mL each.
12. **Protein Confirmation**: Confirm the presence of the correct-sized proteins in the collected fractions by running 5–10 μL of each fraction on SDS-PAGE gels, followed by Coomassie staining.
13. **Protein Pooling and Concentration**: Pool the fractions containing the proteins of interest. Perform buffer exchange (to remove maltose) using binding buffer and Amicon filter tubes (10 kDa cut-off) according to the manufacturer's instructions. After buffer exchange, measure the protein concentration using a Nanodrop, and store the concentrated proteins in 100 μL aliquots at −80°C for use in enzyme reactions.
14. The purification process should yield 2–4 mg of MBP-Spe3, MBP-Spe4, and MBP-PfSPDS per liter of bacterial culture (Fig. 1).

d) **Preparation of a standard curve for spermidine synthase reaction**
   I) **Detection buffer recipe**
      i) 20 mM Sodium borate buffer (pH 9.6) – 6.5 mL (final 13 mM)
      ii) 1 mM Potassium phosphate buffer (pH 7.4) – 2.2 mL (final 0.22 mM)
      iii) 72 mM beta-mercaptoethanol – 243 μL (final 1.75 mM)
      iv) 15 mM Triton X-100 – 830 μL (1.25 mM)
      v) 61 mM DAB – 242.5 μL (1.48 mM)

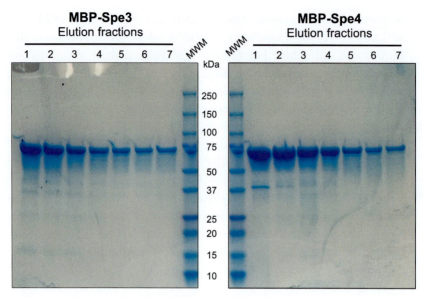

Fig.1 Coomassie stained 4-20% SDS-PAGE of fractions eluted from amylose affinity column for MBP-tagged Spe3 and Spe4 (Singh et al., 2024).

II) Set up the following standards (120 µL total volume)

	Putrescine	dcSAM	Spermidine	Methylthioadenosine	APT reaction buffer
1	6 µL from 10 mM Stock (0.5 mM final)	6 µL from 10 mM Stock (0.5 mM final)	0 µL from 10 mM Stock (0.0 mM final)	0 µL from 10 mM Stock (0.0 mM final)	108 µL
2	4.8 µL from 10 mM Stock (0.4 mM final)	4.8 µL from 10 mM Stock (0.4 mM final)	1.2 µL from 10 mM Stock (0.1 mM final)	1.2 µL from 10 mM Stock (0.1 mM final)	108 µL
3	3.6 µL from 10 mM Stock (0.3 mM final)	3.6 µL from 10 mM Stock (0.3 mM final)	2.4 µL from 10 mM Stock (0.2 mM final)	2.4 µL from 10 mM Stock (0.2 mM final)	108 µL
4	2.4 µL from 10 mM Stock (0.2 mM final)	2.4 µL from 10 mM Stock (0.2 mM final)	3.6 µL from 10 mM Stock (0.3 mM final)	3.6 µL from 10 mM Stock (0.3 mM final)	108 µL
5	1.2 µL from 10 mM Stock (0.1 mM final)	1.2 µL from 10 mM Stock (0.1 mM final)	4.8 µL from 10 mM Stock (0.4 mM final)	4.8 µL from 10 mM Stock (0.4 mM final)	108 µL
6	0 µL from 10 mM Stock (0.0 mM final)	0 µL from 10 mM Stock (0.0 mM final)	6 µL from 10 mM Stock (0.5 mM final)	6 µL from 10 mM Stock (0.5 mM final)	108 µL

i) In a 96 Well Black plate with a clear bottom, add 35 µl of each of the six samples to their respective well in triplicate.

ii) Add 85 µL of detection buffer (recipe in Section 5.d.I) to each well and incubate for 60 minutes at room temperature in the dark.

   Note: Include three additional well with 120 µL of detection buffer alone to serve as background fluorescence controls.

iii) Using a plate reader, read the plates with excitation at 364 nm and emission at 425 nm (λ) (Fig. 2).

e) Preparation of a standard curve for spermine synthase reaction

  i) Set up the following standards (120 µL total volume):

	Spermidine	dcSAM	Spermine	Methylthioadenosine	APT reaction buffer
1	6 µL from 10 mM Stock (0.5 mM final)	6 µL from 10 mM Stock (0.5 mM final)	0 µL from 10 mM Stock (0.0 mM final)	0 µL from 10 mM Stock (0.0 mM final)	108 µL
2	4.8 µL from 10 mM Stock (0.4 mM final)	4.8 µL from 10 mM Stock (0.4 mM final)	1.2 µL from 10 mM Stock (0.1 mM final)	1.2 µL from 10 mM Stock (0.1 mM final)	108 µL
3	3.6 µL from 10 mM Stock (0.3 mM final)	3.6 µL from 10 mM Stock (0.3 mM final)	2.4 µL from 10 mM Stock (0.2 mM final)	2.4 µL from 10 mM Stock (0.2 mM final)	108 µL
4	2.4 µL from 10 mM Stock (0.2 mM final)	2.4 µL from 10 mM Stock (0.2 mM final)	3.6 µL from 10 mM Stock (0.3 mM final)	3.6 µL from 10 mM Stock (0.3 mM final)	108 µL
5	1.2 µL from 10 mM Stock (0.1 mM final)	1.2 µL from 10 mM Stock (0.1 mM final)	4.8 µL from 10 mM Stock (0.4 mM final)	4.8 µL from 10 mM Stock (0.4 mM final)	108 µL
6	0 µL from 10 mM Stock (0.0 mM final)	0 µL from 10 mM Stock (0.0 mM final)	6 µL from 10 mM Stock (0.5 mM final)	6 µL from 10 mM Stock (0.5 mM final)	108 µL

  ii) In a 96-well black plate with a clear bottom, add 35 µL of each of the six samples to their respective wells in triplicate.

  iii) Add 85 µL of detection buffer (recipe in Section 5.d.I) to each well, and incubate for 60 minutes at room temperature in the dark.

   Note: In three additional wells, add 120 µL of detection buffer alone to serve as background fluorescence controls.

  iv) Measure the fluorescence using a plate reader with excitation at 364 nm and emission at 425 nm (λ) (Fig. 3).

6. Setting up spermidine and spermine synthase reactions and detection using the DAB-APT assay

a) Assessing the Activity of Spermidine Synthase (MBP-Spe3, MBP-PfSPDS) and Spermine Synthase (MBP-Spe4)

The activity of freshly purified enzymes can be evaluated using the DAB-APT assay and thin-layer chromatography (TLC) analysis (see Section 7) of the reaction products. This is done by conducting enzyme reactions at 37°C for 60 minutes, using putrescine and dcSAM as substrates for MBP-Spe3 and MBP-PfSPDS, or spermidine and dcSAM as substrates for MBP-Spe4.

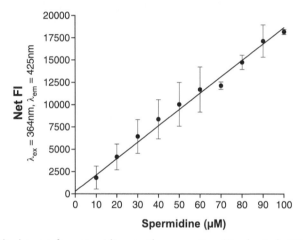

Fig. 2 Standard curve for spermidine synthase reaction (Singh et al., 2024).

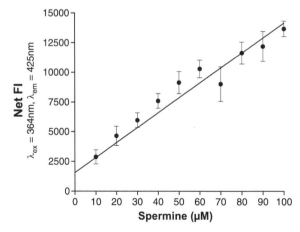

Fig. 3 Standard curve for spermine synthase reaction (Singh et al., 2024).

Reactions are terminated by heat denaturation at 80°C for 15 minutes. The resulting reaction mixtures can be analyzed using the DAB-APT assay and simultaneously run on TLC plates, along with appropriate controls. TLC plates are then stained with ninhydrin (as detailed in Section 7.c) for further analysis.

I) Aminopropyl transferase (APT) enzyme reaction buffer
   i) 50 mM potassium phosphate buffer (pH 7)
   ii) 1 mM DTT
   iii) 1 mM EDTA
   iv) 0.01% bovine serum albumin

II) Preparation of solutions
   i) 10 mM Putrescine (PUT) – Dissolve 1.61 mg of PUT powder in 1 mL water
   ii) 10 mM Spermidine (SPD) – Take 3 µL of 6.7 M SPD stock (liquid) and add to 997 µL water
   iii) 10 mM Spermine (SPM) – Dissolve 20.23 mg of SPM powder in 1 mL water
   iv) 10 mM Decarboxylated S-adenosyl methionine (dcSAM) – Dissolve 10 mg of dcSAM in 2.07 mL water
      Note: make small aliquots of 10 mM dcSAM and store in −20°C.
   v) 10 mM Methylthioadenosine (MTA) – Dissolve 3 mg of MTA in 1 mL of ethanol

III) Setting up spermidine synthase reaction
   For setting up 100 µL of spermidine synthase reaction for DAB-APT assay, add the components listed below in the sequential order:
   i) APT enzyme buffer – 79 µL
   ii) 1 mM putrescine – 10 µL (100 µM final)
   iii) 1 mM dcSAM – 10 µL (100 µM final)
   iv) Spe3 or PfSPDS enzyme – 1 µL (1 µg)

   For setting up 100 µL of spermidine synthase reaction for TLC analysis, add the components listed below in the sequential order:
   i) APT enzyme buffer – 86 µL
   ii) 10 mM putrescine – 5 µL (500 µM final)
   iii) 10 mM dcSAM – 5 µL (500 µM final)
   iv) Spe3 or PfSPDS enzyme – 4 µL (4 µg)

IV) Setting up spermine synthase reaction
   For setting up 100 µL of spermine synthase reaction for DAB-APT assay, add the components listed below in the sequential order:
   i) APT enzyme buffer – 79 µL
   ii) 1 mM spermidine – 10 µL (100 µM final)

iii) 1 mM dcSAM – 10 μL (100 μM final)
iv) Spe4 – 1 μL (1 μg)

For setting up 100 μL of spermine synthase reaction for TLC analysis, add the components listed below in the sequential order:

i) APT enzyme buffer – 86 μL
ii) 10 mM spermidine – 5 μL (500 μM final)
iii) 10 mM dcSAM – 5 μL (500 μM final)
iv) Spe4 – 4 μL (4 μg)

To measure the activity of spermidine synthase and spermine synthase reactions performed in Section 6.**a.III** and **6.a.IV** follow these steps using a 96-well plate format:

1. Add 35 μL of either the spermidine synthase or spermine synthase reaction mixtures to separate wells of a 96-well plate in triplicate.
2. Include control reactions (35 μL per well, in triplicate) for enzyme-only, substrate-only, and product-only conditions.

    Note: If generating standard curves alongside the enzyme reactions, add 35 μL of each standard to separate wells in triplicate.

3. Add 85 μL of detection buffer (recipe in Section 5.**d.I.**) to each well and incubate the plate in the dark at room temperature for 60 minutes.
4. Measure fluorescence using a plate reader with excitation at 364 nm and emission at 425 nm (λ).

b) Calculating kinetic parameters of APT enzymes using DAB-APT assay

The DAB-APT assay can also be used to determine the kinetic parameters (K_m and V_{max}) of substrates for spermidine or spermine synthase enzymes. Below is an example setup for calculating the kinetic parameters of the Spe3 enzyme (determining K_m and V_{max} for putrescine):

Enzyme reactions:

	Spe3 enzyme	Putrescine (1 mM stock)	dcSAM (1 mM stock)	APT Enzyme buffer	Total volume
1	1.2 μL (1.2 μg)	0 μL (0 μM)	14.4 μL (120 μM)	104.4 μL	120 μL
2	1.2 μL (1.2 μg)	1.2 μL (10 μM)	14.4 μL (120 μM)	103.2 μL	120 μL
3	1.2 μL (1.2 μg)	2.4 μL (20 μM)	14.4 μL (120 μM)	102 μL	120 μL
4	1.2 μL (1.2 μg)	3.6 μL (30 μM)	14.4 μL (120 μM)	100.8 μL	120 μL
5	1.2 μL (1.2 μg)	4.8 μL (40 μM)	14.4 μL (120 μM)	99.6 μL	120 μL
6	1.2 μL (1.2 μg)	6 μL (50 μM)	14.4 μL (120 μM)	98.4 μL	120 μL

7	1.2 μL (1.2 μg)	7.2 μL (60 μM)	14.4 μL (120 μM)	97.2 μL	120 μL
8	1.2 μL (1.2 μg)	8.4 μL (70 μM)	14.4 μL (120 μM)	96 μL	120 μL
9	1.2 μL (1.2 μg)	9.6 μL (80 μM)	14.4 μL (120 μM)	94.8 μL	120 μL
10	1.2 μL (1.2 μg)	10.8 μL (90 μM)	14.4 μL (120 μM)	93.6 μL	120 μL
11	1.2 μL (1.2 μg)	12 μL (100 μM)	14.4 μL (120 μM)	92.4 μL	120 μL
12	1.2 μL (1.2 μg)	14.4 μL (120 μM)	14.4 μL (120 μM)	90 μL	120 μL

Controls:

	Putrescine (1 mM stock)	dcSAM (1 mM stock)	APT Enzyme buffer	Total volume
1	0 μL (0 μM)	14.4 μL (120 μM)	105.6 μL	120 μL
2	1.2 μL (10 μM)	14.4 μL (120 μM)	104.4 μL	120 μL
3	2.4 μL (20 μM)	14.4 μL (120 μM)	103.2 μL	120 μL
4	3.6 μL (30 μM)	14.4 μL (120 μM)	102 μL	120 μL
5	4.8 μL (40 μM)	14.4 μL (120 μM)	100.8 μL	120 μL
6	6 μL (50 μM)	14.4 μL (120 μM)	99.6 μL	120 μL
7	7.2 μL (60 μM)	14.4 μL (120 μM)	98.4 μL	120 μL
8	8.4 μL (70 μM)	14.4 μL (120 μM)	97.2 μL	120 μL
9	9.6 μL (80 μM)	14.4 μL (120 μM)	96 μL	120 μL
10	10.8 μL (90 μM)	14.4 μL (120 μM)	94.8 μL	120 μL
11	12 μL (100 μM)	14.4 μL (120 μM)	93.6 μL	120 μL
12	14.4 μL (120 μM)	14.4 μL (120 μM)	91.2 μL	120 μL

	Spermidine (1 mM stock)	MTA (1 mM stock)	APT Enzyme buffer	Total volume
1	0 μL (0 μM)	14.4 μL (120 μM)	105.6 μL	120 μL
2	1.2 μL (10 μM)	14.4 μL (120 μM)	104.4 μL	120 μL
3	2.4 μL (20 μM)	14.4 μL (120 μM)	103.2 μL	120 μL
4	3.6 μL (30 μM)	14.4 μL (120 μM)	102 μL	120 μL
5	4.8 μL (40 μM)	14.4 μL (120 μM)	100.8 μL	120 μL
6	6 μL (50 μM)	14.4 μL (120 μM)	99.6 μL	120 μL

7	7.2 μL (60 μM)	14.4 μL (120 μM)	98.4 μL	120 μL
8	8.4 μL (70 μM)	14.4 μL (120 μM)	97.2 μL	120 μL
9	9.6 μL (80 μM)	14.4 μL (120 μM)	96 μL	120 μL
10	10.8 μL (90 μM)	14.4 μL (120 μM)	94.8 μL	120 μL
11	12 μL (100 μM)	14.4 μL (120 μM)	93.6 μL	120 μL
12	14.4 μL (120 μM)	14.4 μL (120 μM)	91.2 μL	120 μL

- Incubate all enzyme reactions and controls at 37°C for 60 minutes. After incubation, transfer 35 μL of each reaction and control (in triplicate) into a clear-bottom 96-well plate. Add 85 μL of detection buffer to each well, then incubate the plate in the dark at room temperature for 60 minutes. Measure the fluorescence using a plate reader with excitation at 364 nm and emission at 435 nm.
- To determine the Km and Vmax for putrescine with the Spe3 enzyme, maintain a fixed concentration of dcSAM at 120 μM, while varying the concentration of putrescine from 0 to 120 μM. Include controls with varying concentrations of putrescine alone (0−120 μM) and spermidine alone (0-120 μM).
- Perform data analysis as described in Section 8.c.

7. Validation of APT reactions using thin-layer chromatography (TLC) and mass spectrometry

a) **Preparation of solvent solution for running TLC**
   i) n-butanol − 50 mL
   ii) Acetic acid − 50 mL
   iii) Pyridine − 33.3 mL
   iv) Water − 16.7 mL

b) **Preparation of ninhydrin solution for staining TLC plates**
   i) Ninhydrin − 0.2 g
   ii) Ethanol − 99.5 mL
   iii) Acetic acid − 0.5 mL

c) **Step by step procedure for running APT reactions on TLC plates**
   i) Inside a chemical hood, place Whatman blotting papers inside the TLC tank to cover the inner edges of the tank. Put 100 mL of methanol in the tank and subsequently place a TLC plate inside it. Cover the tank and allow the methanol to run through the plate.

ii) Once the methanol has run to the top of the plate, take out the TLC plate and let it dry inside the chemical hood.
iii) After the plate is dry, place it on a white light transilluminator.
iv) Spot 20 μL of the APT enzyme reactions in horizontal lanes (~0.8 cm to 1 cm in width) along with standards (500 μM each of putrescine, dcSAM, spermidine and spermine) on the TLC plate.
v) Let the plate dry at room temperature for 20 minutes, then carefully place it into the TLC tank containing the solvent solution described in recipe in Section 7.a, ensuring no solvent splashes on the plate.
vi) Allow the solvent to move up the plate for ~4 hours, until the solvent front reaches about $3/4^{th}$ up the plate. Afterward, remove the plate from the tank and allow it to dry in the chemical hood for 8-10 hours.
vii) Once completely dry, spray the plate with ninhydrin solution (Section 7.b.) and let it stand for 5 minutes in the chemical hood. Thereafter, place the TLC plate in an oven preheated to 110°C and dry for 4-6 minutes until the color develops (Fig. 4).
viii) Capture an image of the plate for qualitative or quantitative analysis. Quantitative analysis can be performed using ImageJ software, with the percentage of enzyme catalysis determined by the intensity of the substrate band's disappearance or the product band's formation on the TLC plate.

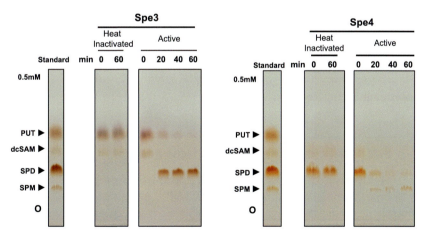

Fig. 4 TLC analysis to show enzyme reactions catalyzed by Spe3 and Spe4 reactions (left and right, respectively) (Singh et al., 2024).

d) Validation of the enzyme reactions by LC-MS

Note: While TLC confirmation of product formation in the APT reactions is typically sufficient, LC-MS can be used as an additional method to confirm product formation if needed.

The products of APT reactions, namely spermidine (from spermidine synthase reactions) and spermine (from spermine synthase reactions), can be confirmed using LC-MS. Quantification of substrates (putrescine or spermidine) and products (spermidine or spermine) can be performed using an API 4000 QTrap® mass spectrometer (Applied Biosystems Sciex, Toronto, Canada) coupled with an Agilent HPI1200 HPLC system (Agilent Technologies, Santa Clara, CA). A targeted Multiple Reaction Monitoring (MRM) method can be applied for polyamine quantification.

Polyamine standards (putrescine, spermidine, and spermine) should be prepared in a concentration range of 25–2500 ng/mL, and quality controls (QCs) should be prepared at 100, 500, and 2000 ng/mL. For liquid chromatography separation, an Agilent C18 guard column should be used at a column temperature of 50°C.

The gradient elution should be set as follows:
- Start with 98% 0.1% formic acid in water (A) and 2% 0.1% formic acid in acetonitrile (B) for 0.2 minutes.
- Increase to 30% B over 4.3 minutes, then to 50% B over 1.5 minutes, and further to 85% B over 0.5 minutes.
- Maintain 85% B for 1.5 minutes, then reduce back to 2% B over 0.5 minutes.
- Allow equilibration at 2% B for 3.5 minutes before the next injection.

The API 4000 QTrap mass spectrometer should be operated with an electrospray ionization (ESI) source in positive ion mode. Data acquisition and analysis can be performed using Analyst 1.7 software (Applied Biosystems Sciex, Toronto, Canada).

8. Quantification and statistical analysis

a) Plotting the data to create a standard curve

i) Subtract the average value of the blank wells (containing detection buffer only) from all sample readings to correct for background fluorescence.

ii) Plot the corrected fluorescence values against the concentration of the polyamines (PUT + dcSAM + SPD + MTA for spermidine synthase reactions, or SPD + dcSAM + SPM + MTA for spermine synthase reactions) in GraphPad Prism.

iii) Use GraphPad Prism to determine the linear fit of the data. This will allow you to calculate the concentration of the product (spermidine or spermine) based on the equation generated by the software.

b) **Calculating the conversion rate of spermidine synthase and spermine synthase catalyzed reactions**

i) To calculate the percent conversion of substrate (putrescine, 100 μM) into the product (spermidine), first calculate net fluorescence readings by subtracting the fluorescence value of the substrate alone (100 μM putrescine after 60 minutes in DAB reaction buffer) from total fluorescence.

ii) The fluorescence value of the product alone (100 μM spermidine after 60 minutes in DAB reaction buffer) is set as 100% for reference.

iii) Calculate the percent conversion of putrescine to spermidine by active Spe3 and control (heat inactivated Spe3) over time (0, 20, 40, 60 minutes) (Fig. 5) using the formula below:

$$\left[\frac{\text{Net fluorescence intensity from enzyme reaction at X minute}}{\text{Net fluorescence intensity of 100 μM spermidine control at 60 minute}} \right] * 100$$

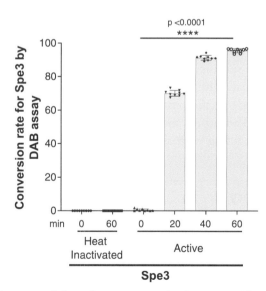

Fig. 5 Conversion rate of the substrate putrescine into spermidine by Spe3 (Singh et al., 2024).

c) **Calculating the kinetic parameters of APT enzymes (example: Spe3)**
   i) Subtract the average fluorescence value of the blank wells (containing detection buffer only) from all readings to corrects for background fluorescence.
   ii) Calculate the average fluorescence readings of the putrescine + dcSAM controls at each concentration, ranging from 0 μM to 120 μM putrescine (dcSAM is at a fixed concentration of 120 μM).
   iii) From the Spe3-catalyzed reactions, where dcSAM is held constant at 120 μM and putrescine varies (0–120 μM), subtract the reading of the corresponding control reading (putrescine + dcSAM without enzyme) to normalize the values.
   iv) Example: For Spe3 reaction performed with 10 μM of putrescine and 120 μM dcSAM, subtract the fluorescence readings of the 10 μM putrescine + 120 μM dcSAM control (no enzyme) from the reaction reading.
   v) Calculate the moles of spermidine formed at different substrate concentrations using the following formula:
   vi) $\left[\frac{nmoles\ of\ spermidine}{FI\ at\ X\ \mu M\ of\ spermidine\ control} \right] * FI\ in\ Spe3\ reaction\ at\ X\mu M$ of putrescine substrate

   vii) The specific activity of Spe3 at different substrate concentrations can be calculated using the formula:
   viii) *Specific activity = [nmoles of spermidine formed/μg of enzyme used/minute]*
   ix) Enter the specific activity values into GraphPad prism in an XY table format, with the substrate (putrescine) concentration on the X-axis and specific activity (spermidine formed) on the Y-axis.
   x) In GraphPad prism, analyze the data and generate a graph using nonlinear fitting with allosteric sigmoidal curve (Fig. 6).

9. Summary

The DAB-APT assay developed by Singh et al. (2024), offers a robust and straightforward fluorescence-based approach for the detailed biochemical characterization of APT enzymes. This assay is particularly well-suited for high-throughput screening of chemical libraries, making it invaluable for identifying novel inhibitors across various therapeutics areas, including cancer, neuroprotection and infectious diseases. By

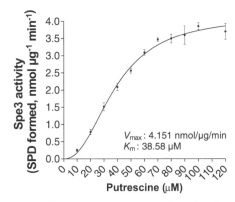

Fig. 6 Kinetic parameters of Spe3 enzyme. V_{max} and K_m for the substrate putrescine are shown (Singh et al., 2024).

facilitating the discovery of targeted therapies, the DAB-APT assay paves the way for combating a wide range of pathogens and diseases. For protozoan targets, the assay could help identify inhibitors that disrupt polyamine metabolism in *Plasmodium* species, potentially leading to novel treatments for malaria. It may also aid in developing therapeutic strategies against *Toxoplasma gondii, Trypanosoma species*, and *Leishmania* species by disrupting polyamine synthesis in these parasites. In fungal pathogens, the DAB-APT assay can be used to identify specific inhibitors that impair polyamine metabolism in important human pathogens, *such as Candida albicans, Candida auris, Aspergillus fumigatus, Histoplasma capsulatum, Cryptosporidium parvum and Mucorales*, further enhancing our arsenal of antifungal therapies. Beyond infectious diseases, the DAB-APT assay holds considerable promise in oncology and neurodegenerative conditions. Dysregulation of polyamine biosynthesis is a hallmark of many cancers, and inhibiting APT enzymes may hinder tumor growth and promote apoptosis in cancer cells, potentially leading to novel anticancer drugs (Casero et al., 2018; Murray-Stewart et al., 2016; Novita Sari et al., 2021). In neurodegenerative diseases, where polyamine levels are often disrupted, APT inhibitors could restore balance and provide neuroprotective benefits (Minois, Carmona-Gutierrez, & Madeo, 2011; Morrison & Kish, 1995).

Overall, the DAB-APT assay represents a promising platform that could transform therapeutic strategies against a wide array of protozoan, fungal, and chronic diseases, ultimately enhancing our capabilities to combat some of the most challenging health threats.

References

Blazquez, M. A. (2024). Polyamines: Their role in plant development and stress. *Annual Review of Plant Biology, 75*(1), 95–117.

Bowman, W. H., Tabor, C. W., & Tabor, H. (1973). Spermidine biosynthesis. Purification and properties of propylamine transferase from Escherichia coli. *Journal of Biological Chemistry, 248*(7), 2480–2486.

Brooks, W. H. (2013). Increased polyamines alter chromatin and stabilize autoantigens in autoimmune diseases. *Frontiers in Immunology, 4*, 91.

Brooks, W. H. (2024). Polyamine dysregulation and nucleolar disruption in Alzheimer's disease. *Journal of Alzheimer's Disease: JAD, 98*(3), 837–857.

Casero, R. A., Jr., Stewart, T. M., & Pegg, A. E. (2018). Polyamine metabolism and cancer: Treatments, challenges and opportunities. *Nature Reviews. Cancer, 18*(11), 681–695.

Cason, A. L., Ikeguchi, Y., Skinner, C., Wood, T. C., Holden, K. R., Lubs, H. A., ... Schwartz, C. E. (2003). X-linked spermine synthase gene (SMS) defect: The first polyamine deficiency syndrome. *European Journal of Human Genetics: EJHG, 11*(12), 937–944.

Cervelli, M., Averna, M., Vergani, L., Pedrazzi, M., Amato, S., Fiorucci, C., ... Marcoli, M. (2022). The involvement of polyamines catabolism in the crosstalk between neurons and astrocytes in neurodegeneration. *Biomedicines, 10*(7).

Choi, J. Y., Black, R., 3rd, Lee, H., Di Giovanni, J., Murphy, R. C., ... Voelker, D. R. (2020). An improved and highly selective fluorescence assay for measuring phosphatidylserine decarboxylase activity. *The Journal of Biological Chemistry, 295*(27), 9211–9222.

Dudley, H. W., Rosenheim, O., & Starling, W. W. (1926). The chemical constitution of spermine: Structure and synthesis. *The Biochemical Journal, 20*(5), 1082–1094.

Enomoto, K., Nagasaki, T., Yamauchi, A., Onoda, J., Sakai, K., Yoshida, T., ... Nagata, K. (2006). Development of high-throughput spermidine synthase activity assay using homogeneous time-resolved fluorescence. *Analytical Biochemistry, 351*(2), 229–240.

Firpo, M. R., Mastrodomenico, V., Hawkins, G. M., Prot, M., Levillayer, L., Gallagher, T., ... Mounce, B. C. (2021). Targeting polyamines inhibits coronavirus infection by reducing cellular attachment and entry. *ACS Infectious Diseases, 7*(6), 1423–1432.

Haider, N., Eschbach, M. L., Dias Sde, S., Gilberger, T. W., Walter, R. D., & Luersen, K. (2005). The spermidine synthase of the malaria parasite Plasmodium falciparum: Molecular and biochemical characterisation of the polyamine synthesis enzyme. *Molecular and Biochemical Parasitology, 142*(2), 224–236.

Han, X., Wang, D., Yang, L., Wang, N., Shen, J., Wang, J., ... Wang, Y. (2024). Activation of polyamine catabolism promotes glutamine metabolism and creates a targetable vulnerability in lung cancer. *Proceedings of the National Academy of Sciences of the United States of America, 121*(13), e2319429121.

Huang, M., Zhang, W., Chen, H., & Zeng, J. (2020). Targeting polyamine metabolism for control of human viral diseases. *Infection and Drug Resistance, 13*, 4335–4346.

Ikeguchi, Y., Bewley, M. C., & Pegg, A. E. (2006). Aminopropyltransferases: Function, structure and genetics. *Journal of Biochemistry, 139*(1), 1–9.

Kaiser, A. (2023). The role of spermidine and its key metabolites in important, pathogenic human viruses and in parasitic infections caused by Plasmodium falciparum and Trypanosoma brucei. *Biomolecules, 13*(5).

Lefevre, P. L., Palin, M. F., & Murphy, B. D. (2011). Polyamines on the reproductive landscape. *Endocrine Reviews, 32*(5), 694–712.

Médici, R., de Maria, P. D., Otten, L. G., & Straathof, A. J. J. (2011). A high-throughput screening assay for amino acid decarboxylase activity. *Advanced Synthesis & Catalysis, 353*(13), 2369–2376.

Mendez, R., Kesh, K., Arora, N., Di Martino, L., McAllister, F., Merchant, N., ... Banerjee, S. (2020). Microbial dysbiosis and polyamine metabolism as predictive markers for early detection of pancreatic cancer. *Carcinogenesis, 41*(5), 561–570.

Michael, A. J. (2018). Polyamine function in archaea and bacteria. *The Journal of Biological Chemistry, 293*(48), 18693–18701.

Minois, N., Carmona-Gutierrez, D., & Madeo, F. (2011). Polyamines in aging and disease. *Aging (Albany NY), 3*(8), 716–732.

Morrison, L. D., & Kish, S. J. (1995). Brain polyamine levels are altered in Alzheimer's disease. *Neuroscience Letters, 197*(1), 5–8.

Morris, D. R., & Pardee, A. B. (1965). A biosynthetic ornithine decarboxylase in Escherichia coli. *Biochemical and Biophysical Research Communications, 20*(6), 697–702.

Murray-Stewart, T. R., Woster, P. M., & Casero, R. A., Jr. (2016). Targeting polyamine metabolism for cancer therapy and prevention. *The Biochemical Journal, 473*(19), 2937–2953.

Novita Sari, I., Setiawan, T., Seock Kim, K., Toni Wijaya, Y., Won Cho, K., & Young Kwon, H. (2021). Metabolism and function of polyamines in cancer progression. *Cancer Letters, 519*, 91–104.

Pegg, A. E., & Michael, A. J. (2010). Spermine synthase. *Cellular and Molecular Life Sciences: CMLS, 67*(1), 113–121.

Pegg, A. E. (2016). Functions of polyamines in mammals. *The Journal of Biological Chemistry, 291*(29), 14904–14912.

Raina, A., & Janne, J. (1968). Biosynthesis of putrescine: Characterization of ornithine decarboxylase from regenerating rat liver. *Acta Chem Scand, 22*(7), 2375–2378.

Rhee, H. J., Kim, E. J., & Lee, J. K. (2007). Physiological polyamines: Simple primordial stress molecules. *Journal of Cellular and Molecular Medicine, 11*(4), 685–703.

Saeki, Y., Uehara, N., & Shirakawa, S. (1978). Sensitive fluorimetric method for the determination of putrescine, spermidine and spermine by high-performance liquid chromatography and its application to human blood. *Journal of Chromatography, 145*(2), 221–229.

Sano, M., & Nishino, I. (2007). Assay for spermidine synthase activity by micellar electrokinetic chromatography with laser-induced fluorescence detection. *Journal of Chromatography. B, Analytical Technologies in the Biomedical and Life Sciences, 845*(1), 80–83.

Schibalski, R. S., Shulha, A. S., Tsao, B. P., Palygin, O., & Ilatovskaya, D. V. (2024). The role of polyamine metabolism in cellular function and physiology. *American Journal of Physiology. Cell Physiology, 327*(2), C341–C356.

Seiler, N., & Raul, F. (2005). Polyamines and apoptosis. *Journal of Cellular and Molecular Medicine, 9*(3), 623–642.

Shah, P., & Swiatlo, E. (2008). A multifaceted role for polyamines in bacterial pathogens. *Molecular Microbiology, 68*(1), 4–16.

Singh, P., Choi, J. Y., Wang, W., Lam, T., Lechner, P., Vanderwal, C. D., ... Ben Mamoun, C. (2024). A fluorescence-based assay for measuring polyamine biosynthesis aminopropyl transferase-mediated catalysis. *The Journal of Biological Chemistry*, 107832.

Solmi, L., Rossi, F. R., Romero, F. M., Bach-Pages, M., Preston, G. M., Ruiz, O. A., & Garriz, A. (2023). Polyamine-mediated mechanisms contribute to oxidative stress tolerance in Pseudomonas syringae. *Scientific Reports, 13*(1), 4279.

Tang, G., Xia, H., Liang, J., Ma, Z., & Liu, W. (2021). Spermidine is critical for growth, development, environmental adaptation, and virulence in Fusarium graminearum. *Frontiers in Microbiology, 12*, 765398.

Terui, Y., Ohnuma, M., Hiraga, K., Kawashima, E., & Oshima, T. (2005). Stabilization of nucleic acids by unusual polyamines produced by an extreme thermophile, Thermus thermophilus. *The Biochemical Journal, 388*(Pt 2), 427–433.

Thongbhubate, K., Nakafuji, Y., Matsuoka, R., Kakegawa, S., & Suzuki, H. (2021). Effect of spermidine on biofilm formation in Escherichia coli K-12. *Journal of Bacteriology, 203*(10).
Tofalo, R., Cocchi, S., & Suzzi, G. (2019). Polyamines and gut microbiota. *Frontiers in Nutrition, 6*, 16.
Valdes-Santiago, L., & Ruiz-Herrera, J. (2013). Stress and polyamine metabolism in fungi. *Frontiers in Chemistry, 1*, 42.
Zappia, V., Cacciapuoti, G., Pontoni, G., & Oliva, A. (1980). Mechanism of propylamine-transfer reactions. Kinetic and inhibition studies on spermidine synthase from Escherichia coli. *The Journal of Biological Chemistry, 255*(15), 7276–7280.
Zhang, H., & Au, S. W. N. (2017). Helicobacter pylori does not use spermidine synthase to produce spermidine. *Biochemical and Biophysical Research Communications, 490*(3), 861–867.

CHAPTER TWENTY TWO

Isolation of *Hafnia paralvei* FB315, a polyamine-high-producing bacterium, from aged cheese

Yuta Ami, Emi Sugiwaka, and Shin Kurihara*
Faculty of Biology-Oriented Science and Technology, Kindai University, Kinokawa, Wakayama, Japan
*Corresponding author. e-mail address: skurihara@waka.kindai.ac.jp

Contents

1. Introduction	390
2. Isolation of bacteria from cheese	391
2.1 Equipment	391
2.2 Reagents	392
2.3 Procedure	392
3. Bacterial culture	393
3.1 Equipment	393
3.2 Reagents	393
3.3 Procedure	393
4. Purification of PuO derived from *Rhodococcus erythropolis* NCIMB 11540	393
4.1 Equipment	393
4.2 Reagents	393
4.3 Procedure	394
5. Simplified quantification of putrescine concentration in culture supernatants using the PuO-POD-4AA-TOPS method (Sugiyama et al., 2020)	395
5.1 Equipment	395
5.2 Reagents	395
5.3 Procedure	395
5.4 Note	397
6. Quantification of polyamine concentration by HPLC	397
6.1 Equipment	397
6.2 Reagents	397
6.3 Procedure	398
7. 16S rRNA gene analysis	398
7.1 Equipment	398
7.2 Reagents	398
7.3 Procedure	399
8. Cultivation in MRS medium supplemented with precursors of putrescine	399
8.1 Equipment	399

8.2 Reagents	399
8.3 Procedure	400
9. Results	400
9.1 Putrescine concentration in aged cheeses	400
9.2 Isolation of bacteria from aged cheeses and screening for polyamine-producing strain	400
9.3 Detailed analysis of the polyamine-producing candidate strain	401
9.4 Effects of putrescine precursors on putrescine production by *Hafnia paralvei* FB315	401
9.5 Growth and Putrescine Production of a High Putrescine-Producing Bacterium Isolated from Aged Cheese	401
10. Discussion	402
Appendix A. Supporting information	404
References	404

Abstract

As previously reported, bacteria in fermented foods produce putrescine, a type of polyamine. In order to produce putrescine, which has been reported to contribute to health promotion, by fermentation, 181 bacterial strains were isolated from six types of aged cheeses (Bleu d'Auvergne, Red Leicester, Langres, Munster, Galet de la Loire, and Mimolette) and screened for strains that secrete high levels of putrescine using the method we developed, termed PuO-POD-4AA-TOPS. This approach resulted in the isolation of *Hafnia paralvei* FB315, a strain producing high levels of putrescine, from Langres cheese. Subsequent analyses revealed that *H. paralvei* FB315 utilizes both agmatine and ornithine, known precursors of putrescine, as substrates for putrescine production, and that the concentration of putrescine in the culture supernatant increases in a time-dependent manner to a maximum of 1840 μM.

1. Introduction

Polyamines are defined as hydrocarbon compounds that contain two or more amino groups in their molecular structure. Due to their positive charge, they bind weakly to negatively charged components such as nucleic acids, phosphorylated proteins, phospholipids, and ATP. The significance of these interactions is evidenced by their role in gene transcription (Ohishi et al., 2008), translation (Yoshida et al., 2004), and cell differentiation (Oka & Perry, 1974). Furthermore, previous studies have reported that polyamine administration extends lifespan (Soda, Dobashi, Kano, Tsujinaka, & Konishi, 2009; Eisenberg et al., 2009; Matsumoto, Kurihara, Kibe, Ashida, & Benno, 2011; Soda, Kano, Chiba, Koizumi, & Miyaki, 2013) improves brain function (Kibe et al., 2014), reduces intestinal inflammation (Nakamura et al., 2021), and enhances anti-tumor immunity (Al-Habsi et al., 2022). It has been demonstrated that humans possess the capacity to synthesize polyamines endogenously (Das & Kanungo, 1982); nevertheless, there is a decline in polyamine production with age (Nishimura, Shiina, Kashiwagi, & Igarashi, 2006). It has been reported that dietary intake of polyamines increases their concentration in human blood (Soda, Kano, et al., 2009) and that putrescine, produced by gut microbiota in the colon lumen, is absorbed into colonic tissues (Nakamura et al., 2021). The present study thus provides a robust foundation for the hypothesis that exogenous polyamine supplementation can effectively counteract the age-related decline in endogenous synthesis, thereby promoting healthy lifespan extension. Nevertheless, levels of polyamines found in foods such as wheat germ (563 μg/g) are not considered high enough to significantly impact the human diet (Nishimura et al., 2006). However, notable concentrations of polyamines have been detected in certain cheeses that undergo maturation by the action of fermentative microorganisms (Nishimura et al., 2006). These polyamines are released extracellularly by the microbes responsible for the maturation process (Kalac & Krausová, 2005; Okamoto, Sugi, Koizumi, Yanagida, & Udaka, 1997). In the event of the successful isolation of high polyamine-producing strains from these cheeses, it may be possible to develop polyamine-rich supplements derived from their culture supernatants. Our research group has investigated bacteria derived from fermented foods that release polyamines extracellularly. To date, the isolation of *Latilactobacillus curvatus* KP 3–4 from kabura-zushi (Hirano et al., 2022), *Staphylococcus epidermidis* FB146 from miso (Shirasawa et al., 2023), and *Levilactobacillus brevis* FB215 from blue cheese (Ami et al., 2023) has been successful.

In the present study, 181 bacterial strains were isolated from six types of aged cheeses (Bleu d'Auvergne, Red Leicester, Langres, Munster, Galet de la Loire, and Mimolette). Subsequently, the strains were evaluated for their polyamine productivity by a simplified quantification of the putrescine concentration in their culture supernatants. This approach led to the identification and isolation of a strain of *Hafnia paralvei* as a high polyamine-producing bacterium.

2. Isolation of bacteria from cheese
2.1 Equipment
1. Kitchen scissors sterilized with 70 % *(v/v)* ethanol.
2. Sterilized beads (Ambion; product number, 10022G).

3. 96-deep-well plates (Thermo Scientific Nunc; product number, 260252).
4. AnaeroPack (Mitsubishi Gas Chemical Company; product number, A-03).
5. Square hermetically-sealed jar (Mitsubishi Gas Chemical Company; product number, A-110).

2.2 Reagents
1. Aged Cheese
2. 70 % ethanol *(v/v)*
3. Sterilized phosphate-buffered saline (PBS)
4. Sterilized beads (Ambion; product number, 10022G)
5. MRS (Becton, Dickinson and Company; product number, 288130) plate
6. MRS (Becton, Dickinson and Company; product number, 288130) broth
7. 50 % *(v/v)* glycerol (Fujifilm Wako; product code 075-00616)

2.3 Procedure
1. The purchased cheese was chopped using kitchen scissors sterilized with 70 % ethanol *(v/v)*.
2. The pieces were then mixed with sterilized phosphate-buffered saline (PBS) and diluted 10-fold.
3. Mixture was stirred until no solid residues remained, resulting in a suspension.
4. The suspension was diluted 10^{-1} *(w/v)*.
5. The 10^{-1} -*(w/v)* suspension was further diluted to 10^{-2}, 10^{-3}, 10^{-4}, and 10^{-5} *(w/v)*.
6. A volume of 100 microliters of the serially diluted suspensions were spread on MRS plates using sterilized beads.
7. The plates were sealed in square hermetically-sealed jar with AnaeroPack.
8. The plates were incubated anaerobically at 37 °C for 48 to 96 h.
9. Single colonies from the MRS plates were transferred to sterilized 96-deep-well plates containing 500 μL of MRS broth.
10. The deep-well plates were sealed in square hermetically-sealed jar with AnaeroPack and incubated anaerobically at 37 °C for 72 to 108 h.
11. The cultured broth was mixed with 50 % *(v/v)* glycerol to achieve a final glycerol concentration of 18 %.
12. The glycerol stocks were stored frozen at −80 °C.

3. Bacterial culture
3.1 Equipment
1. 96-deep-well plates (Thermo Scientific Nunc, Tokyo, Japan)
2. Square hermetically-sealed jar (Mitsubishi Gas Chemical Company; product number, A-110)
3. AnaeroPack (Mitsubishi Gas Chemical Company; product number, A-03)
4. 96-well microtiter plate (WATSON; product number, 195-96 F)

3.2 Reagents
1. MRS (Becton, Dickinson and Company; product number, 288130) broth
2. Phosphate-buffered saline (PBS; sterilized)

3.3 Procedure
1. The frozen glycerol stock of the isolated bacteria was thawed.
2. A volume of 2 μL of the thawed suspension was inoculated into 500 μL of MRS broth in sterilized 96-deep-well plates.
3. The plates were sealed in square hermetically-sealed jar with AnaeroPack and incubated anaerobically at 37 °C for 48 h.
4. After incubation, 50 μL of the cultured broth was mixed with 150 μL of PBS in a 96-well microtiter plate.
5. The optical density at 600 nm (OD_{600}) of the resulting 4-fold dilution was measured to confirm bacterial growth.

4. Purification of PuO derived from *Rhodococcus erythropolis* NCIMB 11540
4.1 Equipment
1. Q500 Sonicator (Qsonica; product number Q500)
2. Open column (Bio-Rad; product number 7370717)
3. Q-Sepharose Fast Flow resin (GE Healthcare; product number 17051001)
4. Amicon® Ultra-15 10k centrifugal filter unit (Millipore; product number UFC901024)

4.2 Reagents
1. TOP10F' competent cells (Thermo Scientific; product number C303003)

2. pBAD-puo_{Re}^{+} plasmid (kindly gifted by Dr. Marco W. Fraaije)
3. LB medium (Becton Dickinson; product number 244620) supplemented with a final concentration of 100 μg/mL ampicillin.
4. TB medium (tryptone, yeast extract, dipotassium hydrogen phosphate, potassium dihydrogen phosphate, and glycerol)
5. 0.02 % *(w/v)* Arabinose (Fujifilm Wako; product code 010-04582)
6. 50 mM Tris-HCl (pH 7.5)

4.3 Procedure

The *puo* gene, derived from *Rhodococcus erythropolis* NCIMB 11540 (puo_{Re}), was overexpressed in genetically engineered *Escherichia coli* and purified using anion exchange chromatography.

1. A plasmid containing the cloned *puo* gene (pBAD-puo_{Re}^{+}) was kindly gifted by Dr. Marco W. Fraaije (University of Groningen, The Netherlands) (Van Hellemond, van Dijk, Heuts, Janssen, & Fraaije, 2008).
2. pBAD-puo_{Re}^{+} plasmid was introduced into TOP10F' competent cells.
3. A transformed strain was inoculated into 5 mL of LB liquid medium supplemented with ampicillin (100 μg/mL).
4. The culture was grown overnight at 37 °C with shaking at 140 rpm.
5. Two milliliters of the overnight culture was inoculated into 200 mL of TB medium in a 500 mL Erlenmeyer flask. The TB medium containing 1.2 % *(w/v)* tryptone, 2.4 % *(w/v)* yeast extract, 0.94 % *(w/v)* dipotassium hydrogen phosphate, 0.22 % *(w/v)* potassium dihydrogen phosphate, 0.8 % *(w/v)* glycerol.
6. Ampicillin (100 μg/mL) and arabinose (0.02 % *(w/v)*) were added to the medium.
7. The culture was grown at 30 °C with shaking at 140 rpm for 24 h.
8. The culture was centrifuged, and the cell pellet was washed twice with 20 mL of 50 mM Tris-HCl (pH 7.5), then resuspended in 8 mL of 50 mM Tris-HCl (pH 7.5).
9. The suspension was sonicated for 5 min on ice using a Q500 Sonicator.
10. The lysate was centrifuged at 21,500g for 10 min, and the supernatant was collected as the cell-free extract.
11. A total of 15.6 mg of protein in the cell-free extract was applied to an open column (Bio-Rad; product number 7370717) packed with 3 mL of Q-Sepharose Fast Flow.
12. The column was washed with 21 mL of 50 mM Tris-HCl (pH 7.5), followed by 40 mL of 50 mM Tris-HCl containing 100 mM KCl.

13. Stepwise elution was performed with 21 mL of 50 mM Tris-HCl containing 200, 300, and 400 mM KCl, respectively.
14. Fractions eluted with 300 mM KCl were analyzed by SDS-PAGE and found to contain PuO$_{Re}$ with high purity.
15. Purified fractions were pooled and concentrated using an Amicon® Ultra-15 10k centrifugal filter unit, which also removed KCl.
16. Concentrated PuO$_{Re}$ was stored at −25 °C until use.

5. Simplified quantification of putrescine concentration in culture supernatants using the PuO-POD-4AA-TOPS method (Sugiyama et al., 2020)

5.1 Equipment
1. Centrifuge (himac; product number CF16RN)
2. Rotor for 96-deep-well plates (himac; product number T5S33)
3. 96-well electronic pipette (INTEGRA; product number 6001)
4. 96-well microtiter plate (WATSON; product number 195-96F)
5. Microplate reader (Thermo Fisher Scientific; product number 51119350)

5.2 Reagents
1. 50 mM Tris-HCl (pH 8.0) (NIPPON GENE; product number 314-90065)
2. 1 mM N-Ethyl-N-(3-sulfopropyl)−3-methylaniline sodium salt (TOPS) (Dojindo Laboratories; product number OC14)
3. 1 mM 4-Aminoantipyrine (4-AA) (TCI; product number A2254)
4. 5 U/mL Peroxidase, from Horseradish Roots (POD) (HRP) (Oriental Yeast; product number 46261003)
5. PuO$_{Re}$ (0.1 μg/mL)
6. 1,4-diaminobutane dihydrochloride (putrescine dihydrochloride) (TCI; product number D0081)

5.3 Procedure
The 96-deep-well plates containing bacterial culture broth were subjected to centrifugation at 1,900g for 20 min, using a centrifuge equipped with a rotor designed for 96-deep-well plates.
1. The resultant supernatants were diluted 4-fold with deionized water.
2. A 50-mL pre-mix solution was prepared, containing: 50 mM Tris-HCl (pH 8.0), 1 mM TOPS, 1 mM 4-AA, 5 U/mL HRP, 0.1 μg/mL PuO$_{Re}$ (Details in Table 1).

Table 1 Reaction mixture composition of the PuO-POD-4AA-TOPS method.

Stock concentration		Reagent	Manufacturer (product number)	Required volume for making 90 mL of pre-mix		The amount required per well		Final concentration	
1	M	Tris–HCl (pH 8.0)	NIPPON GENE (314-90065)	6.0	mL	10	μL	50.0	mM
20	mM	TOPS	Dojindo Laboratories (OC14)	6.0	mL	10	μL	1.0	mM
20	mM	4-AA	TCI (A2254)	6.0	mL	10	μL	1.0	mM
100	U/mL	HRP	Oriental Yeast (46261003)	6.0	mL	10	μL	5.0	U/mL
1	μg/mL	PuO	See materials and methods	12.0	mL	20	μL	0.1	μg/mL
		Deionized water		54.0	mL	90	μL		
4	fold	Diluted culture supernatant				50	μL		
						200	μL		

3. Using a 96-well electronic pipette (VIAFLO 96), 150 μL of the pre-mix solution was dispensed into each well of a 96-well microtiter plate.
4. Subsequently, 50 μL of fourfold diluted culture supernatants from each bacterial strain was added to the wells, thoroughly mixed, and incubated in the dark for 20 min.
5. Absorbance at 550 nm was measured using a microplate reader.
6. Standard curve was constructed using putrescine solutions (final concentrations: 400, 200, 100, 50, and 0 μM). Measured absorbance values from samples were used to calculate putrescine concentrations based on the calibration curve.

5.4 Note

For the purpose of assay controls, *Levilactobacillus brevis* FB215 (Ami et al., 2023) was selected as the positive control, a strain known to accumulate high levels of putrescine in the culture supernatant. In addition, an uninoculated MRS medium that had been subjected to identical conditions served as the negative control.

6. Quantification of polyamine concentration by HPLC

6.1 Equipment

1. Square hermetically-sealed jar (Mitsubishi Gas Chemical Company; product number, A-110)
2. AnaeroPack (Mitsubishi Gas Chemical Company; product number, A-110)
3. Centrifuge (himac; product number CF16RN)
4. Cosmonice Filter W (Merck; product number 06543-04)
5. HPLC system (Chromaster, Hitachi Ltd., Tokyo, Japan)
6. Cation exchange column (Hitachi; product number #2619PH)

6.2 Reagents

1. MRS (Becton, Dickinson and Company; product number, 288130) broth
2. 100 % *(w/v)* trichloroacetic acid (TCA) (Fujifilm Wako; product code 206-16211)
3. PBS
4. 1,4-Diaminobutane Dihydrochloride (TCI; product number D0081)
5. Spermidine Trihydrochloride (Nacalai tesque; product code 32110-41)
6. Spermine Tetrahydrochloride (Nacalai tesque; product code 32113-11)

6.3 Procedure

Bacteria producing high levels of putrescine, identified using the PuO-POD-4AA-TOPS method, were re-inoculated into 500 μL of MRS broth.

1. Cultures was anaerobically incubated at 37 °C for 24 h in square hermetically-sealed jar with AnaeroPack.
2. After incubation, the cultures were centrifuged at 2,700g for 15 min.
3. To 200 μL of the collected supernatant, 20 μL of 100 % *(w/v)* TCA was added to precipitate proteins.
4. The mixture was centrifuged at 18,700g for 5 min to remove precipitated proteins.
5. The supernatant was filtered using a Cosmonice Filter W.
6. Polyamine concentrations were analyzed with an HPLC system equipped with a cation exchange column (Shirasawa et al., 2023).
7. Standard curves were constructed using polyamine standards of known concentrations. Retention times for standard compounds are as follows: Putrescine: 15.2 min, Spermidine: 26.0 min, and Spermine: 39.1 min.
8. Putrescine concentration in cheese, used for bacterial isolation, was measured by suspending the cheese in PBS and analyzing it with the same HPLC method.

7. 16S rRNA gene analysis

7.1 Equipment

1. Centrifuge (himac; product number CF16RN)
2. Beads shocker (SHAKE MASTER NEO ver1.0; Biomedical Science; product code BMS-M10N21)
3. TaKaRa PCR Thermal Cycler Dice® Touch (Takara Bio; product code TP350)

7.2 Reagents

1. Tris-EDTA buffer (TE buffer)
2. KOD FX Neo (TOYOBO product number KFX-201)
3. 10 μM 7 F (Forward PCR primer) (5' AGAGTTTGATYMTGGCTCAG 3')
4. 10 μM 1510 R (Reverse PCR primer) (5' ACGGYTACCTTGTTACGACTT 3')
5. Bacterial disruption beads (Ambion; product number 10022G)

6. ExoSAP-IT™ Express PCR Product Cleanup Reagent (Thermo Fisher Scientific; product number 75001)
7. Wizard® SV Gel and PCR Clean-Up System (Promega; product number A9281)

7.3 Procedure
1. Bacterial culture was centrifuged to form a pellet.
2. Bacterial pellet was resuspended in TE buffer.
3. Approximately 20 mg of bacterial disruption beads were added to the suspension.
4. Cells were disrupted using a beads shocker.
5. The suspension was centrifuged to obtain a crude extract of bacterial chromosome, which served as the template for PCR amplification of the 16S rRNA gene.
6. The template for PCR amplification of 16S rRNA gene was amplified by PCR using KOD FX Neo, 5 F and 1510 R.
7. PCR products were purified using one of the following methods: ExoSAP-IT™ Express PCR Product Cleanup Reagent or Wizard® SV Gel and PCR Clean-Up System.
8. Nucleotide sequences were determined by Sanger sequencing.
9. The obtained sequences were analyzed using BLAST to identify the bacterial species.

8. Cultivation in MRS medium supplemented with precursors of putrescine

8.1 Equipment
1. Square hermetically-sealed jar (Mitsubishi Gas Chemical Company; product number, A-110)
2. AnaeroPack (Mitsubishi Gas Chemical Company; product number, A-03)
3. Centrifuge (himac; product number CF16RN)

8.2 Reagents
1. MRS (Becton, Dickinson and Company; product number, 288130) broth
2. Agmatine sulfate (TCI; product number A0310)
3. L-Ornithine·HCl (Peptide Institute; product number 2716)

8.3 Procedure

1. Bacterial glycerol stocks were streaked onto an MRS plate.
2. The plates were placed into square hermetically-sealed jar containing an AnaeroPack and incubated at 37 °C for 24 h to obtain single colonies.
3. A single colony was then inoculated into 3 mL of MRS broth and incubated in square hermetically-sealed jar with an AnaeroPack at 37 °C for an additional 24 h to obtain pre-culture liquid.
4. 20 µL of the pre-culture liquid was inoculated into 2 mL of MRS broth supplemented with 1 mM of either: agmatine or ornithine.
5. The mixture was incubated in square hermetically-sealed jar equipped with an AnaeroPack at 37 °C for 48 h.
6. The culture was centrifuged at 1,900 g for 20 min.
7. Putrescine concentration of culture supernatant was determined by HPLC.

9. Results

9.1 Putrescine concentration in aged cheeses

Putrescine concentrations were determined in a selection of aged cheeses used as sources for bacterial isolation. The cheeses included in the study were Bleu d'Auvergne, Red Leicester, Langres, Munster, Gaperon de la Loire, and Mimolette. The results demonstrated that the putrescine levels in Bleu d'Auvergne, Langres, and Mimolette were 314, 4136, and 157 µM, respectively. No putrescine was detected in Red Leicester, Munster, and Gaperon de la Loire.

9.2 Isolation of bacteria from aged cheeses and screening for polyamine-producing strain

Suspensions of aged cheeses were plated onto MRS agar plates. The 29, 39, 24, 37, 38, and 14 colonies derived from Bleu d'Auvergne, Red Leicester, Langres, Munster, Gaperon de la Loire, and Mimolette, respectively, were inoculated into MRS broth. The cultures derived from Bleu d'Auvergne were incubated for 78 h, while those derived from the other cheeses were incubated for 108 h. The maximum OD_{600} was 2.29. The putrescine concentration in the culture supernatant was then subjected to simple quantification using the PuO-POD-4AA-TOPS method (Sugiyama et al., 2020), which led to identification of one putrescine-high-producing strain isolated from Langres cheese (Details in Supplemental Table 1).

9.3 Detailed analysis of the polyamine-producing candidate strain

The putrescine-high-producing candidate strain isolated from Langres was streaked to obtain a single colony, designated as FB315, which was used for further analysis. Polyamine concentration in the culture supernatants of FB315 after 24 h of growth was measured using HPLC. The culture supernatants of FB315 contained 221 µM of putrescine, while spermidine and spermine were below the detection limit for the tested strain. Furthermore, a BLAST analysis of the sequenced 16S rDNA of FB315 (1459 bp) revealed complete identity with the corresponding 16S rDNA region of *Hafnia paralvei*.

9.4 Effects of putrescine precursors on putrescine production by *Hafnia paralvei* FB315

As summarized in a previous study (Sugiyama et al., 2017), two pathways have been identified for bacterial putrescine synthesis: the conversion of agmatine to putrescine and the conversion of ornithine to putrescine. In order to assess the effect of these precursors on bacterial putrescine production, *H. paralvei* FB315 and *H. paralvei* ATCC29927^T was cultivated in MRS medium supplemented with 1 mM ornithine or 1 mM agmatine, and the putrescine concentration in the culture supernatants was measured at 24 h after inoculation. In the absence of supplemented precursors, the culture supernatant of FB315 contained 140 µM of putrescine, while the supplementation of the medium with 1 mM agmatine or 1 mM ornithine resulted in putrescine concentrations of 275 and 1025 µM, respectively (Fig. 1A). For *H. paralvei* ATCC29927^T, the culture supernatant contained 205 µM of putrescine in the absence of precursors, while the supplementation of 1 mM agmatine or 1 mM ornithine resulted in putrescine concentrations of 570 µM and 833 µM, respectively (Fig. 1B).

9.5 Growth and Putrescine Production of a High Putrescine-Producing Bacterium Isolated from Aged Cheese

In order to analyze the putrescine production of the high putrescine-producing bacterium *H. paralvei* FB315 in different growth phases, the growth (OD_{600}) was monitored and the putrescine concentration of the culture supernatant was determined by HPLC over time. The results demonstrated that *H. paralvei* FB315 reached an OD_{600} of 2.8 at 48 h of cultivation, followed by a gradual decrease to an OD_{600} of 2.1 by 122 h. Concurrently, the putrescine concentration in the culture supernatant exhibited a steady increase, attaining a maximum of approximately 1840 µM at 122 h of cultivation (Fig. 2).

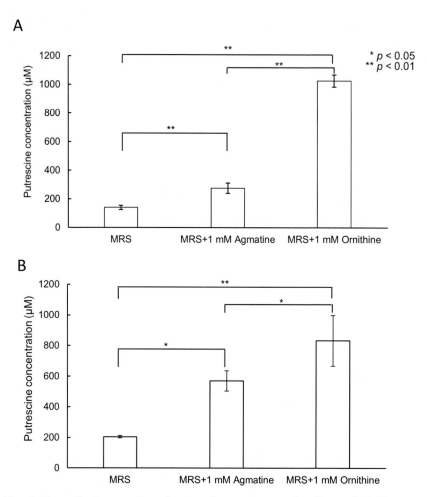

Fig. 1 **Biosynthetic capacity of putrescine precursors by *H. paralvei*.** The concentration of putrescine in the culture supernatant was measured after culturing *H. paralvei* FB215 (A) and ATCC29927^T (B) in media supplemented with the putrescine precursors agmatine or ornithine. Error bars indicate standard deviation ($n = 3$) * = $p < 0.05$, ** = $p < 0.01$.

10. Discussion

In the present study, *Hafnia paralvei* FB315, a bacterium that produces high levels of putrescine, was isolated for the first time from Langres cheese. The Langres cheese was found to have the highest putrescine concentration of all the cheeses analyzed in this study. *Hafnia* species are among the few Gram-negative bacteria used in cheese production (Bourdichon et al., 2012).

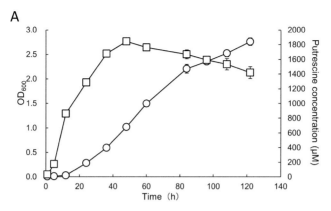

Fig. 2 Putrescine concentration of culture supernatant of *H. paralvei* FB315 in the different growth phases. The growth (OD_{600}) and the concentration of putrescine in the culture supernatant were measured at various time points during the cultivation of *H. paralvei* FB315. Growth and putrescine concentration of culture supernatant is represented by open squares and open circles, respectively. Error bars indicate standard deviation ($n = 3$).

These bacteria have been shown to enhance the final flavor of cheese by degrading proteins and producing aromatic sulfur compounds (Irlinger et al., 2012; Morales, Fernandez-Garcia, & Nunez, 2003). *H. paralvei* has been detected in cheeses made from raw milk in both Japan and France (Unno, Suzuki, Osaki, Matsutani, & Ishikawa, 2022), and it has been reported as the dominant species by the 20th day of ripening in "Torta del Casar" and "Queso de la Serena" cheeses (Merchan et al., 2022). In light of these findings, the high polyamine-producing *H. paralvei* FB315 isolated in this study shows promise for application in the food industry. The type strain *H. paralvei* ATCC29927^T accumulated higher concentrations of putrescine in its culture supernatant than *H. paralvei* FB315 (Fig. 1A and B). However, as its source of isolation is a human clinical specimen (Huys, Cnockaert, Abbott, Janda, & Vandamme, 2010) its application in food is not feasible. Conversely, *H. paralvei* FB315, isolated from a food source in this study, is considered more advantageous than the type strain for food-related applications.

When *H. paralvei* FB315 was cultivated in MRS medium supplemented with 1 mM ornithine or agmatine, the concentration of putrescine in the culture supernatant increased in both media. These results suggest that *H. paralvei* FB315 has the capacity to produce putrescine from both ornithine and agmatine, precursors of putrescine.

Notably, *H. paralvei* FB315 persisted in its putrescine export into the culture supernatant even during the stationary phase and subsequent decrease in OD_{600}, ultimately resulting in the accumulation of approximately 2 mM of putrescine in the culture supernatant at 122 h after inoculation (Fig. 2). In contrast, strains previously reported to accumulate high concentrations of putrescine in the culture supernatant, such as *Latilactobacillus curvatus* KP 3–4, *Staphylococcus epidermidis* FB146, and *Levilactobacillus brevis* FB215, showed a cessation in the increase of putrescine concentration at a certain point (Ami et al., 2023). These observations suggest that *H. paralvei* FB315 possesses a unique putrescine production capability.

Appendix A. Supporting information

Supplementary data associated with this article can be found in the online version at doi:10.1016/bs.mie.2025.02.004.

References

Al-Habsi, M., Chamoto, K., Matsumoto, K., Nomura, N., Zhang, B., Sugiura, Y., ... Honjo, T. (2022). Spermidine activates mitochondrial trifunctional protein and improves antitumor immunity in mice. *Science (New York, N. Y.), 378*(6618), eabj3510. https://doi.org/10.1126/science.abj3510.

Ami, Y., Kodama, N., Umeda, M., Nakamura, H., Shirasawa, H., Koyanagi, T., & Kurihara, S. (2023). *Levilactobacillus brevis* with high production of putrescine isolated from blue cheese and its application. *International Journal of Molecular Sciences, 24*(11), https://doi.org/10.3390/ijms24119668.

Bourdichon, F., Casaregola, S., Farrokh, C., Frisvad, J. C., Gerds, M. L., Hammes, W. P., ... Hansen, E. B. (2012). Food fermentations: Microorganisms with technological beneficial use. *International Journal of Food Microbiology, 154*(3), 87–97. https://doi.org/10.1016/j.ijfoodmicro.2011.12.030.

Das, R., & Kanungo, M. S. (1982). Activity and modulation of ornithine decarboxylase and concentrations of polyamines in various tissues of rats as a function of age. *Experimental Gerontology, 17*(2), 95–103. https://doi.org/10.1016/0531-5565(82)90042-0.

Eisenberg, T., Knauer, H., Schauer, A., Buttner, S., Ruckenstuhl, C., Carmona-Gutierrez, D., ... Madeo, F. (2009). Induction of autophagy by spermidine promotes longevity. *Nature Cell Biology, 11*(11), 1305–1314. https://doi.org/10.1038/ncb1975.

Hirano, R., Kume, A., Nishiyama, C., Honda, R., Shirasawa, H., Ling, Y., ... Kurihara, S. (2022). Putrescine production by *Latilactobacillus curvatus* KP 3-4 isolated from fermented foods. *Microorganisms, 10*(4), https://doi.org/10.3390/microorganisms10040697.

Huys, G., Cnockaert, M., Abbott, S. L., Janda, J. M., & Vandamme, P. (2010). *Hafnia paralvei* sp. nov., formerly known as *Hafnia alvei* hybridization group 2. *International Journal of Systematic and Evolutionary Microbiology, 60*(Pt 8), 1725–1728. https://doi.org/10.1099/ijs.0.018606-0.

Irlinger, F., Yung, S. A., Sarthou, A. S., Delbes-Paus, C., Montel, M. C., Coton, E., ... Helinck, S. (2012). Ecological and aromatic impact of two Gram-negative bacteria (Psychrobacter celer and *Hafnia alvei*) inoculated as part of the whole microbial community of an experimental smear soft cheese. *International Journal of Food Microbiology, 153*(3), 332–338. https://doi.org/10.1016/j.ijfoodmicro.2011.11.022.

Kalac, P., & Krausová, P. (2005). A review of dietary polyamines: Formation, implications for growth and health and occurrence in foods. *Food Chemistry, 90*(1-2), 219–230. https://doi.org/10.1016/j.foodchem.2004.03.044.

Kibe, R., Kurihara, S., Sakai, Y., Suzuki, H., Ooga, T., Sawaki, E., ... Matsumoto, M. (2014). Upregulation of colonic luminal polyamines produced by intestinal microbiota delays senescence in mice. *Scientific Reports, 4*, 4548. https://doi.org/10.1038/srep04548.

Matsumoto, M., Kurihara, S., Kibe, R., Ashida, H., & Benno, Y. (2011). Longevity in mice is promoted by probiotic-induced suppression of colonic senescence dependent on upregulation of gut bacterial polyamine production. *PLoS One, 6*(8), e23652. https://doi.org/10.1371/journal.pone.0023652.

Merchan, A. V., Ruiz-Moyano, S., Hernandez, M. V., Martin, A., Lorenzo, M. J., & Benito, M. J. (2022). Characterization of autochthonal *Hafnia* spp. strains isolated from Spanish soft raw ewe's milk PDO cheeses to be used as adjunct culture. *International Journal of Food Microbiology, 373*, 109703. https://doi.org/10.1016/j.ijfoodmicro.2022.109703.

Morales, P., Fernandez-Garcia, E., & Nunez, M. (2003). Caseinolysis in cheese by *Enterobacteriaceae* strains of dairy origin. *Letters in Applied Microbiology, 37*(5), 410–414. https://doi.org/10.1046/j.1472-765x.2003.01422.x.

Nakamura, A., Kurihara, S., Takahashi, D., Ohashi, W., Nakamura, Y., Kimura, S., ... Hase, K. (2021). Symbiotic polyamine metabolism regulates epithelial proliferation and macrophage differentiation in the colon. *Nature Communications, 12*(1), 2105. https://doi.org/10.1038/s41467-021-22212-1.

Nishimura, K., Shiina, R., Kashiwagi, K., & Igarashi, K. (2006). Decrease in polyamines with aging and their ingestion from food and drink. *Journal of Biochemistry, 139*(1), 81–90. https://doi.org/10.1093/jb/mvj003.

Ohishi, H., Odoko, M., Grzeskowiak, K., Hiyama, Y., Tsukamoto, K., Maezaki, N., ... Nakatani, K. (2008). Polyamines stabilize left-handed Z-DNA: using X-ray crystallographic analysis, we have found a new type of polyamine (PA) that stabilizes left-handed Z-DNA. *Biochemical and Biophysical Research Communications, 366*(2), 275–280. https://doi.org/10.1016/j.bbrc.2007.10.161.

Oka, T., & Perry, J. W. (1974). Spermidine as a possible mediator of glucocorticoid effect on milk protein synthesis in mouse mammary epithelium in vitro. *The Journal of Biological Chemistry, 249*(23), 7647–7652. https://www.ncbi.nlm.nih.gov/pubmed/4436330.

Okamoto, A., Sugi, E., Koizumi, Y., Yanagida, F., & Udaka, S. (1997). Polyamine content of ordinary foodstuffs and various fermented foods. *Bioscience, Biotechnology, and Biochemistry, 61*(9), 1582–1584. https://doi.org/10.1271/bbb.61.1582.

Shirasawa, H., Nishiyama, C., Hirano, R., Koyanagi, T., Okuda, S., Takagi, H., & Kurihara, S. (2023). Isolation of the high polyamine-producing bacterium *Staphylococcus epidermidis* FB146 from fermented foods and identification of polyamine-related genes. *Bioscience of Microbiota, Food and Health, 42*(1), 24–33. https://doi.org/10.12938/bmfh.2022-011.

Soda, K., Dobashi, Y., Kano, Y., Tsujinaka, S., & Konishi, F. (2009). Polyamine-rich food decreases age-associated pathology and mortality in aged mice. *Experimental Gerontology, 44*(11), 727–732. https://doi.org/10.1016/j.exger.2009.08.013.

Soda, K., Kano, Y., Chiba, F., Koizumi, K., & Miyaki, Y. (2013). Increased polyamine intake inhibits age-associated alteration in global DNA methylation and 1,2-dimethylhydrazine-induced tumorigenesis. *PLoS One, 8*(5), e64357. https://doi.org/10.1371/journal.pone.0064357.

Soda, K., Kano, Y., Sakuragi, M., Takao, K., Lefor, A., & Konishi, F. (2009). Long-term oral polyamine intake increases blood polyamine concentrations. *Journal of Nutritional Science and Vitaminology (Tokyo), 55*(4), 361–366. https://doi.org/10.3177/jnsv.55.361.

Sugiyama, Y., Nara, M., Sakanaka, M., Gotoh, A., Kitakata, A., Okuda, S., & Kurihara, S. (2017). Comprehensive analysis of polyamine transport and biosynthesis in the dominant human gut bacteria: Potential presence of novel polyamine metabolism and transport genes. *The International Journal of Biochemistry & Cell Biology, 93*, 52–61. https://doi.org/10.1016/j.biocel.2017.10.015.

Sugiyama, Y., Ohta, H., Hirano, R., Shimokawa, H., Sakanaka, M., Koyanagi, T., & Kurihara, S. (2020). Development of a new chromogenic method for putrescine quantification using coupling reactions involving putrescine oxidase. *Analytical Biochemistry, 593*, 113607. https://doi.org/10.1016/j.ab.2020.113607.

Unno, R., Suzuki, T., Osaki, Y., Matsutani, M., & Ishikawa, M. (2022). Causality verification for the correlation between the presence of nonstarter bacteria and flavor characteristics in soft-type ripened cheeses. *Microbiol Spectr, 10*(6), e0289422. https://doi.org/10.1128/spectrum.02894-22.

Van Hellemond, E. W., van Dijk, M., Heuts, D. P., Janssen, D. B., & Fraaije, M. W. (2008). Discovery and characterization of a putrescine oxidase from *Rhodococcus erythropolis* NCIMB 11540. *Applied Microbiology and Biotechnology, 78*(3), 455–463. https://doi.org/10.1007/s00253-007-1310-4.

Yoshida, M., Kashiwagi, K., Shigemasa, A., Taniguchi, S., Yamamoto, K., Makinoshima, H., ... Igarashi, K. (2004). A unifying model for the role of polyamines in bacterial cell growth, the polyamine modulon. *The Journal of Biological Chemistry, 279*(44), 46008–46013. https://doi.org/10.1074/jbc.M404393200.

CHAPTER TWENTY THREE

Methylated polyamines derivatives and antizyme-related effects

Maxim A. Khomutov, Arthur I. Salikhov, Olga A. Smirnova, Vladimir A. Mitkevich, and Alex R. Khomutov[*]

Engelhardt Institute of Molecular Biology, Russian Academy of Sciences, Moscow, Russia
[*]Corresponding author. e-mail address: alexkhom@list.ru

Contents

1. Introduction	408
2. Rationale	408
2.1 MeSpds differentially modulate (mOAZ1)$_2$-polyamine complex formation	409
2.2 Electrophoresis of mOAZ1-polyamine complex	411
3. Synthesis of research instruments (C-methylated spermidines)	411
3.1 Synthesis of N-(benzyloxycarbonyl)-aminoalcohols (2a-2c)	413
3.2 Synthesis of N-Cbz-protected C-methylated analogs of Spd (4a-4c)	414
3.3 Synthesis of C-methylated spermidine trihydrochloride (5-7)	416
4. Detection of (mOAZ1)$_2$-polyamine complex formation using ITC	418
4.1 Equipment	418
4.2 Reagents	418
4.3 Procedure	418
4.4 Notes	419
5. Electrophoresis of mOAZ1–polyamine complex	419
5.1 Equipment	419
5.2 Reagents	419
5.3 Procedure	419
6. Conclusion	420
Acknowledgements	420
References	421

Abstract

Antizyme is a key regulator of polyamine homeostasis, and the biosynthesis of this short-lived protein is induced in response to the increase of the intracellular polyamine concentration. Once synthesized, antizyme inhibits polyamine transport and directs the ODC subunit to the 26S proteasome, that normalize the polyamine pool in the cell. Here we demonstrated that polyamines induce dimerization of full-length mouse antizyme with the formation of (antizyme)$_2$-polyamine complex. This can be modulated by C-methylated analogues of spermidine and functionally active 2-methylspermidine turned to be a very poor inducer unlike spermidine and its other C-methylated analogues. The protocols for gram-scale synthesis of C-methylated spermidines and for detecting antizyme dimerization using isothermal titration calorimetry and electrophoresis are described.

1. Introduction

The biogenic polyamines spermine (Spm) and spermidine (Spd) are present in all eukaryotic cells in μM–mM concentrations and vitally important for cell growth and viability. Polyamine homeostasis is strictly controlled and one of the key downregulators of the polyamine pool is a short-living protein, ornithine decarboxylase antizyme (OAZ). Antizyme proteins family includes three paralogs – OAZ1, OAZ2, and OAZ3, where OAZ1 is the predominant one (Kahana, 2018). OAZ1 is expressed *via* a polyamine-stimulated +1 ribosomal frameshift of OAZ1 mRNA that is required for the synthesis of functionally active full-length protein. Once synthesized, OAZ1 binds to an ODC subunit and targets it for the ubiquitin-independent degradation by the 26S proteasome (Murakami et al., 1992). In addition, OAZ1 inhibits the uptake of polyamines (Kahana, 2018). Such a unique combination of OAZ1 functions allows to normalize intracellular levels of polyamine rapidly.

OAZ is synthesized and function in the cells with the elevated polyamine pool. However, previous investigations of OAZ and its functions were performed, as a rule, in the absence of polyamines. We have demonstrated for the first time that full-length mouse OAZ1 (mOAZ1) undergoes dimerization in the presence of polyamines with the formation of the (mOAZ1)$_2$-polyamine complex (Hyvönen et al., 2022). The mOAZ1 dimerization can be modulated by functionally active C-methylated Spd mimetics (MeSpds) moving the methyl group along their Spd backbone. 2-MeSpd was a poor inducer of the OAZ1 dimerization on the contrary to 1-MeSpd, 3-MeSpd and Spd. Notably, the efficiency of the (OAZ1)$_2$-polyamine complex formation correlates with the ability of the OAZ1 to inhibit polyamine uptake in DU145 cells (Hyvönen et al., 2022). Respectively, the use of MeSpds as a research tool discloses new avenues for the investigation of OAZ1 and OAZ1-related effects.

2. Rationale

The interaction of full-length mouse OAZ1 (mOAZ1) with natural polyamines and their analogs was studied using isothermal titration calorimetry

(ITC). We found that Spd, Spm and N^1,N^{11}-diethylnorspermine (DENSpm) bind to mOAZ1 with His-tag at the C-terminus (mOAZ1–6XHis-tag), whereas under the conditions used, putrescine (Put) did not. The stoichiometry of polyamine binding to mOAZ1 was 1:2, *i.e.*, the dimeric (mOAZ1)$_2$-polyamine complex was formed (Fig. 1).

The association constants (K_a) for the complexes of Spm and DENSpm with mOAZ1–6XHis-tag were about threefold lower than that for Spd. The thermodynamic parameters of complex formation (ΔH and $T\Delta S$ ratio) were virtually similar for all polyamines, indicating the same character of their binding with the protein (Table 1). To evaluate the effect of His-tag on mOAZ1 dimerization, N-terminally His-tagged mOAZ1 (6XHis-tag-mOAZ1) and mOAZ1 with N-terminal His-tag cleaved with thrombin were used. Interestingly, in the case of 6XHis-tag-mOAZ1 no binding to Spm was observed. This may indicate on the involvement of an unstructured N-terminus of mOAZ1 in the dimerization phenomenon. Moreover, mOAZ1 formed tighter complex with Spm compared to mOAZ1–6XHis-tag.

Hence, it was shown for the first time that polyamines can induce the dimerization of mOAZ1 with the formation of (mOAZ1)$_2$-polyamine complex.

2.1 MeSpds differentially modulate (mOAZ1)$_2$-polyamine complex formation

Having discovered the mOAZ1 dimerization phenomenon, we attempted to modulate the dimerization process using functionally active mimetics of

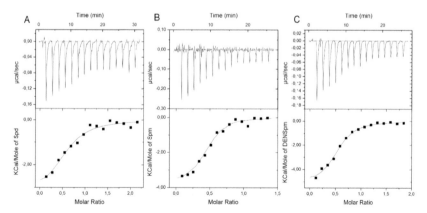

Fig. 1 Interaction of mOAZ1–6XHis-tag with polyamines measured by ITC. Titration curves (upper panels) and binding isotherms (lower panels) for mOAZ1 binding to (A) Spd, (B) Spm, (C) DENSpm. *Data from Hyvönen et al. (2022).*

Table 1 Thermodynamic parameters of the polyamines binding mOAZ1–6XHis-tag, 6XHis-tag-mOAZ1, and mOAZ1 determined by ITC.[a]

Polyamine	Stoichiometry of mOAZ1/polyamine complex	K_a[b], M^{-1}	ΔH[d], kcal/mol	$T\Delta S$[e], kcal/mol	K_d[c], μM
mOAZ1 −6XHis-tag					
Spd	2:1	2.8×10^5	−3.3	4.2	3.6
Spm	2:1	9.2×10^5	−3.7	4.5	1.1
DENSpm	2:1	7.5×10^5	−5.1	3.0	1.3
Put	No binding[f]				
6XHis-tag-mOAZ1					
Spm	No binding[f]				
mOAZ1					
Spm	2:1	2.7×10^6	−3.6	5.5	0.37

[a] All the measurements were performed three to five times.
[b] K_a – affinity constant; standard deviation did not exceed ±25 %.
[c] K_d – dissociation constant; calculated as $1/K_a$ from experimentally determined K_a
[d] ΔH – enthalpy variation; standard deviation did not exceed ±20 %.
[e] $T\Delta S$ – entropy variation; calculated from the equation $T\Delta S = \Delta H + RT \ln K_a$.
[f] No binding means that interaction was not detected in experimental conditions. The concentrations used did not allow to detect $K_d > 50$ μM.
Adapted from Hyvönen et al. (2022); Khomutov et al. (2023).

Spd, namely C-MeSpds. Surprisingly, the position of the methyl group in the Spd analog crucially influenced the ability to bind to and to induce the dimerization of mOAZ1-6XHis-tag (Fig. 2).

The interaction of mOAZ1–6XHis-tag with 1-MeSpd or with 3-MeSpd was characterized by the K_d values, which were close to that observed for Spd, i.e., about 10^{-6} M. However, the binding of 2-MeSpd to such mOAZ1 was weak (the concentrations used did not allow to detect $K_d > 50$ μM) and the formation of (mOAZ1)$_2$-2-MeSpd complex was not detected (Table 2).

Therefore, the formation of (mOAZ1)$_2$-polyamine complex can be regulated by moving the methyl group along the backbone of functionally active C-methylated Spd mimetics.

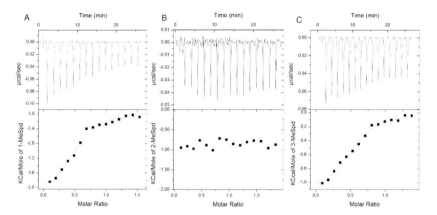

Fig. 2 Interaction of mOAZ1–6XHis-tag with MeSpds measured by ITC. Titration curves (upper panels) and binding isotherms (lower panels) for the binding of the protein to (**A**) 1-MeSpd, (**B**) 2-MeSpd, and (**C**) 3-MeSpd. *Data from Hyvönen et al. (2022).*

2.2 Electrophoresis of mOAZ1-polyamine complex

The formation of (mOAZ1)$_2$-polyamine complexes was also investigated for mOAZ1–6XHis-tag by the electrophoresis on a 12% polyacrylamide gel lacking SDS.

Dimeric and monomeric complexes of mOAZ1–6XHis-tag with Spm, Spd, 1-MeSpd and 3-MeSpd, but not with Put, were detected (Fig. 3A and B). In the case of 2-MeSpd, not even the monomeric (mOAZ1–6XHis-tag)-2-MeSpd complex was detected under these conditions, and only the band corresponding to mOAZ1–6XHis-tag itself was observed (Fig. 3B). These results are in line with ITC data.

3. Synthesis of research instruments (C-methylated spermidines)

Commercially available amino alcohols (**1a-1c**) were converted to the corresponding *N*-(benzyloxycarbonyl) derivatives (**2a-2c**) with excellent yields. Compounds (**2a-2c**) were converted to *O*-methanesulfonates (**3a-3c**), which were used without isolation to alkylate the excess of corresponding diamine yielding *N*-Cbz derivatives of *C*-methylated Spds (**4a-4c**). Hydrogenation of compounds (**4a-4c**) over Pd/C and subsequent recrystallization provided target *C*-methylated Spds (**5–7**) in gram scale with overall yields of 40–70% (Scheme 1).

Table 2 Thermodynamic parameters of the MeSpds binding to mOAZ1–6XHis-tag determined by ITC.[a]

Polyamine	Stoichiometry of mOAZ1/polyamine complex	K_a[b], M^{-1}	ΔH[d], kcal/mol	$T\Delta S$[e], kcal/mol	K_d[c], µM
mOAZ1-6XHis-tag					
H$_2$N~~~N(H)~~~NH$_2$ Spd	2:1	2.8×10^5	−3.3	4.2	3.6
H$_2$N~~~N(H)~~~NH$_2$ 1-MeSpd	2:1	5.4×10^5	−2.2	5.8	1.9
H$_2$N~~~N(H)~~~NH$_2$ 2-MeSpd	No binding[f]				
H$_2$N~~~N(H)~~~NH$_2$ 3-MeSpd	2:1	4.0×10^5	−1.2	6.6	2.5

[a] All the measurements were performed three to five times in buffer containing 25 mM potassium phosphate, pH 7.5, 150 mM KCl, 25 % (v/v) glycerol, 1 mM 2-mercaptoethanol, and 1 mM PMSF.
[b] K_a – affinity constant; standard deviation did not exceed ±25 %.
[c] K_d – dissociation constant; calculated as $1/K_a$ from experimentally determined K_a.
[d] ΔH – enthalpy variation; standard deviation did not exceed ±20 %.
[e] $T\Delta S$ – entropy variation; calculated from the equation $T\Delta S = \Delta H + RT \ln K_a$.
[f] No binding means that interaction was not detected in experimental conditions. The concentrations used did not allow to detect $K_d > 50$ µM.
Adapted from Hyvönen et al. (2022).

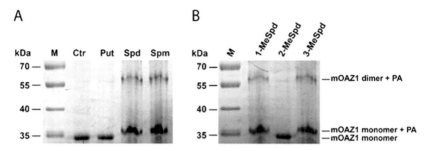

Fig. 3 Formation of (mOAZ1–6XHis-tag)₂-polyamine complex. (A) Spd and Spm induce dimerization of mOAZ1–6XHis-tag. (B) 1-MeSpd and 3-MeSpd, but not 2-MeSpd, bind to the protein and induce its dimerization. *Data from Hyvönen et al. (2022).*

Scheme 1 *i*-CbzCl/NaHCO₃/Na₂CO₃/THF/H₂O; *ii*-MsCl/Et₃N/THF; *iii*-NH₂(CH₂)₄NH₂/THF for 1-MeSpd and 3-MeSpd; NH₂CH₂CH(CH₃)CH₂NH₂/THF for 2-MeSpd; *iv*-H₂/Pd/MeOH/AcOH; *v*-HCl/EtOH.

3.1 Synthesis of *N*-(benzyloxycarbonyl)-aminoalcohols (2a-2c)

3.1.1 Equipment
- Overhead stirrer (Heidolph, Schwabach, Germany)
- Rotary evaporator (Buchi, Flawil, Switzerland)
- TLC plates Kieselgel 60 F$_{254}$ (Merck, Darmstadt, Germany)

3.1.2 Reagents
- 3-Aminobutan-1-ol (**1a**) (Santa Cruz Biotechnology, Dallas, TX, USA)
- Chemicals, all available from Sigma Aldrich, St. Louis, MO, USA:
- 4-Aminobutan-1-ol (**1b**)
- 4-Aminobutan-2-ol (**1c**)
- Benzyl chloroformate (Cbz-Cl)
- Tetrahydrofuran (THF)
- Chloroform (CHCl₃)
- Diethyl ether (Et₂O)
- Hexane

- Na$_2$CO$_3$
- NaHCO$_3$
- NaCl
- MgSO$_4$

3.1.3 General procedure

- Prepare a suspension of 2.65 g (25 mmol) Na$_2$CO$_3$, 2.1 g (25 mmol) NaHCO$_3$ and 25 mmol of the corresponding aminoalcohol (**1a-1c**) in a mixture of 25 mL H$_2$O and 12 mL of abs. THF. Cool the resulting suspension to 4 °C.
- Add 3.5 mL of Cbz-Cl in five equal portions every 20 min to the vigorously overhead stirred suspension. After adding the final portion of Cbz-Cl continue stirring for 1 h at 4 °C and then 3 h at room temperature.
- Filter off the precipitate and separate the organic phase. Extract the aqueous phase with abs. CHCl$_3$ (2 × 20 mL) and concentrate the combined organic extracts *in vacuo*.
- Dissolve the residue in 75 mL of CHCl$_3$ and wash sequentially with 1 M HCl (2 × 20 mL), H$_2$O (2 × 10 mL), 1 M NaHCO$_3$ (10 mL), H$_2$O (10 mL), brine (2 × 15 mL) and dry the solution over MgSO$_4$.
- Filter off MgSO$_4$ and evaporate the filtrate to dryness *in vacuo*. Triturate the residue with 80 mL of an Et$_2$O/hexane (1:3) mixture and leave the resulting suspension overnight at 4 °C.
- Filter the solid off and dry it *in vacuo* at 1 Torr to obtain *N*-Cbz-aminoalcohols (**2a-2c**) as colorless crystals.

N-(benzyloxycarbonyl)-3-amino-butanol-1 (**2a**) was obtained as described above starting from **1a** to give **2a** (4.79 g, 86 %) as colorless crystals. R_f 0.29 (CH$_2$Cl$_2$-MeOH, 98:2), mp 60–61 °C.

N-(benzyloxycarbonyl)-4-amino-butanol-1 (**2b**) was obtained as described above starting from **1b** to give **2b** (4.68 g, 84 %) as colorless crystals. R_f 0.23 (CH$_2$Cl$_2$-MeOH, 98:2), mp 82–84 °C.

N-(benzyloxycarbonyl)-4-amino-butanol-2 (**2c**) was obtained as described above starting from **1c** to give **2c** (4.9 g, 88 %) as colorless oil that solidifies at 4 °C. R_f 0.24, (CH$_2$Cl$_2$-MeOH, 98:2).

3.2 Synthesis of *N*-Cbz-protected *C*-methylated analogs of Spd (4a-4c)

3.2.1 Equipment
- Magnetic stirrer (Heidolph, Schwabach, Germany)
- Rotary evaporator (Buchi, Flawil, Switzerland)

- Glass column for flash chromatography (Ø 6.2 cm)
- UVICORD
- TLC plates Kieselgel 60 F$_{254}$ (Merck, Darmstadt, Germany)

3.2.2 Reagents
- 2-methyl-1,3-diaminopropane (TCI, Japan)
- N-Cbz-aminoalcohols **2a-2c**
- Chemicals, all available from Sigma Aldrich, St. Louis, MO, USA:
- Methanesulfonyl chloride (Ms-Cl)
- Tetrahydrofuran (THF)
- Triethylamine (Et$_3$N)
- Toluene
- 1,4-diaminobutane (freshly distilled, bp 79 °C/37 Torr)
- 2-methyl-1,3-diaminopropane (freshly distilled, bp 59 °C/25 Torr)
- Dichloromethane (CH$_2$Cl$_2$)
- 1,4-dioxane
- Silica gel Kieselgel (40–63 µm)
- NaOH
- NaCl
- K$_2$CO$_3$
- 25 % NH$_4$OH
- P$_2$O$_5$

3.2.3 General procedure
- Dissolve 20 mmol of the corresponding N-Cbz-aminoalcohol (**2a-2c**) and 2.5 mL (25 mmol) of Et$_3$N in 50 mL of abs. THF.
- Equip the flask with a stirring bar, thermometer and a pressure-equalizing dropping funnel and place the flask in an ice bath. Stir the solution until it cools to 4 °C.
- Dissolve 1.55 mL (20 mmol) of Ms-Cl in 10 mL of abs. THF and fill the dropping funnel with it. Add the Ms-Cl solution dropwise within 20 min maintaining temp. below 7 °C, stir the reaction mixture for 1 h at 4 °C and then for 3 h at room temperature.
- Filter off the precipitate and evaporate the filtrate to dryness *in vacuo*. Co-evaporate the residue with toluene (2 × 20 mL) to remove the excess of Et$_3$N.
- Dissolve the residue in 20 mL of abs. THF, cool to 0 °C and add cooled to 0 °C solution of 0.2 mol of corresponding diamine in 20 mL of abs. THF.

- Keep the reaction mixture for 12 h at 4 °C and then for 24 h at room temperature. In the case of **3c** keep the reaction mixture for an additional 16 h at 37 °C.
- Filter off the precipitate and concentrate the filtrate *in vacuo*. Mix the residue with 20 mL of 2 M NaOH, separate the oil and extract the aq. phase with CH_2Cl_2 (3 ×20 mL). Wash combined organic extracts sequentially with H_2O (10 mL), brine (2 ×10 mL) and dry over K_2CO_3.
- Filter off K_2CO_3 and evaporate filtrate to dryness *in vacuo*. Dissolve the residue in 10 mL of a mixture of 1,4-dioxane/25 % NH_4OH (95:5), divide the resulting solution into two equal parts and purify each portion on 180 g of Kieselgel (40–63 µm) eluting with 1,4-dioxane/25 % NH_4OH (95:5).
- Concentrate the fractions containing the product *in vacuo* and dry the residue *in vacuo* over P_2O_5 to obtain compounds **4a-4c**.

N^2-*(benzyloxycarbonyl)-2,9-diamino-5-azanonane* (**4a**) was obtained as described above starting from **2a** (2.67 g, 12 mmol) affording **4a** (2.56 g, 73 %) as a colorless viscous oil, R_f 0.14 (1,4-dioxane-25 % NH4OH = 95:5).

N^8-*(benzyloxycarbonyl)-1,8-diamino-2-methyl-4-azaoctane* (**4b**) was obtained as described above starting from **2b** (3.4 g, 15 mmol) affording **4b** (3.12 g, 71 %) as a colorless viscous oil, R_f 0.14 (1,4-dioxane-25 % NH_4OH = 95:5).

N^1-*(benzyloxycarbonyl)-1,8-diamino-3-methyl-4-azaoctane* (**4c**) was obtained as described above starting from **2c** (2.9 g, 13 mmol) affording **4c** (2.41 g, 63 %) as a colorless viscous oil, R_f 0.26 (1,4-dioxane-25 % NH_4OH = 95:5).

3.3 Synthesis of *C*-methylated spermidine trihydrochloride (5-7)

3.3.1 Equipment
- Magnetic stirrer (Heidolph, Schwabach, Germany)
- Rotary evaporator (Buchi, Flawil, Switzerland)
- TLC plates Kieselgel 60 F_{254} (Merck, Darmstadt, Germany)

3.3.2 Reagents
- *N*-Cbz-spermidine derivatives (**4a-4c**)
- Chemicals, all available from Sigma Aldrich, St. Louis, MO, USA:
- Methanol (MeOH)
- Glacial acetic acid (AcOH)
- Palladium on carbon (Pd/C)
- Ethanol (EtOH)
- 37 % HCl

3.3.3 General procedure

- Dissolve 10 mmol of corresponding N-(benzyloxycarbonyl)-protected triamine **4a–4c** in 35 mL of a mixture of AcOH-MeOH (1:1), add 1 mL of suspension of Pd/C in abs. MeOH and carry out hydrogenation at atmospheric pressure.
- Filter off the catalyst and evaporate the filtrate to dryness *in vacuo*. Dissolve the residue in 20 mL of abs. EtOH and dilute with 5 mL of 37 % HCl. Evaporate the resulting solution to dryness then co-evaporate the residue twice with 20 mL of EtOH.
- Recrystallize residue from a MeOH-EtOH mixture to get target compounds **5–7**.

2,9-Diamino-5-azanonane trihydrochloride (**1-MeSpd**) (**5**) was obtained as described above starting from **4a** (2.0 g, 6.8 mmol) that gave **5** (1.62 g, 89 %) as colorless crystals: mp. 190–193 °C, R_f 0.35 (n-BuOH-AcOH-Py-H$_2$O, 4:2:1:2). ^{1}H NMR (300.13 MHz, D$_2$O) δ: 3.53 (m, 1 H, CH$_3$CH); 3.21 (m, 2 H, CH$_2$NH$_2$); 3.15 (m, 2 H, NHCH$_2$); 3.07 (m, 2 H, NHCH$_2$); 2.16 (m, 1 H, CH(CH$_3$)CH$_2$); 2.02 (m, 1 H, CH(CH$_3$)CH$_2$); 1.86–1.74 (m, 4 H, CH$_2$CH$_2$CH$_2$CH$_2$); 1.36 (d, 3 H, *J* 6.5 Hz, CH$_3$). ^{13}C NMR (75.43 MHz, D$_2$O) δ: 50.03; 48.36; 46.89; 41.85; 33.34; 26.85; 25.69; 20.38. Found, %: C 35.86, H 9.08, N 15.73. C$_8$H$_{24}$N$_3$Cl$_3$. Calculated, %: C 35.76, H 9.00, N 15.64.

1,8-Diamino-2-methyl-4-azaoctane trihydrochloride (**2-MeSpd**) (**6**) was obtained as described above starting from **4b** (2.5 g, 8.53 mmol) that gave **6** (2.05 g, 89 %) as colorless crystals: mp. 197–199 °C (lit.: 189–192 °C), R_f 0.22 (n-butanol-1-AcOH-Py-H$_2$O = 4:2:1:2). ^{1}H NMR (300.13 MHz, D$_2$O) δ: 3.20–3.09 (m, 4 H, CHCH$_2$NH$_2$ + CH$_2$NH$_2$) 3.08–2.87 (m, 4 H, CH$_2$NHCH$_2$); 2.42–2.24 (m, 1 H, CH(CH$_3$)); 1.88–1.68 (m, 4 H, CH$_2$CH$_2$CH$_2$NH$_2$); 1.15 (d, 3 H, *J* 6.8 Hz, CH$_3$). ^{13}C NMR (75.43 MHz, D$_2$O) δ: 51.28, 48.38, 43.03, 39.47, 30.06, 24.58, 23.26, 14.85. Found, %: C 35.76; H 9.17; N 15.80. C$_8$H$_{24}$Cl$_3$N$_3$. Calculated, %: C 35.77; H 9.00; N 15.64.

1,8-Diamino-3-methyl-4-azaoctane trihydrochloride (**3-MeSpd**) (**7**) was obtained as described above starting from **4c** (2.0 g, 6.8 mmol) that gave **7** (1.61 g, 88 %) as colorless crystals: mp. 233–235 °C (lit.: 231–232 °C), R_f 0.24 (n-butanol-1-AcOH-Py-H$_2$O, 4:2:1:2). ^{1}H NMR (300.13 MHz, D$_2$O) δ: 3.50–3.37 (m, 1 H, CH(CH$_3$)); 3.22–2.99 (m, 6 H, H$_2$NCH$_2$(CH$_2$)$_2$CH$_2$NH + CH$_2$NH$_2$); 2.27–2.11 (m, 1 H, CHCH$_2$); 2.03–1.86 (m, 1 H, CHCH$_2$); 1.83–1.70 (m, 4 H, CH$_2$CH$_2$CH$_2$CH$_2$);

1.36 (d, 3 H, J 6.7 Hz, CH(C$\underline{H}_3$)). ^{13}C NMR (75.43 MHz, D$_2$O) δ: 52.69, 44.90, 39.49, 36.56, 30.88, 24.63, 23.62, 15.75. Found, %: C 35.57; H 9.06; N 15.53. C$_8$H$_{24}$Cl$_3$N$_3$. Calculated, %: C 35.77; H 9.00; N 15.64.

4. Detection of (mOAZ1)$_2$-polyamine complex formation using ITC

4.1 Equipment
- MicroCal iTC200 (GE Healthcare, Chicago, IL, USA)

4.2 Reagents
- mOAZ1-6XHis-tag (Hyvönen et al., 2022)
- 6XHis-tag-mOAZ1 (Khomutov et al., 2023)
- mOAZ1 (Khomutov et al., 2023)
- Polyamine ligands (1-MeSpd, 2-MeSpd or 3-MeSpd, Section 3, Scheme 1)
- Chemicals, all available from Sigma Aldrich, St. Louis, MO, USA:
- 1,4-diaminobutane dihydrochloride (Put)
- Spermidine trihydrochloride (Spd)
- Spermine tetrahydrochloride (Spm)
- N^1,N^{11}-Diethylnorspermine tetrahydrochloride (DENSpm)
- Glycerol
- 2-mercaptoethanol
- Phenylmethane sulfonyl fluoride (PMSF)
- KH$_2$PO$_4$
- K$_2$HPO$_4$
- KCl

4.3 Procedure
- Prepare the buffer containing 25 mM potassium phosphate (pH 7.5), 150 mM KCl, 50 % (v/v) glycerol, 1 mM 2-mercaptoethanol and 1 mM PMSF
- Prepare 20 μM solution of the protein and 200 μM solution of the corresponding polyamine ligand in the above buffer
- Inject aliquots (2.5 μL) of ligands into the cell containing 0.2 mL of the protein solution in to achieve a complete binding isotherm. Experiment is performed at 31 °C.
- Calculate the thermodynamic parameters of polyamines binding to OAZ1 as described (Petrushanko et al., 2014).

4.4 Notes
- For the precise isothermal titration calorimetry (ITC) experiments only freshly purified samples turned out to be suitable.
- The heat of dilution was measured by injection of the ligand into the buffer solution or by additional injections of ligand after saturation; the values obtained were subtracted from the heat of reaction to obtain the effective heat of binding.

5. Electrophoresis of mOAZ1-polyamine complex
5.1 Equipment
- Vertical gel electrophoresis apparatus: GibcoBRL, model V16 (Thermo Fisher Scientific, Waltham, MA, USA)
- Power supply: model PowerPac 3000 (Bio-Rad, CA, USA)

5.2 Reagents
- mOAZ1-6XHis-tag (Hyvönen et al., 2022)
- Polyamine ligands (1-MeSpd, 2-MeSpd or 3-MeSpd, Section 3, Scheme 1)
- Chemicals, all available from Sigma Aldrich, St. Louis, MO, USA:
- 1,4-diaminobutane dihydrochloride (Put)
- Spermidine trihydrochloride (Spd)
- Spermine tetrahydrochloride (Spm)
- Coomassie R-250
- Bromophenol blue
- Glycerol
- Acrylamide
- Bis-acrylamide
- TEMED
- Ammonium persulfate
- Tris-base
- Glycine
- 37 % HCl

5.3 Procedure
- Incubate the mOAZ1-6XHis-tag (0.4 mg/1 mL) with 200 µM of the corresponding polyamine ligand in 25 mM potassium phosphate buffer (pH 7.5), containing 150 mM KCl, 50 % (v/v) glycerol, 1 mM 2-mercaptoethanol and 1 mM PMSF at room temperature for 30 min

- Quench 2.5 μL of the sample, containing 1 μg of the protein, with 2 μL of loading buffer (200 mM Tris-HCl, pH 6.8; 0.01 % bromophenol blue; 40 % (v/v) glycerol), total volume of the sample 15 μL.
- Apply the quenched sample onto polyacrylamide gel (375 mM Tris-HCl, pH 8.8 for 12 % resolving gel, 62.5 mM Tris-HCl, pH 6.8, 5 % for stacking gel) lacking SDS.
- Carry out the electrophoresis in a running upper buffer (25 mM Tris base, 250 mM glycine and 0.01 % SDS) and lower buffer (25 mM Tris base, 250 mM glycine) for the first 30 min at 50 V, then at 70 V.
- Stain the gel with Coomassie R-250 according standard protocol.

6. Conclusion

It was demonstrated that natural polyamines Spm and Spd, but not Put, induce dimerization of mouse recombinant full-length OAZ1, forming an (OAZ1)$_2$-polyamine complex. It is essential that the N-terminally His-tagged mOAZ1 doesn't dimerize in the presence of Spm, on the contrary with mOAZ1 and C-terminally His-tagged mOAZ1. This may indicate on importance of the mOAZ1 N-terminus for polyamine induced dimerization. The formation of (OAZ1)$_2$-polyamine complex was confirmed with ITC and electrophoresis.

The formation of the (OAZ1)$_2$-polyamine complex can be modulated by functionally active C-methylated spermidine analogs (MeSpds) by changing the position of the methyl group along the Spd backbone. 2-MeSpd was a poor inducer as opposed to Spd, 1-MeSpd, and 3-MeSpd which were good inducers. This might be an indication of the importance of the second position of the Spd chain for the interaction with OAZ1.

The formation of the (OAZ1)$_2$-polyamine complex correlates with cellular functions of OAZ1, *i.e.*, inhibition of polyamine uptake in DU145 cells (Hyvönen et al., 2022). Interestingly, the efficiency of +1 frameshift of OAZ1 mRNA, which is required for the synthesis of full-length protein, can be also modulated by moving the methyl group along the Spd backbone (Hyvönen et al., 2022). These findings offer a new insight into the OAZ1-mediated regulation of the polyamine homeostasis and provide chemical tools to study it.

Acknowledgements

This work was supported by Russian Science Foundation, Grant #25-24-00374.

References

Hyvönen, M. T., Smirnova, O. A., Mitkevich, V. A., Tunitskaya, V. L., Khomutov, M., Karpov, D. S., et al. (2022). Role of polyamine-induced dimerization of antizyme in its cellular functions. *International Journal of Molecular Sciences, 23*, 4614.

Kahana, C. (2018). The antizyme family for regulating polyamines. *Journal of Biological Chemistry, 293*, 18730–18735.

Khomutov, M. A., Salikhov, A. I., Mitkevich, V. A., Tunitskaya, V. L., Smirnova, O. A., Korolev, S. P., et al. (2023). C-Methylated spermidine derivatives: Convenient syntheses and antizyme-related effects. *Biomolecules, 13*, 916.

Murakami, Y., Matsufuji, S., Kameji, T., Hayashi, S., Igarashi, K., Tamura, T., et al. (1992). Ornithine decarboxylase is degraded by the 26S proteasome without ubiquitination. *Nature, 360*, 597–599.

Petrushanko, I. Y., Mitkevich, V. A., Anashkina, A. A., Klimanova, E. A., Dergousova, E. A., Lopina, O. D., et al. (2014). Critical role of gamma-phosphate in structural transition of Na,K-ATPase upon ATP binding. *Scientific Reports, 4*, 5165.

CHAPTER TWENTY FOUR

Novel LC-MS/MS assay to quantify D,L-alpha-difluoromethylornithine (DFMO) in mouse plasma

Matthew A. Swanson[a], Carlye Szarowicz[a], Schuyler T. Pike[a], Chad R. Schultz[b], André S. Bachmann[b], and Thomas C. Dowling[c],*

[a]Department of Biological Sciences, College of Arts and Sciences, Ferris State University, Big Rapids, MI, United States
[b]Department of Pediatrics and Human Development, College of Human Medicine, Michigan State University, Grand Rapids, MI, United States
[c]Department of Pharmaceutical Sciences, College of Pharmacy, Ferris State University, Big Rapids, MI, United States
*Corresponding author. e-mail address: thomasdowling@ferris.edu

Contents

1. Introduction	424
2. Rationale	425
3. Materials and methods	425
3.1 Chemicals and reagents	425
3.2 Standard solutions preparation	426
3.3 Calibration curve, QCs and study sample preparation	426
3.4 Chromatographic and mass spectrometric conditions	426
3.5 Method validation	427
3.6 Application of the method to *in vivo* pharmacokinetic study	428
4. Results	428
4.1 LC-MS/MS method development	428
4.2 Method validation	429
4.3 Assay stability	432
4.4 Application to *in vivo* pharmacokinetic study	433
5. Conclusion	433
Acknowledgments	433
Declaration of competing interest	433
CRediT authorship contribution statement	433
References	434

Abstract

D,L-alpha-difluoromethylornithine (DFMO) is an irreversible inhibitor of ornithine decarboxylase (ODC) that is being investigated to treat cancers such as pediatric neuroblastoma. A novel and sensitive LC-MS/MS assay was developed and validated to quantify DFMO concentrations in support of pre-clinical pharmacokinetic studies in mice. The study was performed using a Shimadzu triple quad LC-MS/MS equipped with an Atlantis HILIC Silica 3 μm 2.1 × 100 mm column, and an isocratic mobile phase (75:25 acetonitrile and 0.2 % formic acid) at a flow rate of 0.5 mL/min. Multiple Reaction Monitoring (MRM) was used to identify the precursor ion (183 *m/z*) with quantification of daughter ions at transitions of 183 > 120.10, 183 > 166.10, and 183 > 80.05. Plasma standards and quality control samples (20 μL) were processed using protein precipitation with cold acetonitrile. The lower limit of detection (LLOQ) was 5 ng/mL. Assay performance was determined from multiple runs (n = 10) with standards ranging from 250–50,000 ng/mL and three levels of quality control (500, 4000, and 40,000 ng/mL). Standard curves were linear with r^2 values between 0.9960 and 0.9999. Quality control samples were stable and exhibited maximum inter-day % bias of ≤3 % and CV% of ≤0.7 %. The assay was successfully applied to an *in vivo* study to determine the pharmacokinetics of DFMO in athymic nu/nu mice.

1. Introduction

D,L-alpha-difluoromethylornithine (DFMO) (Fig. 1) is an irreversible inhibitor of ornithine decarboxylase (ODC), a MYC-target gene (Bachmann & Geerts, 2018; Bello-Fernandez, Packham, & Cleveland, 1993; Metcalf et al., 1978; Nilsson et al., 2005). Inhibition of ODC causes decreased levels of metabolites such as putrescine in the polyamine biosynthetic pathway (Wallick, Gamper, Thorne, Feith, & Takasaki, 2005), S-adenosylmethionine, and thymidine (Witherspoon, Chen, Kopelovich, Gross, & Lipkin, 2013), all of which are involved in cell proliferation and tumor promotion (Casero & Marton, 2007; Casero, Murray Stewart, & Pegg, 2018; Gerner & Meyskens, 2004; Pegg & Feith, 2007). Although DFMO has primarily been utilized and FDA-approved as a treatment for second stage human African trypanosomiasis (African sleeping sickness) (Alirol et al., 2013; Priotto et al., 2008; Priotto et al., 2009; Sjoerdsma & Schechter, 1999) and female hirsutism (Blume-Peytavi & Hahn, 2008; Jackson et al., 2007; Wolf et al., 2007), it has also been under investigation for many years as a cancer treatment option in preventative and active tumor suppression (See NCT03794349) due to its potential role in inhibiting cell proliferation (Manni, Washington, Mauger, Hackett, & Verderame, 2004; Witherspoon et al., 2013). Multiple clinical trials have commenced to determine the effectiveness of DFMO in neuroblastoma

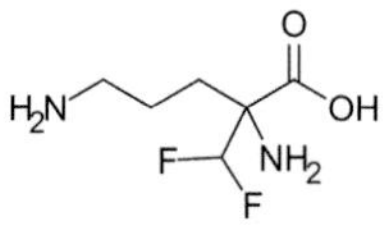

Fig. 1 DFMO molecular structure–molecular weight: 182.17 g/mol.

(Hogarty et al., 2024; Lewis et al., 2020; Moreno, Barone, & DuBois, 2020; Oesterheld et al., 2023; Saulnier Sholler et al., 2015; Sholler et al., 2018) and other cancers including colorectal adenomas (Bachmann & Levin, 2012; Meyskens et al., 2008). Most recently, DFMO was FDA-approved for the treatment and relapse prevention of neuroblastomas (USWN, 2024). Optimized dosing of DFMO using advanced pharmacokinetic approaches is required to further the understanding of its pharmacological actions as well as safe and effective dosing requirements for future treatments of cancer and other diseases, including Bachmann-Bupp syndrome (Bachmann, VanSickle, Michael, Vipond, & Bupp, 2022; Rajasekaran et al., 2021) and diabetes (Sims et al., 2023). Taken together, the recent development DFMO as a therapeutic agent requires a sensitive, reproducible and precise analytic method for quantitation of DFMO content in biological specimens.

2. Rationale

Previously reported DFMO assays have not been optimized for pharmacokinetic analysis. For example, high-performance liquid chromatograph methods have required pre- and postcolumn derivatization and ultraviolet or fluorescence detection (Kumar, Singh, & Kumar, 2013). These techniques are limited by disadvantages including extensive processing time, low recovery, analyte instability, and increased background signal (Cho & Kang, 2018; Yang, Peng, & Wang, 2014). A previously reported liquid chromatography-mass spectrometry (LC-MS/MS) method was limited by the requirement for derivatization using 6aminoquinolyl-N-hydroxysuccinimidyl carbamate (Yang et al., 2014). The present study describes a sensitive and reproducible LC-MS/MS assay to quantify DFMO in mouse plasma without derivatization.

This assay was applied to an *in vivo* pharmacokinetic study of DFMO in an athymic nu/nu mouse model (Schultz, Swanson, Dowling, & Bachmann, 2021).

3. Materials and methods
3.1 Chemicals and reagents

DFMO (analytical grade, Fig. 1) was obtained from Sigma-Aldrich (St. Louis, MO). HPLC grade acetonitrile (ACN), formic acid, and HPLC grade water were obtained from Fisher Scientific. Calibration standards and quality control samples were prepared using blank mouse plasma (sodium citrate) obtained from Valley Biomedical (Winchester, VA).

3.2 Standard solutions preparation

Stock solutions of DFMO were dissolved in water at a base concentration of 1 mg/mL and stored for long-term storage in −80 °C. The stock solution was diluted with blank plasma to prepare an initial 50,000 ng/mL plasma standard of DFMO. Serial dilutions of the 50,000 ng/mL standard were made to achieve final concentrations in plasma: 25,000, 10,000, 5000, 1000, and 250 ng/mL. The quality controls, listed from high to low (QCH, QCM, QCL) were prepared with final concentrations of 40,000 ng/mL, 7500 ng/mL and 500 ng/mL. All standards and quality control (QC) samples were prepared on ice and stored at −80 °C for preservation.

3.3 Calibration curve, QCs and study sample preparation

To prepare for LCMS injection, after thawing, each plasma standard and QC sample was thoroughly vortex-mixed. Sample processing consisted of protein precipitation by adding 300 μL of chilled ACN to 20 μL of spiked plasma. Samples were vortexed for six seconds followed by a centrifugation step using an Eppendorf 5424 for six minutes at 12,000 g. Following centrifugation, 100 μL of supernatant was transferred to an amber glass vial and inserted into the autosampler and maintained at 15° C until LC-MS/MS analysis.

3.4 Chromatographic and mass spectrometric conditions

Quantification of DFMO was achieved using a Shimadzu LC-MS/MS 8040 system with electrospray ionization (ESI) and atmospheric pressure chemical ionization (APCI). The system contained binary pumps (Shimadzu LC-30AD), a degasser unit (Shimadzu DGU-20A$_{5R}$), an autosampler unit (Shimadzu SIL-30AC), a diode array detector (Shimadzu SPD-M30A), a prominence column oven (Shimadzu CTO-20A), and a quantitative electrospray ionization mass spectrometer (Shimadzu LCMS 8040). Shimadzu Lab Solutions® software was used for peak integrations and quantification. The isocratic mobile phase consisted of 75:25 acetonitrile to aqueous with 0.2 % formic acid. Chromatographic separation was accomplished using a Waters Atlantis HILIC Silica 3 μm 2.1 × 100 mm column preceded by an Atlantis HILIC 3 μm VanGuard Cartridge 2.1 × 5 mm. The column was stored in 100 % acetonitrile. The autosampler injection volume was 5 μL per sample with a temperature of 15° C. Needle rinses were performed before and after aspiration using pure acetonitrile, and the rinse pump was washed using reagent alcohols for two seconds.

3.5 Method validation

The method was developed in accordance with the guidances on bioanalytical method validation from the US Food and Drug Administration (FDA) (US Department of Health & Human Services Food & Drug Administration, 2024) and the European Medicines Agency (EMA) (European Medicines Agency, 2022). Validation studies included intra- and inter-day accuracy and precision of DFMO standards and QCs, assay linearity, lower limit of quantitation (LLOQ), lower limit of detection (LLOD), and stability in processing and freeze-thaw cycles.

3.5.1 Linearity

Calibration standards and QCs were processed on different days and analyzed within six hours of processing as described in Section 3.3. Product ions were plotted and combined to form a total ion count (TIC), which was used to quantify the area under the curve. Ten calibration curves were analyzed by the external standard method with weighted linear regression ($1/x^2$). Linear regression fit was determined by the accuracy of the determined concentrations and the correlation coefficient (r^2) value.

3.5.2 Intra- and inter-day precision and accuracy

Intra- and inter-day precision and accuracy was determined by analyzing sets of QCs at each concentration on three different days. Standard deviation, percentage coefficient of variation (CV%) and bias (error %) was determined for each set of QCs. For at least 2/3 of the nominal concentrations reported for samples at each QC level, the acceptable bias was <15 % of the nominal concentration. The acceptable % CV was <15 %, except for the LLOQ, with an acceptable limit of <20 %.

3.5.3 Stability

The stability of DFMO in processed and unprocessed samples was assessed under separate conditions. To assess short-term stability, QCL and QCH samples were processed in triplicate as and either left on the benchtop (room temperature) or refrigerated (4 °C) for 24 h. Freezethaw stability was determined for QCL and QCH samples in triplicate. Here, plasma was subjected to up to three cycles alternating between room temperature and −80 °C. The analyte concentration in all QC samples was determined and compared with the analyte concentration of freshly prepared QC samples.

Stability under each condition was affirmed if the QC concentrations deviated ≤15 % when compared to the nominal value.

3.6 Application of the method to in vivo pharmacokinetic study

The validated assay method was used to quantify DFMO concentrations in samples obtained from an *in vivo* mouse PK study [see reference (Schultz et al., 2021) for additional details]. Briefly, DFMO was administered as a single 500 mg/kg dose *via* oral gavage. Blood was harvested from three animals per group by intracardial puncture at 10 and 30 min, 1, 2, 4, and 6 h post administration of DFMO. Blood samples were centrifuged at 1000 g for 15 min and plasma supernatant was harvested. Plasma samples were immediately stored at −80 °C until processing, as described above. All procedures and animal handling were performed in accordance with the Michigan State University IACUC (approved protocol 07–18-090–00). Plasma samples that exceeded the ULOQ were diluted with blank plasma to quantify within the standard curve concentrations.

4. Results

4.1 LC-MS/MS method development

The method was optimized to obtain maximum DFMO signal. The isocratic mobile phase was delivered at a flow rate of 0.5 mL/min, resulting in a total sample run time of 6 minutes. The collision induced dissociation (CID) gas used was argon with a set pressure of 230 kPa. Nebulizing and drying gas flow was set to 2 and 17 L/min, respectively. The desolvation line (DL) and heating block temperature was set to 250 and 400 °C, respectively. DFMO was ionized in the ESI source in positive MRM mode to obtain the following transitions: 183.00 > 120.10 (Q1 Pre-Bias = −20.0 V, CE = −22.0, Q3 Pre-Bias = −19.0 V); 183.00 > 166.10 (Q1 Pre-Bias = −20.0 V, CE = −15.0, Q3 Pre-Bias = −15.0 V); and 183.00 > 80.05 (Q1 Pre-Bias = −20.0 V, CE = −35.0, Q3 Pre-Bias = −18.0 V). DFMO was initially identified using the first quadrupole (Q1), then product ions were identified using the third quadrupole (Q3) after collision with argon gas. After product ions were identified (Fig. 2), an automated program was utilized to determine precise collision energy and biases for the quadrupoles.

DFMO peaks resolved at approximately 3 min with symmetry and reliable resolution, as shown by a 40,000 ng/mL QCL sample (Fig. 3).

Novel LC-MS/MS assay to quantify D,L-alpha-difluoromethylornithine (DFMO) in mouse plasma **429**

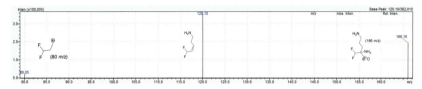

Fig. 2 Mass spectra of product ions formed from DFMO collision with argon gas.

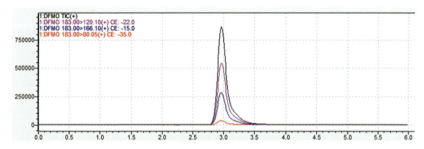

Fig. 3 Representative eluted peak of quality control (HC) sample injection.

Table 1 Performance of DFMO calibration curves (n = 10 replicates).

Slope (mean, SD)	y-Intercept (mean, SD)	r2 (mean, SD)
192.3 (81.7)	−7714 (5233)	0.9985 (0.0014)

Resultant data was utilized to create an MRM to quantify DMFO using product ion quantities between concentrations.

4.2 Method validation

DFMO calibration standards between 1 ng/mL and 50,000 ng/mL were evaluated. The lower limit of quantification (LLOQ) was 250 ng/mL and the lower limit of detection (LLOD) was 5 ng/mL. Six standard concentrations were determined: 250, 1000, 5000, 10,000, 25,000, and 50,000 ng/mL. Fresh calibration curves (n = 10) were prepared from a stock solution using serial dilutions. Linear regression for calibration curves ranged from 0.9960 to 0.9999 using $1/c^2$ weighted values (Table 1, Fig. 4). All plasma standards exhibited inter-day accuracy and precision maximum percent bias and coefficient of variation (CV) values of ± 2.0 % and ± 1.2 %, respectively (Table 2). Quality control (QC) samples were prepared at 500, 7500, and 40,000 ng/mL, and seven replicates were run in tandem with standards to determine the precision

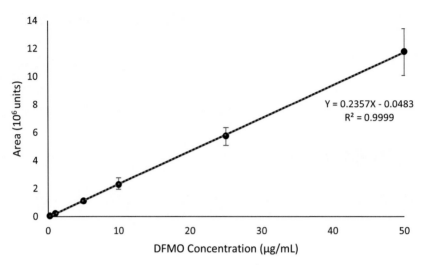

Fig. 4 Representative calibration curve for DFMO in mouse plasma (n = 10).

Table 2 Inter-day accuracy and precision of DFMO calibration standards (n = 20).

Standard concentration (ng/mL)	Mean concentration (ng/mL, SD)	Bias (%)	CV (%)
250	250.4 (2.90)	0.2	1.2
1000	997.8 (9.41)	−0.2	0.9
5000	4899.6 (55.3)	−2.0	1.2
10,000	9933.5 (49.8)	0.7	0.5
25,000	25,113.0 (156)	0.4	0.6
50,000	51,141.0 (125)	2.3	0.2

Table 3 Inter-day accuracy and precision of DFMO quality control samples (n = 70).

QC concentration (ng/mL)	Mean concentration (ng/mL, SD)	Bias (%)	CV (%)
500	506.7 (2.8)	1.3	0.6
7500	7281 (48)	−2.9	0.7
40,000	39,856 (159)	−0.4	0.4

Table 4 24-Hour stability of processed and unprocessed controls at room temperature and 4 °C (n = 3).

Processing method	Condition	Low control (500 ng/mL)		High control (40,000 ng/mL)	
		Mean concentration (ng/mL)(CV%)	Bias (%)	Mean concentration (ng/mL)(CV%)	Bias (%)
Unprocessed	RT	495 (3.0)	−1.0	39,154 (6.5)	−2.1
	4 °C	484 (4.9)	−3.2	39,113 (9.7)	−2.2
Processed	RT	525 (2.7)	5.0	36,559 (9.1)	−8.6
	4 °C	506 (4.0)	1.2	39,127 (3.1)	−2.2

Table 5 Freeze-thaw stability for low and high controls (n = 4).

Freeze/ thaw cycle	Low control (500 ng/mL)		High control (40,000 ng/mL)	
	Mean concentration (ng/mL) (CV%)	Bias (%)	Mean concentration (ng/mL) (CV%)	Bias (%)
1	479 (2.0)	−4.3	39,066 (0.5)	−2.3
2	490 (2.3)	−2.1	38,484 (0.4)	−3.8
3	465 (0.7)	−7.0	36,208 (0.4)	−9.5

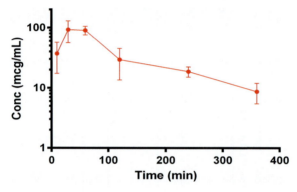

Fig. 5 Plasma-time profile obtained from an *in vivo* pharmacokinetic study (results previously reported in reference (Schultz et al., 2021; figure used with permission)).

and accuracy of the method. Quality control samples displayed maximum interday bias of ± 3% and CV% of ± 0.7% (Table 3). The results are within the acceptable limits for accuracy and precision in accordance with the FDA and EMA guidance documents (European Medicines Agency, 2022; US Department of Health & Human Services Food & Drug Administration, 2024).

4.3 Assay stability

Short-term stability of unprocessed and processed LC and HC samples at bench top (room temperature) and 4° C was determined over 24 h. DFMO was stable under all conditions, with <5% CV and bias % for QCL, and <10% CV and bias % for HCL (Table 4). DFMO was stable after 3 freeze-thaw cycles, with QCL having CV ≤2.4% and bias ≤9.5% (Table 5).

4.4 Application to *in vivo* pharmacokinetic study

A representative plasma-time concentration profile from the single-dose *in vivo* pharmacokinetic study is shown in Fig. 5. Plasma concentrations were detected at 10 min post-dose administration, with the maximum concentration occurring at 60 min. Drug elimination was described by a first-order process (k = 0.0051 min^{-1}) with a terminal phase half-life of approximately 2.3 h.

5. Conclusion

DFMO is a promising anti-cancer agent that is currently in clinical development for several diseases most recently approved by the FDA for neuroblastoma. To further elucidate the pharmacokinetic properties of DFMO, an optimized bioanalytical method was developed using LC-MS/MS. This method employs a simple plasma processing method and allows for a rapid sample analysis (6-minute run time) without the need for pre-column derivatization. The method is applicable to preclinical pharmacokinetic screening in early drug discovery.

Acknowledgments

The project received funding from a grant from the St. Baldrick's Foundation (#591704) to ASB.

Declaration of competing interest

ASB is the sole inventor of U.S. patent (US 9072,778 B2) issued on July 7, 2015, entitled "Treatment regimen for N-Myc, C-Myc, and L-Myc amplified and overexpressed tumors", which discusses the use of DFMO. ASB is one of three inventors of U.S. patent (US 11,273,137 B2) issued on March 15, 2022, entitled "Methods and compositions to prevent and treat disorders associated with mutations in the *ODC1* gene", which discusses the use of DFMO and which was licensed to Orbus Therapeutics, Inc. No competing financial or other interests were disclosed by the other authors.

CRediT authorship contribution statement

Matthew A. Swanson: Formal analysis, Data curation, Investigation, Validation, Methodology, Writing – original draft. **Carlye Szarowicz:** Formal analysis, Data curation, Investigation, Validation, Methodology, Writing – review & editing. **Schuyler T. Pike:** Supervision, Project administration, Writing – review & editing. **Chad R. Schultz:** Investigation, Data curation, Conceptualization, Writing- review & editing. **André S. Bachmann:** Project administration, Funding acquisition, Conceptualization, Writing – review & editing. **Thomas Dowling:** Project administration, Conceptualization, Writing – review & editing.

References

Alirol, E., Schrumpf, D., Amici Heradi, J., Riedel, A., de Patoul, C., Quere, M., & Chappuis, F. (2013). Nifurtimox-Eflornithine combination therapy for second-stage gambiense human African trypanosomiasis: Medecins Sans Frontieres experience in the democratic Republic of the Congo. *Clinical Infectious Diseases: An Official Publication of the Infectious Diseases Society of America, 56*(2), 195–203. https://doi.org/10.1093/cid/cis886.

Bachmann, A. S., & Geerts, D. (2018). Polyamine synthesis as a target of MYC oncogenes. *The Journal of Biological Chemistry, 293*(48), 18757–18769. https://doi.org/10.1074/jbc.TM118.003336.

Bachmann, A. S., & Levin, V. A. (2012). Clinical applications of polyaminebased therapeutics. In P. M. Woster, & R. A. Casero (Eds.). *Polyamine drug discovery* (pp. 257–276). Royal Society of Chemistry.

Bachmann, A. S., VanSickle, E. A., Michael, J., Vipond, M., & Bupp, C. P. (2022). Bachmann-Bupp syndrome and treatment. *Developmental Medicine and Child Neurology, 66*(4), 445455.

Bello-Fernandez, C., Packham, G., & Cleveland, J. L. (1993). The ornithine decarboxylase gene is a transcriptional target of c-Myc. *Proceedings of the National Academy of Sciences of the United States of America, 90*(16), 7804–7808.

Blume-Peytavi, U., & Hahn, S. (2008). Medical treatment of hirsutism. *Dermatologic Therapy, 21*, 329–339.

Casero, R. A., Jr., & Marton, L. J. (2007). Targeting polyamine metabolism and function in cancer and other hyperproliferative diseases. *Nature Reviews. Drug Discovery, 6*(5), 373–390.

Casero, R. A., Jr., Murray Stewart, T., & Pegg, A. E. (2018). Polyamine metabolism and cancer: Treatments, challenges and opportunities. *Nature Reviews. Cancer, 18*, 681–695. https://doi.org/10.1038/s41568-018-0050-3.

Cho, H. E., & Kang, M. H. (2018). PH gradient-liquid chromatography tandem mass spectrometric assay for determination of underivatized polyamines in cancer cells. *Journal of Chromatography B, 1085*, 21–29. https://doi.org/10.1016/j.jchromb.2018.03.043.

European Medicines Agency. ICH guideline M10 on bioanalytical method validation and study sample analysis. 2022. https://www.ema.europa.eu/en/documents/scientific-guideline/ich-guideline-m10bioanalytical-method-validation-step-5_en.pdf. Accessed 23 February 2024.

Gerner, E. W., & Meyskens, F. L., Jr. (2004). Polyamines and cancer: Old molecules, new understanding. *Nature Reviews. Cancer, 4*(10), 781–792.

Hogarty, M. D., Ziegler, D. S., Franson, A., Chi, Y.-Y., Tsao-Wei, D., Liu, K., et al. (2024). Phase 1 study of high-dose DFMO, celecoxib, cyclophosphamide and topotecan for patients with relapsed neuroblastoma: A new approaches to Neuroblastoma therapy trial. *British Journal of Cancer*. https://doi.org/10.1038/s41416-023-02525-2.

Jackson, J., Caro, J. J., Caro, G., Garfield, F., Huber, F., Zhou, W., ... Eflornithine, H. S. G. (2007). The effect of Eflornithine 13.9% cream on the bother and discomfort due to hirsutism. *International Journal of Dermatology, 46*, 976–981.

Kumar, A., Singh, V., & Kumar, P. (2013). A new spectrophotometric method for the determination of Eflornithine hydrochloride in parenteral formulation. *American Journal of Pharmacological Sciences, 1*(3), 38–41. https://doi.org/10.12691/ajps-1-3-2.

Lewis, E. C., Kraveka, J. M., Ferguson, W., Eslin, D., Brown, V. I., Bergendahl, G., ... Saulnier Sholler, G. L. (2020). A subset analysis of a phase II trial evaluating the use of DFMO as maintenance therapy for high-risk neuroblastoma. *International Journal of Cancer, 147*(11), 3152–3159. https://doi.org/10.1002/ijc.33044.

Manni, A., Washington, S., Mauger, D., Hackett, D. A., & Verderame, M. F. (2004). Cellular mechanisms mediating the anti-invasive properties of the ornithine decarboxylase inhibitor alpha-difluoromethylornithine (DFMO) in human breast cancer cells. *Clinical & Experimental Metastasis, 21*(5), 461–467. https://doi.org/10.1007/s10585-004-2724-3.

Metcalf, B. W., Bey, P., Danzin, C., Jung, M. J., Casara, P., & Vevert, J. P. (1978). Catalytic irreversible inhibition of mammalian ornithine decarboxylase (E.C. 4.1.1.17) by substrate and product analogs. *Journal of the American Chemical Society, 100*, 2551–2553.

Meyskens, F. L., Jr, McLaren, C. E., Pelot, D., Fujikawa-Brooks, S., Carpenter, P. M., Hawk, E., et al. (2008). Difluoromethylornithine plus sulindac for the prevention of sporadic colorectal adenomas: A randomized placebo-controlled, double-blind trial. *Cancer Prevention Research (Philadelphia, PA), 1*, 32–38.

Moreno, L., Barone, G., DuBois, S. G., Molenaar, J., Fischer, M., Schulte, J., et al. (2020). Accelerating drug development for Neuroblastoma: Summary of the second Neuroblastoma drug development strategy forum from innovative therapies for children with cancer and international society of paediatric oncology Europe Neuroblastoma. *European Journal of Cancer, 136*, 52–68. https://doi.org/10.1016/j.ejca.2020.05.010.

Nilsson, J. A., Keller, U. B., Baudino, T. A., Yang, C., Norton, S., Old, J. A., et al. (2005). Targeting ornithine decarboxylase in Myc-induced lymphomagenesis prevents tumor formation. *Cancer Cell, 7*(5), 433–444.

Oesterheld, Ferguson, J., Kraveka, W., Bergendahl, J. M., Clinch, G., Lorenzi, T., et al. (2023). Eflornithine as postimmunotherapy maintenance in high-risk Neuroblastoma: Externally controlled, propensity score-matched survival outcome comparisons. *Journal of Clinical Oncology: Official Journal of the American Society of Clinical Oncology*, JCO2202875.

Pegg, A. E., & Feith, D. J. (2007). Polyamines and neoplastic growth. *Biochemical Society Transactions, 35*(Pt 2), 295–299. https://doi.org/10.1042/BST0350295.

Priotto, G., Kasparian, S., Mutombo, W., Ngouama, D., Ghorashian, S., Arnold, U., ... Kande, V. (2009). Nifurtimox-Eflornithine combination therapy for second-stage African Trypanosoma brucei gambiense trypanosomiasis: A multicentre, randomised, phase III, non-inferiority trial. *Lancet, 374*(9683), 56–64. https://doi.org/10.1016/S0140-6736(09)61117-X.

Priotto, G., Pinoges, L., Fursa, I. B., Burke, B., Nicolay, N., Grillet, G., ... Balasegaram, M. (2008). Safety and effectiveness of first line Eflornithine for Trypanosoma brucei gambiense sleeping sickness in Sudan: Cohort study. *BMJ (Clinical Research ed.), 336*(7646), 705–708. https://doi.org/10.1136/bmj.39485.592674.BE.

Rajasekaran, S., Bupp, C. P., Leimanis-Laurens, M., Shukla, A., Russell, C., Junewick, J., et al. (2021). Repurposing Eflornithine to treat a patient with a rare ODC1 gain-of-function variant disease. *Elife, 10*, e67097.

Saulnier Sholler, G. L., Gerner, E. W., Bergendahl, G., MacArthur, R. B., VanderWerff, A., Ashikaga, T., ... Bachmann, A. S. (2015). A phase I trial of DFMO targeting polyamine addiction in patients with relapsed/refractory Neuroblastoma. *PLoS One, 10*(5), e0127246. https://doi.org/10.1371/journal.pone.0127246.

Schultz, C. R., Swanson, M. A., Dowling, T. C., & Bachmann, A. S. (2021). Probenecid increases renal retention and antitumor activity of DFMO in neuroblastoma. *Cancer Chemotherapy and Pharmacology, 88*(4), 607–617.

Sholler, G. L. S., Ferguson, W., Bergendahl, G., Bond, J. P., Neville, K., Eslin, D., ... Kraveka, J. M. (2018). Maintenance DFMO increases survival in high risk Neuroblastoma. *Scientific Reports, 8*(1), 14445. https://doi.org/10.1038/s41598-018-32659-w.

Sims, E. K., Kulkarni, A., Hull, A., Woerner, S. E., Cabrera, S., Mastrandrea, L., et al. (2023). Inhibition of polyamine biosynthesis preserves beta cell function in type 1 diabetes. *Cell Reports Medicine, 4*(11), 101261.

Sjoerdsma, A., & Schechter, P. J. (1999). Eflornithine for African sleeping sickness. *Lancet, 354*(9174), 254. https://doi.org/10.1016/S0140-6736(05)66324-6.

US Department of Health and Human Services Food and Drug Administration. M10 bioanalytical method validation and study sample analysis. https://www.fda.gov/regulatory-information/search-fda-guidance-documents/m10bioanalytical-method-validation-and-study-sample-analysis. Accessed 23 February 2024.

USWN, LLC. Iwilfin (Eflornithine)[package insert]. U.S. Food and Drug Administration website: https://www.accessdata.fda.gov/drugsatfda_docs/label/2023/215500s000lbl.pdf. Accessed 20 February 2024.

Wallick, C. J., Gamper, I., Thorne, M., Feith, D. J., Takasaki, K. Y., et al. (2005). Key role for p27Kip1, retinoblastoma protein Rb, and MYCN in polyamine inhibitor-induced G1 cell cycle arrest in MYCN-amplified human neuroblastoma cells. *Oncogene, 24*(36), 5606–5618.

Witherspoon, M., Chen, Q., Kopelovich, L., Gross, S. S., & Lipkin, S. M. (2013). Unbiased metabolite profiling indicates that a diminished thymidine pool is the underlying mechanism of colon cancer chemoprevention by alpha-difluoromethylornithine. *Cancer Discovery, 3*(9), 1072–1081. https://doi.org/10.1158/2159-8290.CD-12-0305 Epub 2013 Jun 14.

Wolf, J. E., Jr., Shander, D., Huber, F., Jackson, J., Lin, C. S., Mathes, B. M., ... Eflornithine, H. S. G. (2007). Randomized, double-blind clinical evaluation of the efficacy and safety of topical Eflornithine HCl 13.9% cream in the treatment of women with facial hair. *International Journal of Dermatology, 46*, 94–98.

Yang, S., Peng, K., & Wang, M. Z. (2014). A simple and sensitive assay for Eflornithine quantification in rat brain using pre-column derivatization and UPLC-MS/MS detection. *Biomedical Chromatography, 29*(6), 918–924. https://doi.org/10.1002/bmc.3374.

CHAPTER TWENTY FIVE

Polyamine quantitation by LC-MS using isobutyl chloroformate derivatives

Christine Isaguirre[a,1], Megan Gendjar[a,1], Kelsie M. Nauta[b], Nicholas O. Burton[b], and Ryan D. Sheldon[a,*]

[a]Mass Spectrometry Core, Van Andel Institute, Grand Rapids, MI, United States
[b]Department of Metabolism and Nutritional Programming, Van Andel Institute, Grand Rapids, MI, United States
*Corresponding author. e-mail address: ryan.sheldon@vai.org

Contents

1. Introduction	438
2. Methods	439
2.1 Before you begin	439
2.2 Metabolite extraction	443
2.3 Polyamine derivatization	444
2.4 LC-MS analysis	446
2.5 Data analysis	449
3. Expected outcomes	450
3.1 Representative results	450
3.2 Quantitation	452
4. Advantages, limitations, and troubleshooting	452
4.1 Advantages	452
4.2 Limitations	455
4.3 Troubleshooting	456
Acknowledgments	456
References	456

Abstract

Polyamines are an important class of metabolites that are poorly covered in standard metabolomics workflows. Here, we describe a protocol for isobutyl-chloroformate derivatization that can be applied to metabolite extracts following other metabolomics applications. This simple procedure allows for quantitative measurement of thirteen polyamines and two internal standards in a short (15-minute) LC-MS method. We report at least two triple quadrupole mass spectrometer transitions for each compound. Among these are unique transitions for co-eluting isomers N1- and N8-acetylspermidine, enabling quantitation of each isomer individually. We further

[1] These authors contributed equally to this work.

define the linear dynamic range for each compound, and present data in several biological sample types. This simple and robust method enables stand-alone and post-metabolomics polyamine analysis.

1. Introduction

Polyamines are a family of cationic metabolites that contain two or more amino groups. These compounds are ubiquitous in metabolic processes spanning all domains of life and are involved in a variety of human diseases (Sagar, Tarafdar, Agarwal, Tarafdar, & Sharma, 2021). An example is Bachmann-Bupp syndrome, a gain-of-function mutation in ornithine decarboxylase 1, is characterized by elevated levels of plasma polyamines and leads to a spectrum of clinical symptoms including alopecia and several behavioral/neurological disorders (Bupp, Michael, VanSickle, Rajasekaran, & Bachmann, 1993; Bupp, Schultz, Uhl, Rajasekaran, & Bachmann, 2018). Additionally, Snyder Robinson syndrome is an X-linked intellectual disability syndrome caused by a loss of function mutation in the spermine synthase gene (Arena et al., 1996; Schwartz, Peron, & Kutler, 1993; Snyder & Robinson, 1969) leading to altered levels of the central polyamines spermine, spermidine, and putrescine. Moreover, polyamine metabolism is implicated across a spectrum of prevalent human diseases, including cancer (Holbert, Cullen, Casero, & Stewart, 2022), diabetes (Fernandez-Garcia et al., 2019; Kulkarni, Anderson, Mirmira, & Tersey, 2022), and Alzheimer's disease (Ivanov, Atanasova, Dimitrov, & Doytchinova, 2020; Morrison & Kish, 1995). Robust, accessible, and reproducible analytical tools for polyamine measurement are critical for ongoing efforts to define the molecular mechanisms of these and other polyamine associated conditions.

The polycationic nature of polyamines that mediate their biological importance also poses several analytical challenges. Liquid chromatography—mass spectrometry (LC-MS) based metabolomics is a widely-used tool for measuring the relative abundance of metabolites in biological samples. However, polyamines generally do not behave well in typical LC-MS workflows. Low-pH mobile phase hydrophilic interaction or amide chromatography methods can detect some acetylated polyamines, but these methods struggle to consistently retain and resolve unmodified polyamines such as spermine, spermidine, and putrescine. Thus, polyamines are often excluded or only partially covered in typical metabolomics experiments.

Chemical derivatization is an effective strategy to enrich compounds with certain functional groups by improving chromatographic performance, boosting ionization efficiency and, consequently, lowering detection limits (Tang et al., 2023). A chemical derivatization approach for polyamines using isobutyl chloroformate has been previously described (Byun et al., 2008) and refined by our lab. Here, we present a full protocol that can be easily adapted to any sample type and appended to standard metabolomics workflows, enabling addition of polyamines to the metabolomics repertoire.

This protocol includes (1) extraction of metabolites, including polyamines, from tissues, cells, and biofluids, (2) derivatization of metabolite extracts with isobutyl chloroformate, and (3) separation and detection of polyamines via LC-triple quadrupole (QQQ) mass spectrometry. A method summary is presented in Fig. 1. Advantages of this method are (1) compatibility with other upstream metabolomics workflows, (2) simplicity of derivatization, (3) superior retention and separation by a simple reversed phase chromatography, and (4) well-defined and unique transitions, including for co-eluting isomers N1-acetylspermidine and N8-acetylspermidine.

2. Methods
2.1 Before you begin
2.1.1 Stock solutions and standards
1. Prepare 1 M sodium bicarbonate
   a. For 1 M sodium bicarbonate, dissolve 840.07 mg of sodium bicarbonate into 9 mL of LC-MS grade water.
   b. Vortex until sodium bicarbonate goes into solution.
   c. Adjust the pH to 9 ± 0.01 with 1 N sodium hydroxide.
      Tip: Add NaOH slowly, 10-20 μL at a time and routinely test the pH of the solution. Keep track of the volume added during pH adjustment.
   d. Bring the solution up to a final volume of 10 mL with LC-MS grade water.
      Caution! Stock pH may drift over time. Verify stock pH before each use, only use if within +/− 0.2 pH units.
2. Preparation of chemical standards at a stock concentration of 2 mg/mL.

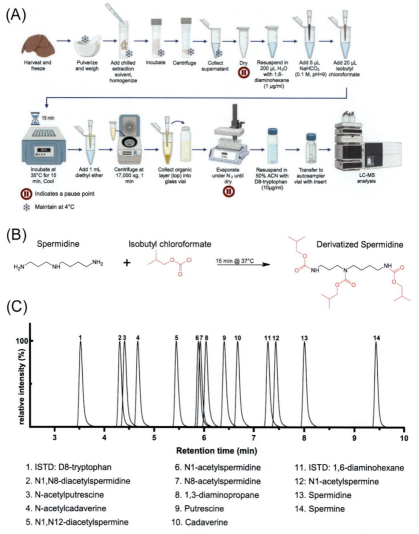

Fig. 1 Workflow overview. (A) Schematic overview of the protocol, with pause points as indicated. (B) Chemical structure of a representative polyamine, spermidine, and the derivatization product formed by reaction with isobutyl chloroformate. (C) Representative chromatogram of derivatized polyamines detected by LC-QQQ.

  a. Using a mass balance, weigh between 10-20 mg of powdered standard into a 15 mL conical tube.
  b. Dissolve the standard into the appropriate amount of LC-MS grade water to achieve a 2 mg/mL final concentration.

Tip: if the standard mass is <10 mg, then resuspend the entire contents by adding 1 mL of water to the standard vial transferring to a 15 mL conical. Repeat these steps 3 times to recover all the compound from the vial. Adjust the final volume to achieve 2 mg/mL concentration.
  c. Further dilute the standard to 2 µg/mL with LC–MS grade water.
      i. Add 10 µL of the 2 mg/mL stock solution to 990 µL of LC–MS grade water to make a 20 µg/mL stock solution.
      ii. Add 100 µL of the 20 µg/mL stock solution to 900 µL of LC–MS grade water to make a 2 µg/mL final concentration.
  d. Add 25 µL of the sodium bicarbonate (NaHCO$_3$; pH 9.0) and transfer 200 µL solution to a new LC autosampler vial. Proceed with polyamine derivatization, beginning with Step 2.3 below.
3. Preparation of internal standards (ISTD).
  a. For 1 µg/mL internal standards, 1,6-diaminohexane and D8-Tryptophan (D8-Trp), dissolve 2 mg of a standard into 1 mL of LC–MS grade water.
  b. Further dilute the standard to 1 µg/mL with LC–MS grade water.
      i. Add 10 µL of the 2 mg/mL stock solution to 990 µL of LC–MS grade water to make a 20 µg/mL stock solution.
      ii. Add 50 µL of the 20 µg/mL stock solution to 950 µL of LC–MS grade water to make a 1 µg/mL final concentration.
      iii. Aliquot and store in −80 °C.
  c. Use the ISTDs as indicated in the polyamine derivatization workflow below.

2.1.2 Materials and equipment
- Spermidine (Sigma 85558)
- Spermine (Sigma 85590)
- Putrescine (Sigma 51799)
- N-acetylputrescine (Sigma A8784)
- N-acetylcadaverine (Toronto Research Chemicals A620505)
- N1,N8-diacetylspermidine (Cayman Chemical 21588)
- N1-acetylspermidine (Toronto Research Chemicals A187845)
- N8-acetylspermidine (Cayman Chemical 37434)
- N1-acetylspermine (Sigma 01467)
- N1,N12-Diacetylspermine (Cayman Chemical 17918)
- N-(3-oxopropyl) acetamide (Sigma PH016963)
- 1,6-diaminohexane Toronto (Research Chemicals D416170)

- N-(5-aminopentyl) acetamide/N-acetylcadaverine (Toronto Research Chemicals A620505)
- Cadaverine (Sigma 33220)
- 1,3-diaminopropane (Sigma D23602)
- D8-tryptophan (Cambridge Isotope Libraries DLM-6903)
- Sodium bicarbonate (VWR BDH9280)
- Isobutyl chloroformate (Sigma 177989)
- Diethyl ether (Sigma 309966)
- LC-MS Grade Acetonitrile (Fisher Scientific A955-4)
- LC-MS Grade Water (Fisher Scientific W6-4)
- 1.5 mL tubes (Eppendorf 22364111)
- Autosampler vials with inserts (Sigma 29391-U)
- Autosampler vials without inserts (Sigma 29651-U)
- Certified polypropylene bonded caps with septa (Sigma 29319-U)
- 1 mL syringes (BD DGW87705)
- 10 mL syringes (MEDLINE SYR110010)
- BD General Use Hypodermic Needles [23G x 1.5″] (Fisher Scientific 14-826-6C)
- 1,6-diaminohexane (Sigma D23602)
- Micro pipette tips
- Formic Acid (Fisher Scientific A117-50)
- CORTECS T3 Column, 120 Å, 1.6 μm, 2.1 mm × 150 mm, 1/pk (Waters 186008500)
- CORTECS T3 VanGuard Pre-column, 120 Å, 1.6 μm, 2.1 mm × 5 mm, 3/pk (Waters 186008508)
- LC-MS Grade Isopropanol (Fisher Scientific A461-4)
- LC-MS Grade Methanol (Fisher Scientific A456-4)
- Heat Block or Incubator
- Vortex
- Refrigerated Centrifuge
- Water batch sonicator
- Rotary vacuum evaporator
- Nitrogen Evaporator
- pH meter
- Fume hood
- LC triple quadrupole mass spectrometer (described here: Agilent 6470 LC-MS)

2.2 Metabolite extraction

Timing: hands-on time = 2 h, total time = 4–6 h.

1. Generate crude metabolite extracts from cells, tissue, or biofluids with organic solvent according to established metabolite extraction methods (See e.g., (Sheldon, Ma, DeCamp, Williams, & Jones, 2021)). Described here is a representative protocol using 40 % acetonitrile, 40 % methanol, and 20 % water for extraction solvent.
   a. For cells, begin with roughly 1-2 million cells. Metabolite abundance per cell is dependent on cell volume, so number of cells required will vary by cell type. 80 % confluency in a 6-well tissue culture plate is a good starting point.
      i. Prepare cells by aspirating media, washing once with 0.9 % NaCl, and snap freezing on dry ice. Do not trypsinize or scrape live cells, as this will affect metabolism. Aspirated, frozen plates should be stored at −80°C until extraction.

      Note: to avoid metabolic changes that happen once cells are removed from the incubator, only process a few replicates (≤6) in small batches. Be cautious to avoid systematic ordering effects by distributing an equal number of replicates from each group into each batch, and processing them in a random order.

      ii. Add ice-cold extraction solvent to the frozen plate and disperse cell contents using a cell scraper.
      iii. Transfer the crude extract to a 1.5 mL conical tube and proceed with the below general metabolite extraction protocol.
   b. For tissues, pulverize and weigh ~40 mg of frozen tissue to a bead mill homogenizer tube.
      i. Add extraction solvent so the final concentration is 40 mg tissue per mL of solvent in each sample.

      Important: tissue to solvent ratio will affect extraction efficiency and should be made equal throughout an experiment by modifying extraction solvent volume based on individual sample mass.

      ii. Use a bead homogenizer to homogenize samples 3 times for 30 s, resting the samples on wet ice in between to prevent sample warming.
      iii. Proceed with the below general metabolite extraction protocol.
   c. For plasma/serum, begin with 40 μL of plasma frozen in a 1.5 mL tube. Add 1000 μL of ice-cold extraction solvent and vortex. Proceed with the below general metabolite extraction protocol.

2. Starting with crude extracts (above):
   a. Pulse vortex sample and extraction solvent 3x for 3 s each.
   b. Sonicate in a waterbath sonicator for 5 min
   c. Incubate on wet ice for at least 1 h.
   d. Centrifuge crude extracts at 14,000 x g for 10 min at 4°C.
   e. Use a 1 mL pipette to decant 800 μL (80%) of metabolite containing supernatant to a fresh 1.5 mL tube.
   f. Dry metabolite extracts in 1.5 mL Eppendorf tubes using a vacuum evaporator or under a stream of nitrogen gas. These dried metabolite extracts are the starting point for polyamine derivatization.

Pause point: Dried metabolite extracts can be capped and stored long-term at −80 °C.

Important: perform the extraction procedure on an empty tube to serve as a process blank. Treat this blank the same as biological samples throughout the entire procedure.

Note: Metabolite extracts may first be used for other non-derivatized LC-MS metabolomics analyses, then dried for polyamine derivatization as described here.

Note: Data shown here are from metabolites extracted using 40% acetonitrile, 40% methanol, and 20% water (v/v) as described previously (Sheldon et al., 2021). Many metabolite extraction approaches exist, and their suitability for polyamine recovery and derivatization should be tested in each case. Best practices should be followed (Alseekh et al., 2021; Lu et al., 2017).

2.3 Polyamine derivatization

Timing: 2–3 h.
1. Resuspend dried metabolite extracts in 200 μL of LC-MS grade H_2O containing 1 μg/mL of 1,6-diaminohexane internal standard (ISTD). This internal standard is important to account for derivatization efficiency and any sample loss during processing.
2. Add 5 μL of 1 M sodium bicarbonate ($NaHCO_3$; pH 9.0) stock solution to each sample.
3. Using a 1 mL syringe in a fume hood, transfer >20 μL per sample (e.g., if working with 10 samples, gather >200 μL) of isobutyl chloroformate to a glass vial for a working stock.
4. In the fume hood, add 20 μL of isobutyl chloroformate to each resuspended metabolite extract and vortex.

    Note: To increase pipetting accuracy, wet the pipette tip by

drawing up and expelling the isobutyl chloroformate several times before aliquoting to the sample.
5. Incubate at 35 °C for 15 min using a dry heat block.
6. Allow samples to cool to room temperature before proceeding.
7. Using a 10 mL syringe in a fume hood, transfer >1 mL per sample of diethyl ether and transfer to a (20 mL or 40 mL) glass vial for a working solution.
8. In the fume hood, add 1 mL of diethyl ether to each sample and pulse vortex.

    Note: To increase pipetting accuracy, wet the pipette tip by drawing up and expelling the diethyl ether several times before aliquoting to the sample.
9. Centrifuge at 17,000 x g for 1 min to induce phase separation.
10. Using a pipettor, transfer 800 μL of the organic (top) layer to a new autosampler vial.

    Note: Minimize sample loss during transfer. Wet each pipette tip in a stock of fresh diethyl ether before collecting the sample
11. Dry the organic layer under a gentle stream of nitrogen gas.
    a. Bring down the needles just on process blank to test flow rate and then add all samples.
    b. Needles should be placed just below the threads of the vials.
       Caution! High nitrogen flow rate and/or getting too close to the sample will force the sample out of the vial.
    c. Dry for 10 min, then visually inspect for dryness. If not dry, leave for another 10 min. Repeat as needed until the sample is dry.
12. Preparing dried samples for analysis.
    a. Resuspend samples in 1:1 (v/v) acetonitrile (ACN) and LC-MS grade H_2O containing 10 μg/mL D8-Tryptophan ISTD. This ISTD will not be derivatized, and is therefore used to identify missed/partial LC injections and LC-MS signal drift over time.
    b. Add resuspension solvent to sample dried in autosampler vials and vortex.
       Suggested resuspension volumes per matrix are as follows:
       i. Cells and Media: 50 μL
       ii. Plasma: 100 μL
       iii. Tissue: 200 μL
          Note: there may be white particulates that do not resuspend. This is normal. These tend to stick to the vial walls. Be careful not to transfer them with the resuspended sample in the next step.

c. Transfer 5-10 µL less than the total resuspension volume to new autosampler vials with inserts for analysis.

Pause point: Resuspended, derivatized polyamines are stable in capped autosampler vials maintained in the dark at 4 °C for at least two weeks. Dried polyamine derivatives may be stored at −80 °C indefinitely.

2.4 LC-MS analysis

LC-MS analysis can be conducted on any electrospray ionization mass spectrometer using reversed phase chromatography, though source parameters, transition settings, and transition time will need to be optimized. Here, we present parameters using an Agilent 6470 LC-QQQ.

Timing: 1–3 h.

2.4.1 Mobile phase preparation

a. Clean mobile phase bottles and graduated cylinders by rinsing twice with LC-MS grade water and once with acetonitrile. Dispose of used solvents in appropriate liquid hazardous waste containers.
b. Mobile Phase A (water with 0.1 % formic acid): add 1 mL of formic acid to 1 L of LC-MS grade water in a clean solvent bottle.
c. Mobile Phase B (99 % acetonitrile with 0.1 % formic acid): combine 990 mL of acetonitrile, 10 mL of LC-MS grade water, and 1 mL of formic acid in a clean solvent bottle.

Tip: Formic acid may degrade over time, which will affect chromatography. Use fresh, single-use ampules. It is also useful to track mobile phase pH.
d. Swirl both mobile phases to thoroughly mix.
e. Degas both mobile phases in a water bath sonicator at room temperature for 10 min.
f. Place mobile phases on LC system, insert solvent lines, and purge each channel for at least 5 min.

2.4.2 LC-MS system preparation.

a. Connect the LC column (Waters CORTECS T3) to the LC system.
b. Set column compartment temperature to 50 °C
c. Turn on the mass spectrometer to allow for system parameters to reach their setpoint.
    i. Gas Temperature (°C): 200
    ii. Gas Flow (L/min): 13
    iii. Nebulizer (psi): 45

    iv. Sheath Gas Temperature (°C): 325
    v. Sheath Gas Flow (L/min): 12
    vi. Capillary Voltage (V): 3000
    vii. Nozzle Voltage (V): 500

d. Begin by flowing 100% Mobile phase B at 100 μL/min until pressure stabilizes (approximately 10 min).

e. Change flow to 100% A at 100 μL/min until pressure stabilizes.

f. When pressure has stabilized, increase flow rate sequentially in 100 μL/min steps, allowing pressure to stabilize between each step, until flow reaches 400 μL/min. When the pressure has equilibrated at 400 μL/min the column is in starting mobile phase condition and is ready to begin analysis.

Tip: track and record LC system pressures during equilibration. These will vary between instruments. High system pressure may be an early indicator of a clogged or aging/deteriorating column, and low system pressure may indicate a leak.

Timing: 20 min of instrument time per sample.

2.4.3 LC-MS aquisition

a. The LC gradient is listed in Table 1.

b. The QQQ transition acquisition list is detailed in Table 2.

c. Utilize an acquisition strategy including appropriate blanks, pooled quality control samples, and column conditioning according to

Table 1 Liquid chromatography gradient.

Time (min)	A (%)	B (%)	Flow (mL/min)
Start. Cond. min	100	0	0.4
0.50 min	100	0	0.4
2.00 min	60	40	0.4
10.00 min	1	99	0.4
15.00 min	1	99	0.4
15.10 min	100	0	0.6
18.00 min	100	0	0.6
18.40 min	100	0	0.4
18.50 min	100	0	0.4

Table 2 LC-QQQ transition list of carbamylated polyamines. The first transition listed for each compound is the quantifier ion, and the others are qualifier ions. Retention time (RT) is provided in minutes based on the chromatography in Fig. 1. Note that RT will vary between labs and should be validated with standards. D8-tryptophan is an underivatized internal standard added at resuspension and used to control for instrument variability. CE, collision energy, Frag, fragmentation voltage.

Compound name	Precursor (m/z)	Product (m/z)	RT (min)	Frag (V)	CE (V)
ISTD: D8-tryptophan	213	195	3.3	135	55
	213	150	3.3	80	20
N1,N8-diacetylspermidine	330	256	4.1	135	10
	330	214	4.1	135	20
	330	155	4.1	135	40
N-acetylputrescine	231	157	4.2	135	10
	231	115	4.2	135	20
N-acetylcadaverine	245	171	4.4	135	10
	245	145	4.4	135	10
	245	129	4.4	135	20
	245	112	4.4	135	25
N1,N12-diacetylspermine	487	371	5.3	135	15
	487	255	5.3	135	30
	487	100	5.3	135	45
N1-acetylspermidine	388	240	5.8	100	20
	388	100	5.8	100	30
N8-acetylspermidine	388	172	5.8	100	20
	388	102	5.8	100	40
1,3-diaminopropane	275	201	6	135	10
	275	101	6	80	15
	275	58	6	80	30

Putrescine	289	215	6.2	90	5
	289	159	6.2	80	10
	289	115	6.2	80	15
Cadaverine	303	229	6.65	80	15
	303	129	6.65	80	15
	303	86	6.65	80	30
ISTD: 1,6-diaminohexane	317	243	7.1	80	10
	317	143	7.1	80	20
	317	100	7.1	135	30
N1-acetylspermine	545.3	471.3	7.3	160	15
	545.3	255	7.3	150	35
	545.3	155	7.3	135	55
Spermidine	446.1	372.1	8	135	15
	446.1	298	8	135	15
	446.1	198	8	135	30
Spermine	603	529.2	9.4	135	15
	603	255	9.4	150	45
	603	155	9.4	50	10

established best practices (Broeckling et al., 2023). For example, see (Sheldon et al., 2021).

Following LC-MS analysis and verifying data quality (below), derivatized polyamines may be dried and kept at −80 °C for at least 6 months.

Note: these samples may take longer than usual to dry.

2.5 Data analysis

1. Immediately after LC-MS analysis, verify data quality by reviewing the following:

a. D8-Tryptophan ISTD, added during final resuspension, as it is useful for identifying missed injections and instrument signal drift over time. RSD should be less than 10%.
b. 1,6-Diaminohexane ISTD, added prior to derivatization, allows for identification sample loss and variance introduced during processing. RSD should be less than 20%.

Note: High ISTD variance indicates a technical problem with the experimental workflow. The presented strategy allows the user to isolate the source of variance. Ideally, this will lead the user to take corrective actions and either repeat or reanalyze the experiment. In some cases, especially when samples are limited, normalizing to the internal standard in each sample can correct variance. This should be done with caution.

2. Peak picking and integration can be completed using any vendor or open-source software. Data presented here were analyzed in Skyline (v.24.0).
a. Discard peaks with <3-fold peak area over blank
b. Ensure quantifier and qualifier ions co-elute and occur at the proper ion ratio (this will vary between instrument platforms). Discard compounds where this does not occur.333

3. Expected outcomes
3.1 Representative results

An overview of the protocol, a representative chemical structure of a derivatized polyamine, and a representative chromatography are presented in Fig. 1. The protocol is applicable to any biological sample type that contains polyamines. To demonstrate this, we analyzed the polyamine content of three common sample types: HEK293 cells, mouse liver, and mouse blood plasma (Fig. 2A). Metabolites were extracted using acetonitrile:methanol:water (40%:40%:20%, v/v), according to (House et al., 2024) and subjected to isobutyl chloroformate derivatization as described above. Peak areas of each polyamine detected are presented relative to the abundance detected in HEK293 cells in Fig. 2A. The relative standard deviation of the D8-tryptophan ISTD, added at final resuspension, was 1.2%, which indicates excellent inter-run LC-MS stability. The relative standard deviation of the 1,6-diaminohexane ISTD, added prior to derivatization, was higher (8.0%) due to variance accumulated during

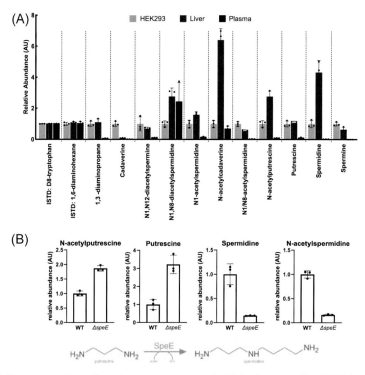

Fig. 2 Representative polyamine measurements in biological samples. (A) Polyamines detected in HEK293 cells, mouse liver, and mouse plasma. Signal intensity is normalized to the intensity detected in HEK293 cells for each compound. (B) Abundances of acetyl-putrescine, putrescine, spermidine, and acetyl-spermidine in *Bacillus Subtills* lacking the spermidine synthase gene (ΔspeE) and wild-type controls.

derivatization, though was within the acceptable range of <20 %. Data here are intended to demonstrate the polyamine coverage and low technical variance across several sample types, but matrix effects and starting material amount differences preclude direct comparison of polyamine abundances between the sample types.

To demonstrate ability of the assay to the detect biological phenotypes, we cultured bacteria (*Bacillus subtilis*) lacking spermidine synthase(ΔspeE), which catalyzes the production of spermidine and methyl-thioladenosine from putrescine and decarboxylated S-adenosylmethionine (dSAM) (Sekowska, Bertin, & Danchin, 1998). When compared to wild-type bacteria (WT), the ΔspeE mutant demonstrated a 7-fold decrease in spermidine, consistent with a loss of SpeE activity (Fig. 2B). The ΔspeE mutant also demonstrated a 3-fold increase in putrescine, which has not been previously reported in this system, likely due to the limited sensitivity or coverage of historical methods (Fig. 2B) (Sekowska et al., 1998). Accordingly, the acetylated cognate of each polyamine was increased and decreased, respectively.

3.2 Quantitation

We next determined the lower limit of quantitation (LLOQ) and linear dynamic range (LDR) of the method. Polyamine standards were prepared in a stock mix containing 20 μg/mL of each compound. The stock mix was serially diluted in half-log steps to 0.2 ng/mL, dried, and derivatized as described. LC-MS peak areas and expected concentrations were log_{10}transformed and plotted in Fig. 3. The LLOQ for each compound was between ~10–30 nM, and the upper bound of the LDR was ~2–10 μM. Importantly, quantitation above this upper bound is possible, but will require fit to a quadratic regression. Note, that while a quadratic fit may improve quantitative accuracy above the LDR, it may impair accuracy in low-abundant samples. Alternatively, the sample and standard curve may be diluted to bring unknowns within the linear range. Note, LLOQ and LDR will vary between instruments and should be verified in each case.

4. Advantages, limitations, and troubleshooting
4.1 Advantages

The current protocol has several advantages over other polyamine analysis workflows, as: (1) It can be applied to metabolite extracts previously used for other metabolomics studies, (2) the derivatization protocol is simple, (3) can be executed in a few hours, and (4) provides full coverage of central polyamine metabolism. Additional compounds may be added to the method through derivatization of chemical standards to determine transitions and RT of each compound.

Another advantage of this method is the novel approach to achieve specificity of N1 versus N8-acetylspermidine (AcSpd) (Fig. 4). Most commonly, this is achieved through polyamine derivatization with dansyl-chloride, followed by separation and detection by UV-HPLC. However, even when using a long 45-min gradient, only partial, non-baseline chromatographic separation of these two isomers is achieved (Kabra, Lee, Lubich, & Marton, 1986; Stewart et al., 2022). Here, using LC coupled mass-spectrometry we identified two unique transitions for each compound, allowing specific quantification of each even in the

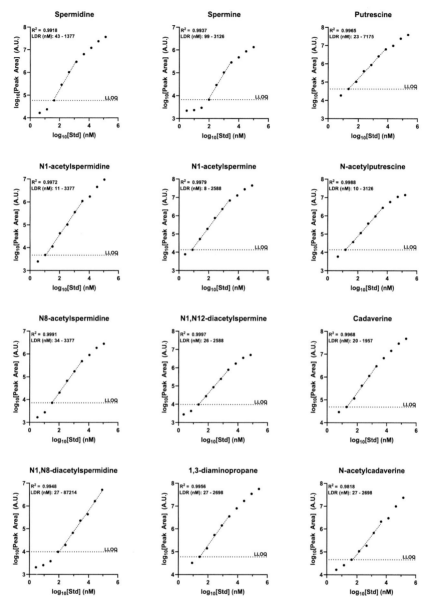

Fig. 3 Calibration curves of polyamine standards. Serial dilutions of polyamine analytical standards were performed from 20 μg/mL to 0.2 ng/mL in half-log steps. Standards were derivatized and detected by LC-MS as described. Shown is molar concentration of each compound. Concentrations and peak areas are log-transformed and fit via linear regression. The lower limit of quantitation (LLOQ) was defined as the lowest curve point to be at least threefold over process blank samples. The linear dynamic range (LDR) was defined as the nM concentration from the LLOQ to upper linear point.

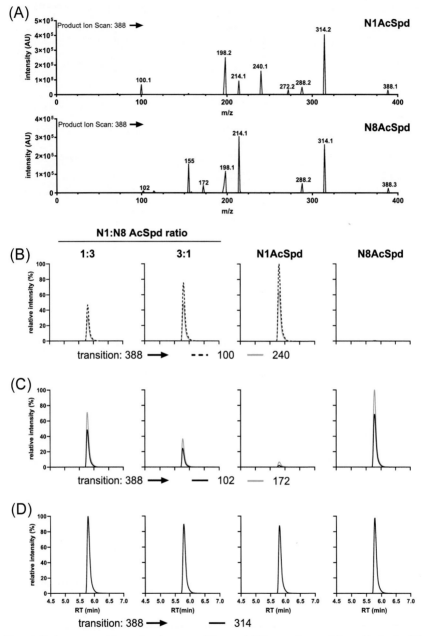

Fig. 4 Unique transitions for isomers N1- and N8-acetylspermidine (AcSpd). (A) Product ion scans of neat derivatized standards for N1-AcSpd (top) and N8-AcSpd (bottom) reveal unique daughter ions to each compound. (B–D) Transition specificity was evaluated by mixing N1- and N8-AcSpd at 1:3 and 3:1 molar ratios (right two

absence of chromatographic separation (Fig. 4). Thus, the current method uses compound-specific transitions to distinguish N1-AcSpd and N8-AcSpd in roughly one-third of the time per sample as traditional LC alone approaches.

4.2 Limitations

1. Derivatization is irreversible. If metabolite extracts are used for upstream metabolomics analysis, then it is important to verify data quality (e.g. missed injection requiring a sample to be reran) for each sample before proceeding with derivatization.
2. Incomplete or multiple-derivatized species. As with any derivatization approach, it is possible that target compounds will form several derivatized forms. This will limit sensitivity, as the compound is distributed across several peaks, and quantitative accuracy. Full scan high-resolution mass spectrometry of derivatized chemical standards provides a mechanism to screen for unintentional derivatives.
3. Derivatization is not specific for polyamines. We have analyzed derivatized biological samples with untargeted, high-resolution LC-MS and found thousands of features not present in blanks. These are likely other metabolites that are derivatized or otherwise carried through the derivatization process. Even if they are not included in the transition list, these compounds will still create ion-suppression, which may affect detected intensity of the polyamines.
4. Co-ion suppression of N1 and N8 acetylspermidine. While the current protocol identifies unique MS transitions for each isomer, their co-elution may cause an ion-supressive effect that may obscure interpretation in some cases. For example, if N1-acetylspermidine is significantly higher in group A than group B, this could ion suppress N8-acetylspermidine specifically in group A. Thus, lower observed N8-acetylspermidine in group A would be a false-positive phenotypic finding caused by a true positive increase in N1-acetylspermidine. Caution is warranted when a divergent phenotype such as this is observed.

---

columns, respectively) and individually as neat standards (left two columns, respectively). Total N-AcSpd content was the same in each condition. (B) N1-AcSpd specific transitions 388→100 (black) and 388→240 (gray) showed increasing dose dependency from 1:3 to 3:1 N1:N8-AcSpd, and was completely absent from N8-AcSpd standard. (C) N8-AcSpd specific transitions 388→102 (black) and 388→172 (gray) showed decreasing dose dependency from 1:3 to 3:1 N1:N8-AcSpd, and was markedly decreased (< 5%) in the N1-AcSpd standard. (D) Shared N-AcSpd transition 388→314 was present at the same level in all conditions, regardless of N1:N8 AcSpd ratio.

5. Lack of compound-specific internal standards. Unfortunately, isotopically labeled standards for each polyamine are either very expensive or not commercially available. Inclusion of isotopically labeled cognates would improve quantitative accuracy and alleviate the concerns of ion suppression listed above.

4.3 Troubleshooting

1. Problem: low mass spectrometry signal. Assuming a properly functioning instrument, low signal can indicate non-optimal source and/or transition settings. These should be tuned to maximize signal. This may also indicate a problem with derivatization. Inclusion of a post-derivatization internal standard (D8-tryptophan in the current protocol) is useful to determine whether this is an instrument or derivatization problem.
2. Problem: poor peak shape. This is likely a chromatography problem. Remake mobile phases ensuring that all components are added and mixed completely. Ensure that the column is sufficiently re-equilibrated before each sample injection. If not resolved, switch to a new column.
3. Problem: drifting or unstable retention time. If the retention time steps earlier with each sample injection, this may indicate that the column is not fully reconditioning between each sample. Increase the time and/or flow rate of mobile phase A between injections. If this does not resolve the problem, replace the column.

Acknowledgments

This protocol was developed and data collected in the Van Andel Institute Mass Spectrometry Core (RRID:SCR_024903). The authors thank Dr. Martha Escobar Galvis for her assistance with manuscript formatting and copy-editing. The authors have no conflicts of interest to disclose. This project was funded by the Van Andel Institute's MeNu (Metabolism and Nutrition) program (RDS) and NIH grant DP2DK139569 (NOB).

References

Alseekh, S., Aharoni, A., Brotman, Y., Contrepois, K., D'Auria, J., Ewald, J., ... Fernie, A. R. (2021). Mass spectrometry-based metabolomics: A guide for annotation, quantification and best reporting practices. *Nature Methods, 18*(7), 747–756. https://doi.org/10.1038/s41592-02101197-1.

Arena, J. F., Schwartz, C., Ouzts, L., Stevenson, R., Miller, M., Garza, J., ... Lubs, H. (1996). X-linked mental retardation with thin habitus, osteoporosis, and kyphoscoliosis: Linkage to Xp21.3-p22.12. *American Journal of Medical Genetics, 64*(1), 50–58. https://doi.org/10.1002/(SICI)1096-8628(19960712)64:1<50::AID-AJMG7>3.0.CO;2-V.

Broeckling, C. D., Beger, R. D., Cheng, L. L., Cumeras, R., Cuthbertson, D. J., Dasari, S., ... Mosley, J. D. (2023). Current practices in LC-MS untargeted. Metabolomics: A

scoping review on the use of pooled quality control samples. *Analytical Chemistry, 95*(51), 18645–18654. https://doi.org/10.1021/acs.analchem.3c02924.
Bupp, C., Michael, J., VanSickle, E., Rajasekaran, S., & Bachmann, A. S. (1993). Bachmann-Bupp syndrome. In M. P. Adam, J. Feldman, G. M. Mirzaa, R. A. Pagon, S. E. Wallace, & A. Amemiya (Eds.). *GeneReviews(R)*. Seattle, WA: University of Washington.
Bupp, C. P., Schultz, C. R., Uhl, K. L., Rajasekaran, S., & Bachmann, A. S. (2018). Novel de novo pathogenic variant in the ODC1 gene in a girl with developmental delay, alopecia, and dysmorphic features. *American Journal of Medical Genetics. Part A, 176*(12), 2548–2553. https://doi.org/10.1002/ajmg.a.40523.
Byun, J. A., Lee, S. H., Jung, B. H., Choi, M. H., Moon, M. H., & Chung, B. C. (2008). Analysis of polyamines as carbamoyl derivatives in urine and serum by liquid chromatographytandem mass spectrometry. *Biomedical Chromatography: BMC, 22*(1), 73–80. https://doi.org/10.1002/bmc.898.
Fernandez-Garcia, J. C., Delpino-Rius, A., Samarra, I., Castellano-Castillo, D., Munoz-Garach, A., Bernal-Lopez, M. R., ... Tinahones, F. J. (2019). Type 2 diabetes Is associated with a different pattern of serum polyamines: A case(-)control study from the PREDIMED-plus trial. *Journal of Clinical Medicine, 8*(1), https://doi.org/10.3390/jcm8010071.
Holbert, C. E., Cullen, M. T., Casero, R. A., Jr., & Stewart, T. M. (2022). Polyamines in cancer: Integrating organismal metabolism and antitumour immunity. *Nature Reviews. Cancer, 22*(8), 467–480. https://doi.org/10.1038/s41568-022-00473-2.
House, R. R. J., Soper-Hopper, M. T., Vincent, M. P., Ellis, A. E., Capan, C. D., Madaj, Z. B., ... Sheldon, R. D. (2024). A diverse proteome is present and enzymatically active in metabolite extracts. *Nature Communications, 15*(1), 5796. https://doi.org/10.1038/s41467-024-50128-z.
Ivanov, S. M., Atanasova, M., Dimitrov, I., & Doytchinova, I. A. (2020). Cellular polyamines condense hyperphosphorylated Tau, triggering Alzheimer's disease. *Scientific Reports, 10*(1), 10098. https://doi.org/10.1038/s41598-020-67119-x.
Kabra, P. M., Lee, H. K., Lubich, W. P., & Marton, L. J. (1986). Solid-phase extraction and determination of dansyl derivatives of unconjugated and acetylated polyamines by reversed-phase liquid chromatography: Improved separation systems for polyamines in cerebrospinal fluid, urine and tissue. *Journal of Chromatography, 380*(1), 19–32. https://doi.org/10.1016/s0378-4347(00)83621-x.
Kulkarni, A., Anderson, C. M., Mirmira, R. G., & Tersey, S. A. (2022). Role of polyamines and hypusine in beta cells and diabetes pathogenesis. *Metabolites, 12*(4), https://doi.org/10.3390/metabo12040344.
Lu, W., Su, X., Klein, M. S., Lewis, I. A., Fiehn, O., & Rabinowitz, J. D. (2017). Metabolite measurement: Pitfalls to avoid and practices to follow. *Annual Review of Biochemistry, 86*, 277–304. https://doi.org/10.1146/annurev-biochem-061516-044952.
Morrison, L. D., & Kish, S. J. (1995). Brain polyamine levels are altered in Alzheimer's disease. *Neuroscience Letters, 197*(1), 5–8. https://doi.org/10.1016/0304-3940(95)11881-v.
Sagar, N. A., Tarafdar, S., Agarwal, S., Tarafdar, A., & Sharma, S. (2021). Polyamines: Functions, metabolism, and role in human disease management. *Medical Sciences, 9*(2), https://doi.org/10.3390/medsci9020044.
Schwartz, C. E., Peron, A., & Kutler, M. J. (1993). Snyder-Robinson syndrome. In M. P. Adam, J. Feldman, G. M. Mirzaa, R. A. Pagon, S. E. Wallace, & A. Amemiya (Eds.). *GeneReviews(R)*Seattle (WA): University of Washington. https://www.ncbi.nlm.nih.gov/pubmed/23805436.

Sekowska, A., Bertin, P., & Danchin, A. (1998). Characterization of polyamine synthesis pathway in Bacillus subtilis 168. *Molecular Microbiology, 29*(3), 851–858. https://doi.org/10.1046/j.1365-2958.1998.00979.x.

Sheldon, R. D., Ma, E. H., DeCamp, L. M., Williams, K. S., & Jones, R. G. (2021). Interrogating in vivo T-cell metabolism in mice using stable isotope labeling metabolomics and rapid cell sorting. *Nature Protocols, 16*(9), 4494–4521. https://doi.org/10.1038/s41596-021-00586-2.

Snyder, R. D., & Robinson, A. (1969). Recessive sex-linked mental retardation in the absence of other recognizable abnormalities. Report of a family. *Clinical Pediatrics, 8*(11), 669–674. https://doi.org/10.1177/000992286900801114.

Stewart, T. M., Foley, J. R., Holbert, C. E., Klinke, G., Poschet, G., Steimbach, R. R., ... Casero, R. A., Jr. (2022). Histone deacetylase-10 liberates spermidine to support polyamine homeostasis and tumor cell growth. *The Journal of Biological Chemistry, 298*(10), 102407. https://doi.org/10.1016/j.jbc.2022.102407.

Tang, S., Zhang, P., Gao, M., Xiao, Q., Li, Z., Dong, H., ... hang, Y. (2023). A chemical derivatization-based pseudotargeted LC-MS/MS method for high coverage determination of dipeptides. *Analytica Chimica Acta, 1274,* 341570. https://doi.org/10.1016/j.aca.2023.341570.

Printed in the United States
by Baker & Taylor Publisher Services